Fundamentals of Automatic Process Control

CHEMICAL INDUSTRIES

A Series of Reference Books and Textbooks

Founding Editor

HEINZ HEINEMANN

Berkeley, California

Series Editor

JAMES G. SPEIGHT

CD & W, Inc.

Laramie, Wyoming

MOST RECENTLY PUBLISHED

Fundamentals of Automatic Process Control, Uttam Ray Chaudhuri and Utpal Ray Chaudhuri

The Chemistry and Technology of Coal, Third Edition, James G. Speight

Practical Handbook on Biodiesel Production and Properties, Mushtaq Ahmad, Mir Ajab Khan, Muhammad Zafar, and Shazia Sultana

Introduction to Process Control, Second Edition, Jose A. Romagnoli and Ahmet Palazoglu

Fundamentals of Petroleum and Petrochemical Engineering, Uttam Ray Chaudhuri

Advances in Fluid Catalytic Cracking: Testing, Characterization, and Environmental Regulations, edited by Mario L. Occelli

Advances in Fischer-Tropsch Synthesis, Catalysts, and Catalysis, edited by Burtron H. Davis and Mario L. Occelli

Transport Phenomena Fundamentals, Second Edition, Joel Plawsky

Asphaltenes: Chemical Transformation during Hydroprocessing of Heavy Oils, Jorge Ancheyta, Fernando Trejo, and Mohan Singh Rana

Chemical Reaction Engineering and Reactor Technology, Tapio O. Salmi, Jyri-Pekka Mikkola, and Johan P. Warna

Lubricant Additives: Chemistry and Applications, Second Edition, edited by Leslie R. Rudnick

Catalysis of Organic Reactions, edited by Michael L. Prunier

The Scientist or Engineer as an Expert Witness, James G. Speight

Process Chemistry of Petroleum Macromolecules, Irwin A. Wiehe

Interfacial Properties of Petroleum Products, Lilianna Z. Pillon

Clathrate Hydrates of Natural Gases, Third Edition, E. Dendy Sloan and Carolyn Koh

Chemical Process Performance Evaluation, Ali Cinar, Ahmet Palazoglu, and Ferhan Kayihan

Fundamentals of Automatic Process Control

Uttam Ray Chaudhuri • Utpal Ray Chaudhuri

CRC Press
Taylor & Francis Group
Boca Raton London New York

CRC Press is an imprint of the
Taylor & Francis Group, an **informa** business

CRC Press
Taylor & Francis Group
6000 Broken Sound Parkway NW, Suite 300
Boca Raton, FL 33487-2742

First issued in paperback 2019

ISBN-13: 978-1-4665-1420-1 (hbk)
ISBN-13: 978-0-367-38072-4 (pbk)

Library of Congress Cataloging-in-Publication Data

Chaudhuri, Uttam Ray.
Fundamentals of automatic process control / Uttam Ray Chaudhuri, Utpal Ray Chaudhuri.
p. cm. -- (Chemical industries ; 134)
Includes bibliographical references and index.
ISBN 978-1-4665-1420-1 (hardback)
1. Chemical process control. I. Chaudhuri, Utpal Ray. II. Title.

TP155.75.C425 2012
660'.281--dc23 2012015433

Visit the Taylor & Francis Web site at
http://www.taylorandfrancis.com

and the CRC Press Web site at
http://www.crcpress.com

Dedication

This book is dedicated to the memory of
the late Kartick Ch Ray Chaudhuri (father)
and
Mrs. Kamala Ray Chaudhuri (mother)

Contents

Preface

Processing plants engage in various activities, involving mass- and energy-transfer operations, thermal and chemical conversions, storage, filling, packaging, conveying operations, etc. Modern plants use automatic process control operations. Hence, the modern-day process engineer should be well conversant about automatic control instruments and their use. In fact, manufacturing and maintenance of such instruments are the handiwork of the instrumentation engineers. A chemical engineer has to understand the necessary control strategy to achieve control of the desired quality and flow rate of products by maintaining operating parameters, such as temperature, pressure, flow rate, or level in each and every piece of equipment present in a plant. The cost of controlling instruments at a plant is about 40%–60% of the initial investment of the plant. Because of the complex nature of a processing plant, the most accurate control strategy is necessary to avoid loss of materials, degradation of quality, accidents, and environmental pollution. Hence, no compromise should be made because of the high cost of automatic process control systems as far as the cost of accidents, loss, and environmental destruction are concerned. Though modern-day plants use computer control systems or distributed control systems, a proper control strategy should be used as decided by the process engineer of the plant. The basic knowledge of hardware and software must be acquired by the engineer who will have to operate and supervise the control system for day-to-day operation. Hence, a strong theoretical and practical knowledge of process control are essential for both plant engineers and operators. In this book, theoretical analysis of process dynamics and control have been done in more detail with a large number of problems and solutions spread throughout the book. In addition, the book is accompanied by a CD-ROM containing the Virtual Laboratory software discussed in Chapter 8. (The CD-ROM can be found at www.routledge.com/9781466514201.) We hope this book will be useful to students and practicing engineers

Acknowledgments

The authors acknowledge the following family members for their constant support and cooperation in the completion of this book:

Mrs. Sampa Ray Chaudhuri
Ms. Aratrika Ray Chaudhuri
Mr. Ashish Sengupta
Mrs. Ashima Sengupta

Authors

Uttam Ray Chaudhuri is an associate professor in the Department of Chemical Technology at Calcutta University. He holds a doctoral degree in chemical engineering from the Indian Institute of Technology, Kharagpur. He received his graduate and postgraduate degrees in chemical engineering from Jadavpur University. He has more than 30 years of experience in industry, research, and teaching in the field of chemical engineering. He has to his credit quite a good number of research publications in foreign and Indian journals.

Utpal Ray Chaudhuri is a professor in the Food Technology and Biochemical Engineering Department of Jadavpur University, Kolkata. He has more than 30 years' experience in industry and teaching in the field of chemical engineering and food technology. He received his graduate and postgraduate degrees at the chemical engineering department at Jadavpur University.

1 An Introduction to Automatic Process Control

1.1 OBJECTIVE OF AUTOMATION

The word "automatic" indicates work done without the help of any human intervention, i.e., recording, understanding, taking necessary action, etc. As we know, a single person cannot watch recorders or machines continuously for 24 hours a day. He must be relieved by another person periodically, usually in eight-hour shifts. Besides these limitations, a human operator has difficulty monitoring machines because of tiredness resulting from the drudgery and other natural human reactions. All these limitations lead to fluctuations and degradation in the quality and flow of products and may even lead to accidents. In addition to these problems, increasing demand for higher salaries and other expenses included in human operators' cost of production also increase over time. In some plants, where explosive or radiation hazards are present, remote operations are possible only with automatic-control instruments. Hence, automatic process control should achieve the following benefits:

(i) Reduced human operations
(ii) Closely monitored product quality
(iii) Increased rate of production
(iv) Reduced wastage of materials and energy consumption
(v) Safe operations and reduced accidents
(vi) Decreased operational costs
(vii) Remote operation envisaged

1.2 HARDWARE AND SOFTWARE

There are five basic elements of control hardware. These are the primary element or sensor, the secondary element or signal-generating element (transducer, transmitter, converter, etc.), the controlling element or controller, the transmission cable, and the final control element.

Process parameters and variables are measured by the primary elements, which are the measuring instruments or sensors. Measurement of the variables or properties are based on certain unique phenomena, such as physical, chemical, or thermoelectrical factors. Examples of temperature sensors include gas- or liquid-filled bulbs, thermocouple joints, bimetallic joints, resistance elements, etc. An orifice plate is a unique example of a flow sensor; whereas, a floating body and submerged displacer unit are examples of level sensors. Because a signal, usually an electrical current or voltage measurement, is not directly available from all types of sensors, a secondary element is necessary to convert the primary measurement into an electrical signal. Of course, there are primary elements, such as thermocouples, which generate electrical voltage unaided by a secondary element.

The process variable as sensed by the primary element cannot be transmitted unless converted to an electrical (or pneumatic) signal by a secondary element, known as a transducer, for recording, transmitting, and processing by the controller. For example, a pressure drop across an orifice plate is converted to a pneumatic (air pressure) signal or electrical voltage/current with the help of a differential pressure cell (DPC). Thus, the combination of the sensor (orifice plate) and the DPC jointly perform the function of a flow transducer. Other examples of transducers are thermocouples, strain gauge, turbine meters, etc., which are the sensors for temperature, pressure, and flow rate, respectively. Instruments, such as amplifiers, potentiometers, wheatstone bridges, etc., are also required for the conversion of electric signals to a desired range (usually in the 4–20 mA range).

Electrical transmission cables, usually copper or aluminum wires, are used to conduct electrical signals; whereas a pneumatic signal (air pressure in the range of 3–15 psi) is transmitted through copper or aluminum tubes. The signal is then carried to the controller located at the point of measurement or in a control room away from the point of measurement (field). Optical fiber cables are also used in transmission, especially for the transmission of digital signals.

The controlling element instrument is known as the controller, which performs appropriate functions for maintaining the desired level (set point) of parameters to restore quality and rate of production. The output signal is generated according to the equation or logic that manipulates the appropriate flow of certain streams. Thus, a controller acts as the brain of the controlled system, mimicking the activities of a human operator, such as reading and comparing the operating parameters or variables, in order to make decisions for correcting actions that manipulate the flow of material or energy with the help of final control elements.

The most common and successful control logic is a proportional–integral–derivative (PID) relationship. Controllers are classified according to the type of signal: pneumatic (handling air signals), electrical (handling electrical current or voltage signals), and microprocessor or digital (handling both electric and digital signals). Today, controllers are mainly microprocessor or digital controllers. These are increasingly replacing the older analog (pneumatic or electrical) controllers because of their lower cost and size, higher speed, memory, and communication capabilities.

Final control elements are the valves or switches that are capable of increasing or decreasing the flow of material or energy passing through any process. Control valves manipulate the flow rate of gas or liquid; whereas, the control switches manipulate the electrical energy entering a system.

Software (or programs) are the equations and logic that mimic the expertise of a intelligent human operator. Software is accessible, storable (and also erasable in

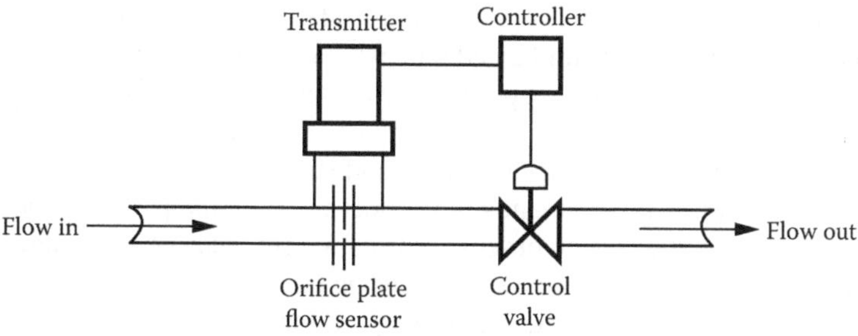

FIGURE 1.1 Flow rate control in a pipeline.

many controllers), and can be manipulated as desired. It resides in the controller or computer written in the form of a program. In the early controllers, the control equations were hardwired programs with the help of resistance–inductance–capacitor–amplifier circuits (analog circuits). Few of the parameters of these equations could be changed by rescrewing and changing wiring connections. Modern controllers use microprocessor chips, and software is a digital program where equation parameters or even the entire equation can be altered by pressing front panel keys. Software generates output signals that ultimately actuate the appropriate control valves or switches to manipulate the flow rate of material or energy to achieve the quality and desired rate of production. A traditionally used control software is a PID equation available in a program residing in the controller memory. The parameters of this program can be varied with appropriate buttons on the front panel of the controller. This will be discussed in more detail elsewhere in this book.

A flow-control system in a pipeline with all the elements of control system is shown in Figure 1.1.

1.3 PROCESS AND VARIABLES

In the eyes of control engineers, a process refers to equipment where certain activity is carried out to achieve a certain target. For example, a motor car is a process, and the activity is the motion. In this case, a control engineer will look to the instruments to control the motion of the car. Whereas, in a chemical plant, the process is a piece of equipment, a group of pieces, or a plant unit wherein certain operation(s) or activity is carried out. Instruments are required to control the activities of such a process. Examples are storage tanks, heat exchangers, distillation columns, extraction columns, reactors, filtration units, pumps, compressors, etc.

Operations are defined by the values of variables. These are the properties and flow rates of input and output streams entering and leaving a process. Properties of the streams are density, viscosity, chemical composition, etc., which are the functions of operating parameters, such as temperature, pressure, flow rate, level, etc. Thus, variation in the values of these parameters will change the properties of the product.

An example of temperature control in a steam-heated tank is shown in Figure 1.2. A cold water stream enters a steam-heated tank where a steam coil heats the liquid and hot water leaves the tank. The purpose of the control system is to deliver hot water at a particular temperature at the given fixed rate. The input streams are cold water and steam. The output streams are hot water and steam condensate. Variation in the rates and properties of these streams will affect the tank temperature and the exit temperature of the hot water. In order to maintain a set value of the temperature of the hot water leaving, the steam flow rate must be manipulated.

For example, as the rate of cold water increases (or as the cold water temperature further falls), the steam rate has to be increased to maintain the hot water temperature and vice versa. Thus, the cold water flow rate or its temperature (or both flow rate and temperature) is called the load variable. The temperature of the hot water is the controlled variable, and the steam flow rate is the manipulated variable. The desired temperature of the hot water is the set point variable or simply the set point. The difference between the set point and the control variable at any time is called the error or deviation variable.

If the cold fluid at a temperature of T_i°C enters the tank (process) at a rate of W kg/hr, the hot fluid leaves at the same rate but at the temperature of T°C where the desired temperature of the hot water is T_{set}°C, the volume of liquid in the tank is V m^3, and the heating rate is q kW. The variables then can be listed as below:

Load or input variable: T_i
Control variable: T
Set point: T_{set}
Manipulated variable: q
Deviation or error variable (ε): $T_{set} - T$

It is imperative to note that the error or deviation ε is always obtained by deducting the control variable from the set point, i.e., $(T_{set} - T)$ and not $(T - T_{set})$. The goal of the control system is to get the control temperature to reach set point (T_{set}) and to maintain it at that level during all events of fluctuation of the load (influent rate or

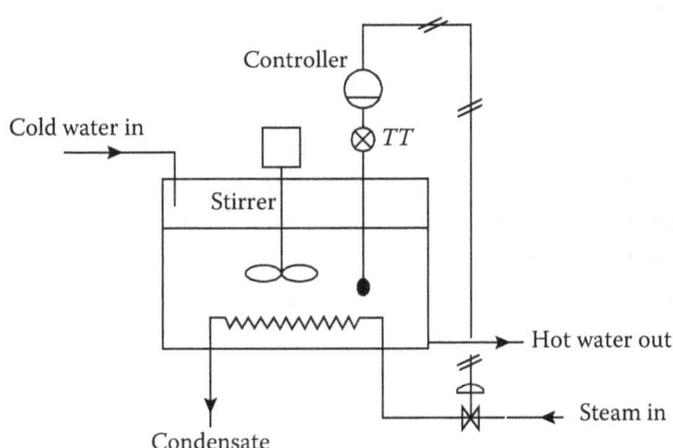

FIGURE 1.2 Temperature control in a steam-heated tank.

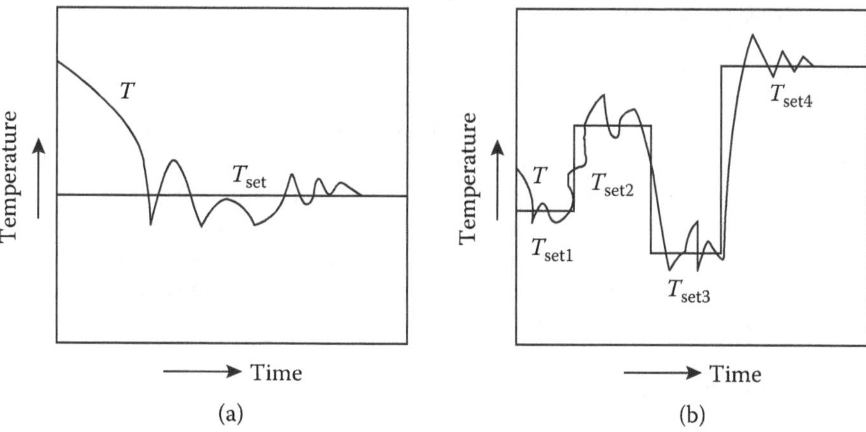

FIGURE 1.3 Control temperature and set point. Horizontal lines indicate set points (a) when the set points is unchanged and (b) when the set point is changed.

its temperature). If the set point is varied, the controller will help the control variable (T) to achieve the new set point. Thus, the error or deviation becomes zero (or near zero) such that $T = T_{set}$. The qualitative explanation of the desired goal for a controlled system is presented in Figure 1.3.

1.4 COMMON TRANSDUCERS

Name of the Instrument	Sensor	Secondary Element	Output
Orifice meter	Orifice plate	DP cell	Air pressure or current
Venturi meter	Venturi	DP cell	Air pressure or current
Piot tube velocity meter	Pitot	DP cell	Air pressure or current
Magnetic turbine meter	Magnetic turbine	Dc potential converter	Electric voltage
Vortex shedding meter	Bluff body	Peizoelectric element	Electric voltage
Doppler meter	Ultrasonic wave	Sound to electric Voltage converter	
Thermocouple	Bimetallic joints	None	Voltage
RTD	Resistance element	Wheatson bridge	Voltage
Bimetal	Bimetallic joints	Peizoelectric element	Voltage
Optoelectric	Semiconductor element	Amplifier	Current
Bourdon Tube	Elastic tube	Strain to electricity converter	Voltage
Strain gage	Resistance element	Wheatson bridge	Voltage
LVDT gage	Ferro-magnetic material	Voltage	Voltage
Capacitance gage	Plate or disk	Capacitance to voltage converter	Voltage
Float-tape	Float	Potentiometer	Voltage
Displacer	Submerged body	strain to electricity converter	Voltage
Ultrasonic	Ultra sound wave	Voltage	Voltage
Static pressure gage	Pressure gage	Pressure to electric volt	Voltage

1.5 CONTROLLERS

Controllers are the brains of the control system. Depending on the type of signal (electrical, pneumatic, digital), three types of controllers are constructed. The oldest type of controllers were pneumatic controllers, which could access and deliver pneumatic signals made of air flow at a pressure of 3 to 15 psi (or 0.2 to 1.02 kg/cm^2). With this instrument, air should be clean and dry generated within the process to avoid moisture and particulate matter that may otherwise damage controllers and connecting instruments. After the development of electrical devices, electrical controllers have replaced pneumatic controllers. These controllers access and deliver electrical signals, usually as electric current in the range of 4 to 20 milliamperes (direct current, or DC, signal). Electrical signals are much faster in action as compared to pneumatic controllers. Both pneumatic and electrical signals are continuous or "analog" signals. Later, with the development of computers and microprocessors, digital controllers have replaced analog controllers. Today's controllers are capable of accessing and processing both electrical and digital signals. These are much smaller in size, much faster, and cheaper as compared to older pneumatic and electrical controllers. A modern controller is shown in Figure 1.4.

Ports or slots for screwing in cables for sensors (RTD, thermocouple, etc.), output signals to the final control element, power supply cables (+ and – ports), communication signal cables, etc. are located at the rear of the controller shown in Figure 1.4. The process variable or parameter value and set point are displayed on the digital display screen in the front panel. The set point can be increased or decreased by the "up" and "down" keys. The "selection/tune" key is a toggle switch, which is pressed as desired for setting controller parameters other than set point and can be increased or decreased by the "up" and "down" keys. With the "selection/tune" key, the address of the controller and digital communication parameters, such as baud

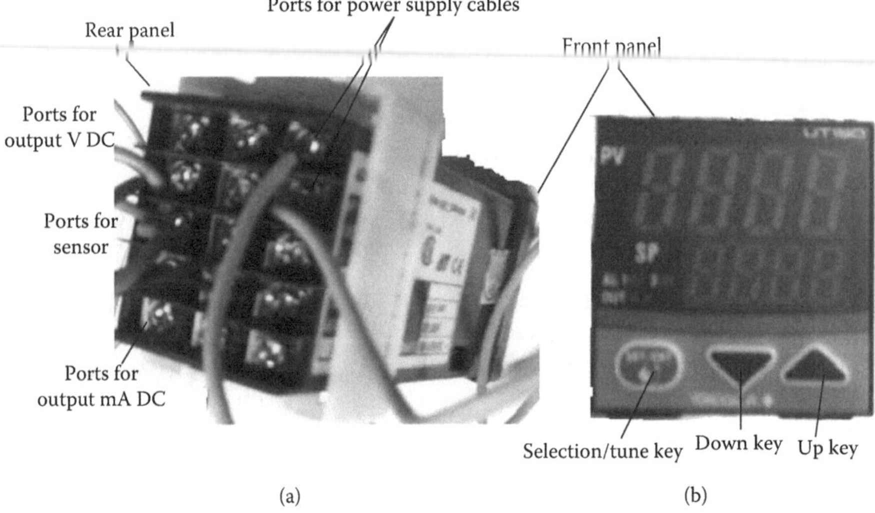

FIGURE 1.4 View of a modern temperature controller: (a) rear view (b) front panel.

rate, parity, sampling time, etc., can be changed and saved into the memory of the digital communication card of the controller. The equation of such a controller is given as

$$O_c = A + K_c\varepsilon + \frac{K_c}{\tau_i}\int \varepsilon dt + K_c\tau_d \frac{d\varepsilon}{dt} \tag{1.1}$$

where ε is the deviation of control variable from set point, O_c is the output signal of the controller, K_c is the proportional gain of the controller, τ_i is the integration time constant, τ_d is the derivative time constant, A is the bias value of the output signal of the controller.

These entities will be discussed in more detail in subsequent chapters.

1.6 CONTROL VALVES

Control valves are the final control element in the control loop. Varieties of control valves are used. Most commonly, these are globe valves with the actuator of the valve connected through the stem. The construction of a common pneumatic control valve is shown in Figure 1.5. Varieties of control valves differing in actuators are available.

1.6.1 Control Valve Sizing

The maximum flow through a valve is related to the area of cross section of the valve and the pressure drop. Area is variable when the relative position of the plug and seat changes with the stem travel. Hence, a factor C_v, known as the valve sizing factor, is used to determine the maximum flow through the valve as

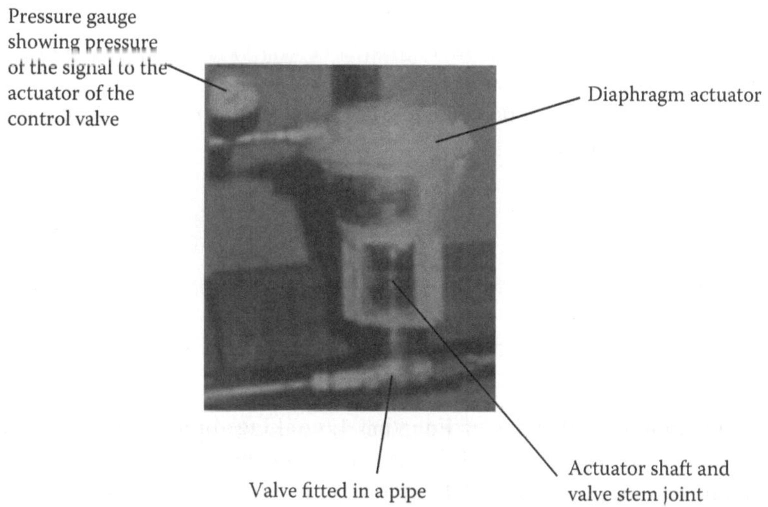

FIGURE 1.5 A pneumatic control valve with a diaphragm actuator.

$$Q = C_v \sqrt{(\Delta P/G)} \tag{1.2}$$

where

Q: maximum flow rate in gallons per minute (gpm) when the valve is fully open
ΔP: pressure drop in pounds per square inch (psi)
G: specific gravity of the fluid flowing

1.6.2 Control Valve Characteristics

Flow through the valve is determined by the length traveled (L) by the stem (stem travel), which is actuated by the input pressure (or electrical signal in the case of a motor-operated valve). Thus, the relationship between flow rate and stem travel is the valve characteristic relationship. If m is the ratio of the flow rate (q) and the maximum flow rate (q_{max}) through the valve, and x is the ratio of stem travel (L) for the rated flow (q) to the maximum stem travel (L_{max}) for the maximum flow rate (q_{max}), there will be varieties of characteristic relationships as listed below:

(a) Linear valve characteristics:

$$m = x. \tag{1.3}$$

While the valve is fully closed, i.e., when $x = 0$, $q = 0$, and $m = 0$, and when the valve is fully open, i.e., $x = 1$, $q = q_{max}$, or $m = 1$.

Thus, for a linear relation, $m = \beta x$, i.e., $\beta = 1$. So the the linear equation is $m = x$.

(b) Equal percentage valve characteristics:

$$m = m_0 \, e^{\beta x} \tag{1.4}$$

where m_0 is the value of m at $x = 0$.

By definition, equal percentage change in the flow rate will have proportional stem travel, i.e.,

$$\Delta m/m = \beta \, \Delta x. \tag{1.5}$$

The differential form is

$$dm/m = \beta \, dx. \tag{1.6}$$

Integrating both sides of Equation 1.6, taking limits at $x = 0$ and $x = 1$, and considering that the flow rate at $x = 0$ is m_0 (non-zero), the valve is assumed to be passing even when the valve is fully shut. This, however, may be considered as true because, practically, the control valve is desired to be neither fully open nor fully shut during control action.

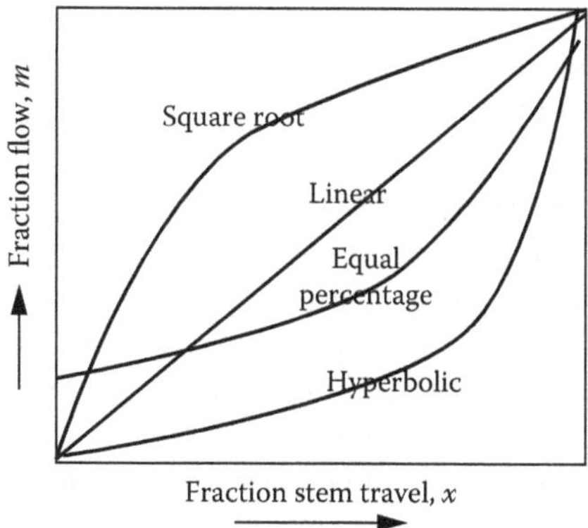

FIGURE 1.6 Control valve characteristic curves.

Thus, integrating Equation 1.6, we obtain

$$\ln m/m_0 = \beta x \text{ or } m = m_0\, e^{\beta x}. \tag{1.7}$$

From this relationship, it is clear that at $x = 0$ (valve fully shut), $m = m_0$, and $x = 1$ (valve fully open), $m = 1$, so β can be evaluated from the following relationship:

$$\beta = \ln (1/m_0) \tag{1.8}$$

(c) Square root control valve characteristics:

$$m = ax^{0.5} \tag{1.9}$$

where a is a constant specific for the size of the valve, and mass fraction m is proportional to the square root of the fraction stem travel x.

(d) Hyperbolic control valve characteristics:

$$m = ax^n \tag{1.10}$$

where a and n are constants specific for the control valve.

The various control valve characteristic curves are presented in Figure 1.6.

1.7 OPEN- AND CLOSED-LOOP SYSTEMS

In a process control system, open loop means while no control action is available in the process. An open-loop system may occur as a result of any one of the following common causes:

(a) There is no controller installed in the process.
(b) The controller is not connected with the control valve.

(c) The control valve is stuck at any position.
(d) Breakdown of the sensor.
(e) Breakdown of the cables.

Open-loop systems as described elsewhere in this book will be considered as a process without control action.

A process with controlling instruments where the controller takes the action is a closed-loop system. In this book, a closed-loop system will mean a negative feedback control loop, which will be discussed in more detail in subsequent chapters. In such a system, the controller is fed with the signal from the transducer, which gives information about the process variable to be controlled. Figures 1.1 and 1.2 are typical closed-loop systems.

1.8 PROCESS PIPING AND INSTRUMENTATION DIAGRAM

A controlled process plant consists of a number of process vessels connected to various fluid-moving machineries (such as pumps, compressors, blowers, etc.) and other processing vessels through pipelines. Instruments are installed at appropriate positions for necessary control of the variables. A comprehensive idea about the control

TABLE 1.1
Symbols of Instruments in a PI Diagram

Instrument	Symbol	Instrument	Symbol
Orifice plate	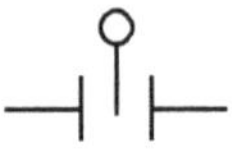	Thermocouple	
Pressure gauge		Level gauge	
Transmitter		Field-mounted controller	
Pneumatic control valve	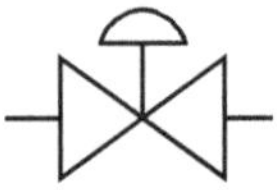	Panel-mounted controller	
Motor-operated control valve (MOV)	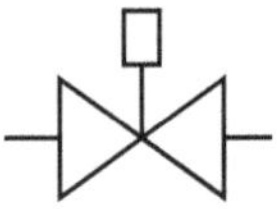	Pressure safety valve	
Pneumatic signal (3–15 psi)		Current (4–20 mA) or voltage signal	
Digital signal		Piping	
Capillary connection		Manually operated valve	

system is best understood by a diagram depicting all the vessels, machinery, and controlling instruments with their corresponding standard symbols. This is known as the process piping and instrumentation diagram or PI diagram. Examples of such PI diagrams are presented in Figures 1.1 and 1.2. Some of the common symbols for the control elements are presented in Table 1.1.

1.9 CONTROL PANEL

Traditionally, indicators, recorders, and controllers are located in a housing known as a control room, which is away (at a distance of a few feet to a few kilometers) from the process. The area in which the indicators, recorders, and controllers are located in the control room is known as the control panel. In modern plants (with computer-controlled processes), separate control panels are absent; rather, these are available instead on the monitors of the computers. The pictorial representation of the PI diagram of a process plant is available on-screen through which the plant operation is monitored by operators in the same way a driver monitors a car through the dashboard. A modern control room is presented in Figure 1.7.

1.10 QUESTIONS AND ANSWERS

EXERCISE 1.1

(a) What are the objectives behind automatic process control?
Answer: See Section 1.1

(b) What are the differences between a transducer and an ordinary measuring instrument?
Answer: An ordinary measuring instrument like a mercury-in-glass thermometer cannot generate any signal (electrical, pneumatic, etc.). A transducer is an instrument that can generate a signal (electrical, pneumatic, etc.) that can be transmitted, recorded, and processed for control action.

FIGURE 1.7 A modern control room.

(c) Name the basic hardware instruments required for process control.
Answer: Sensor, transducer, controller, and final control element. Also see Section 1.2.

(d) Distinguish between a process and variables.
Answer: A process is a piece of equipment or a group of pieces of equipment where certain unit operations or a unit process is carried out. A unit operation involves mass or energy transfer in absence of any chemical reaction or conversion. Examples of processes involving unit operations only are heating a liquid in a tank, heat transfer in a heat exchanger or a furnace, distillation, extraction, etc. A unit process involves an operation with a chemical conversion or reaction, e.g., cracking, reforming, fermentation, etc. A variable is the entity that determines the quality and rate of production or delivery. For example, temperature is a variable that determines not only the degree of heat of a liquid being heated in a tank heater but also determines the density of a liquid or product composition in a chemical reaction.

(e) Distinguish between open- and closed-loop systems.
Answer: See Section 1.7.

(f) Draw a schematic representation of a flow control system.
Answer: See Figure 1.1.

EXERCISE 1.2

In a pressure vessel as shown in Figure 1.8, a gas is entering and leaving. Suggest a pressure control system. Show all the basic instruments required for such a system in a neat sketch.

Answer:

The pressure in the vessel first must be measured by a pressure sensor capable of producing an electrical (or pneumatic) signal. A controller will access this signal and the output of the controller will actuate a control valve. The control strategy is to vary or manipulate the outflow of the gas while pressure increases above or decreases below the set point. In addition to above, the vessel must be fitted with a pressure safety valve, which will allow release of the gas when the pressure in the vessel is too high, i.e., greater than the maximum operating pressure. The high

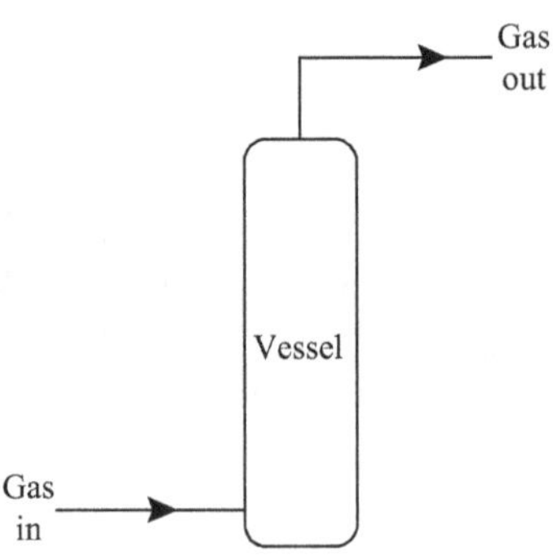

FIGURE 1.8 Open-loop pressure vessel.

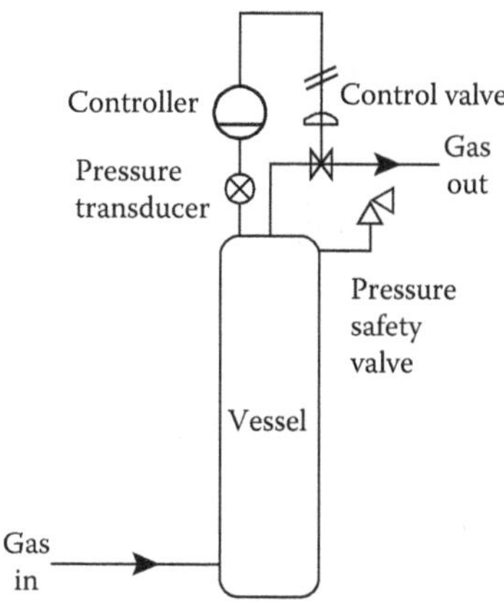

FIGURE 1.9 Pressure control system with all instruments.

pressure otherwise may damage the vessel. The controlled system along with the instruments are shown in Figure 1.9.

EXERCISE 1.3

(a) What are the various primary elements, secondary elements, and signals available?
Answer: Vide Table 1.1.
(b) What is the usual output signal from a pneumatic transducer?
Answer: Air pressure from 3 to 15 psi.
(c) What is the usual signal from an electrical transducer?
Answer: Electric current from 4 to 20 milliamperes (DC).
(d) Why is the signal from an electrical transducer 4 milliamperes as the minimum rather than 0 milliamperes?
Answer: If the transducer sends no signal (i.e., 0 milliamperes) while the process variable is at its minimum level and also when the transducer is damaged or defective, it will be difficult to identify the fault of the transducer. Usually the minimum signal is, therefore, 4 milliamperes to help identify the breakdown of the transducer while the signal falls to zero.

EXERCISE 1.4

In a tank, a liquid level has to be maintained where a liquid enters and leaves as shown in Figure 1.10. Present a control system indicating the instruments in the diagram.
Answer:

The control system has to have a level transducer that will send a signal to a controller. The controller will then actuate a control valve at the exit pipe of the tank so that the effluent flow will increase if the level is higher than the desired set point and vice versa. The PI diagram is shown in Figure 1.11.

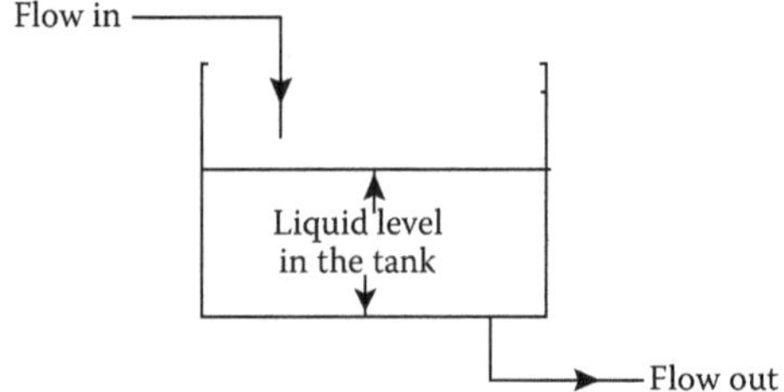

FIGURE 1.10 Open-loop process of the liquid level in a tank.

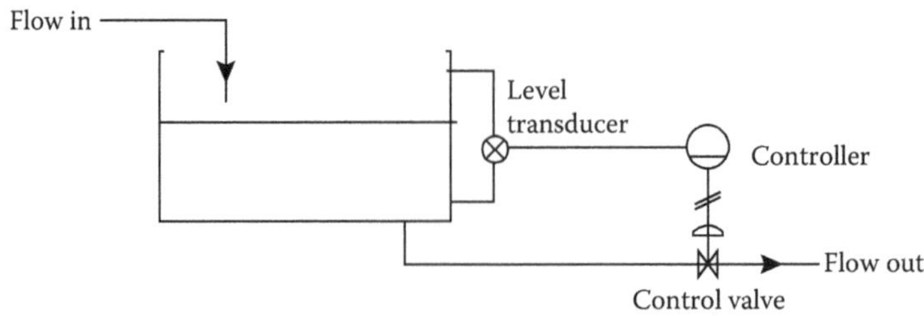

FIGURE 1.11 Level control system with instruments.

EXERCISE 1.5

In a tank, an acidic effluent is neutralized by a stream of alkali, and the neutralized effluent leaves in a continuous manner as shown in Figure 1.12. Present a control system such that the pH of the effluent is controlled.

Answer:

The control system will require a pH transducer, a controller, and a control valve at the alkali flow stream. This is shown in Figure 1.13.

EXERCISE 1.6

An orifice meter has been installed to measure the flow rate of a liquid with a density of 892.7 kg/cum in a schedule 40 4-inch-diameter pipe. The pressure drop across the orifice is found to be 762 mm of mercury. The diameter of the orifice is 57 mm. Determine the flow rate given that the coefficient of the orifice is 0.61.

Solution:

From the steel pipe standard chart of schedule 40 4-inch pipe, the inside diameter (D) is 4.026 inches, i.e., 102.26 mm. The flow rate through the orifice is given as

$$q = \frac{A_o C_o \sqrt{(2g\Delta p/\rho)}}{\sqrt{(1-\beta^4)}} \tag{1.11}$$

where

A_o: cross-sectional area of the orifice = $3.14/4 \times (5.7)^2 \times 10^{-4}$ m^2

C_o: coefficient of discharge through orifice = 0.61

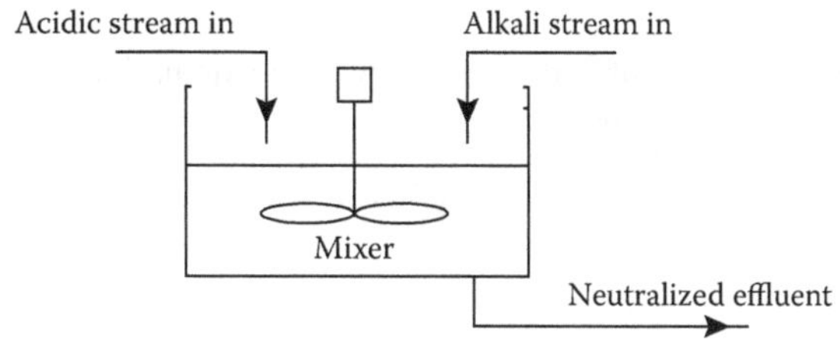

FIGURE 1.12 Open-loop neutralization process.

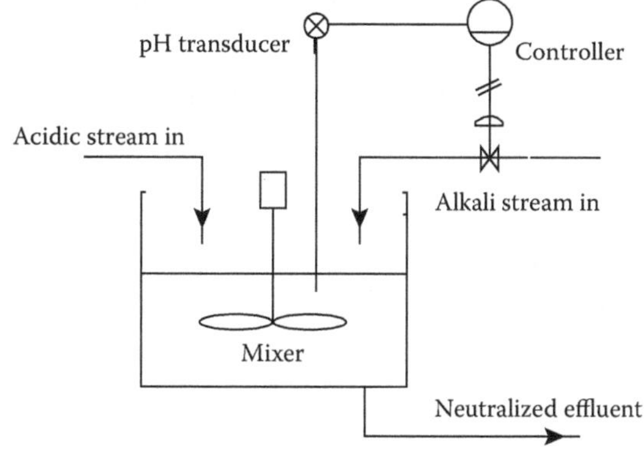

FIGURE 1.13 A pH control system in a continuous neutralizing process.

g: acceleration due to gravity = 9.81 m/sec^2

Δp: pressure drop across the orifice = 762 mm mercury = 762/760 × 1.0133 × 10^4 kg/m^2

ρ: density of the liquid flowing = 8927 kg/m^3

β: the ratio of orifice diameter to pipe diameter = d_0/D = 57/102.26 = 0.5574

so

q: flow rate through the orifice

$$= \frac{25.5\times10^{-4}\times0.61\times\sqrt{(2\times9.81\times1.002\times1.0133\times10^{4})}}{\sqrt{8927\times(1-0.5574^{4})}}$$

$$= 0.00773\ \text{m}^3/\text{sec}.$$

EXERCISE 1.7

Repeat the problem (2) with air as the fluid for which flow rate has to be measured using a similar orifice meter for the same pressure drop where the upstream pressure at the orifice is 1000 mm of mercury at a temperature of 30°C.

Solution:

Because air is a compressible fluid, the flow rate calculation will be given by the following modified equation as

$$q = A_o C_o Y \sqrt{(2g\Delta p)}\ \rho_a \tag{1.12}$$

where

A_o = 3.14/4 × $(5.7)^2 \times 10^{-4}$ m^2

C_o = 0.61

g = 9.81 m/sec^2

Δp = 762 mm mercury = 762/760 × 1.0133 × 10^4 kg/m^2

$$\rho_a = \frac{P_a M \text{ (assuming ideal gas law)}}{RT}$$

P_a = upstream pressure = 1000 mm of Hg absolute

M = mol wt of air = 28.88

T = 273 + 30 = 303 K

R = 0.0832 lit-atm/gmole k

so

ρ_a = 1.507 kg/m^3 and Y = expansion factor given as

$$Y = 1 - \frac{(0.41 + 0.35 \times \beta^4)(1 - P_b/P_a)}{\gamma} \tag{1.13}$$

where

γ = cp/cv of air = 1.44

β = 0.5574 or Y = 0.7676

P_b = downstream pressure = 1000 − 762 = 238 mm absolute

so

q = 0.654 kg/sec

(note that β is included in the factor Y, and q is the mass flow rate).

EXERCISE 1.8

Flue gas is flowing through a one-meter-diameter circular duct of a furnace stack. A pitot tube is used to measure velocity exactly at the center of the duct at a certain height along the stack and reads 14 mm of water. The same pitot measures a pressure of 388 mm of water at the wall of the duct at the same height. If the coefficient of the pitot is 0.98, calculate the flow rate of gas at 15°C and at a pressure of 760 mm of mercury. Assume viscosity of air at 15°C is 2.2 × 10^{-5} Pa·s.

Solution:

A pitot tube measures the pressure at the center and at the wall surface of the duct at the same height, which are impact and static pressure respectively. Hence, the maximum velocity of gas that occurs at the center is given by the relation as

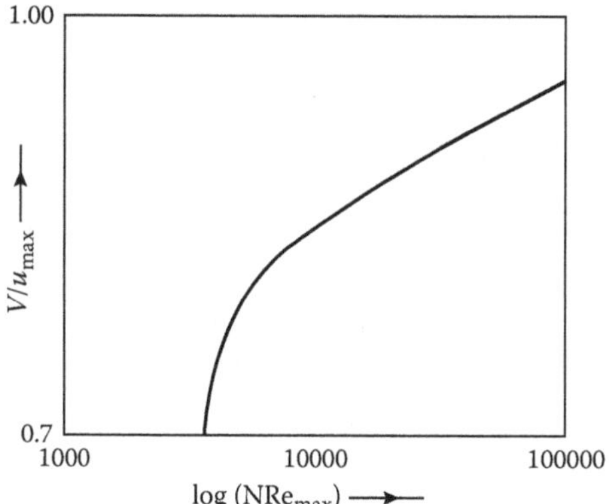

FIGURE 1.14 Experimental plot of V/u_{max} vs. log (NRe_{max}).

$$u_{max} = C_o\sqrt{(2g\Delta p/\rho)} \tag{1.14}$$

where

$C_o = 0.98$

Δp = 14 mm of water = 14 kg/m^2

$\rho = \dfrac{P_a M \text{ (assuming ideal gas law)}}{RT}$

where

P_a = 1 atm + 388 mm water = 1.0521 kg/cm^2

so

ρ = 1.25 kg/m^3.

Hence, u_{max} = 14.5 m/sec.

While u at the wall = 0, that at the other positions between the center and wall will vary between 145 and 0 m/sec. Hence, to calculate flow rate through the duct, average velocity (V) throughout the cross section needs to be calculated. The average velocity may be evaluated from the graph of V/u_{max} vs. NRe_{max} where

$$NRe_{max} = \frac{Du_{max}\rho}{\mu} \tag{1.15}$$

$$= \frac{1 \times 14.5 \times 1.25}{2.2 \times 10^{-5}} = 8.238 \times 10^{5}$$

from Figure 1.14, V/u_{max} = 0.8.

Hence, $q = 3.14 \times 1^2 \times (0.8 \times 14.5)/4 = 9.106$ m^3/sec.

EXERCISE 1.9

A venturi meter is installed in a schedule 40 4-inch pipe to measure the flow rate of water. The pressure drop across the venturi meter is 1.27 m of mercury. Throat diameter of the venturi is 38 mm. Take the coefficient of discharge as 0.98. Determine the flow rate of water.

Solution:

From the standard steel pipe data table, the inside diameter of a schedule 40 4-inch pipe is 4.026 in = 0.10065 m.

The flow rate of water is given by the same equation as in Equation 1.11.

$$q = \frac{A_o C_o \sqrt{(2g\Delta p/\rho)}}{\sqrt{(1-\beta^4)}} \tag{1.11}$$

where

A_o = 3.14/4 × (0.10065)^2 m^2

C_o = 0.98

g = 9.81 m/sec^2

Δp = 1270 mm mercury = 1.69 × 10^4 kg/m^2

ρ = 1000 kg/m^3

β = 38/100.65 = 0.38

so

q = 0.143 m$^{3/}$sec.

EXERCISE 1.10

As shown in Figure 1.15, a control valve has been installed in a pipeline having a 25 mm inside diameter and is 10 meters in length. The control valve has a coefficient of 4 (C_v). Determine the flow rate of water through the pipe when the control valve is 50% open.

It is given that water has a viscosity of 1.5 milli Pa·s. Assume the valve characteristic follows linearity between y and x where $y = q/q_{max}$ and $x = L/L_{max}$. L and q are the stem travel and flow rate, respectively.

Solution:

Maximum flow rate is given by the valve as

$$\begin{aligned} q &= C_v \sqrt{(\Delta p_v/G)} \\ &= 4\sqrt{(\Delta p_v/1)} \text{ gpm} \end{aligned} \tag{1.16}$$

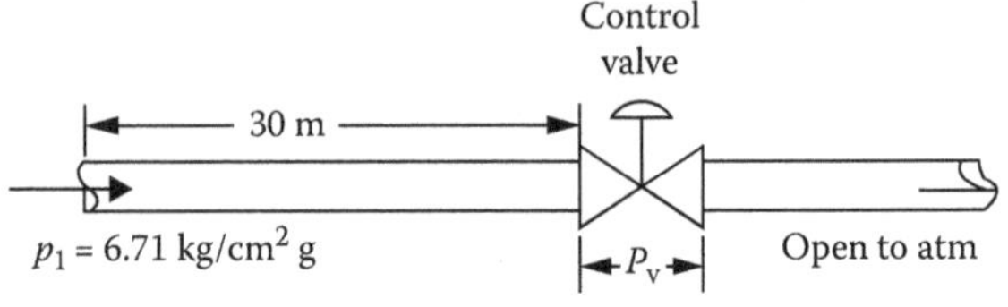

FIGURE 1.15 Control valve installed over a pipe.

where Δp_v = pressure drop across the valve = p_1 – pressure drop in the pipe

$$= p_1 - 2f\,u^2\,L_p\,\rho/Dg \tag{1.17}$$

where

f = is the Fanning friction factor
u = is the average velocity
L_p = 30 m
ρ = 1000 kg/m^3
D = 0.025 m
g = 9.81 m/sec^2.

Assume flow rate = 20 gpm.

So

$$u = \frac{20 \times 3.78 \times 4 \times 10^{-3}}{60 \times \pi \times (0.025)^2}$$

$$= 2.52 \text{ m/sec (as 1 gallon} = 3.78 \times 10^{-3}\,\text{m}^3)$$

$$\text{NRe} = \frac{Du\rho}{\mu} = \frac{0.025 \times 2.52 \times 1000}{1.5 \times 0.001} = 42000$$

from friction factor chart for Newtonian fluid (available from any handbook on fluid dynamics), f = 0.0052.

Hence, pressure drop in the pipe = $2f\,u^2\,L_p\,\rho/Dg$

$$= 2 \times 0.0052 \times (2.52)^2 \times 30 \times 1000/0.025 \times 9.81 = 8078 \text{ kg/m}^2 = 11.7 \text{ psi}$$

so

$$\Delta p_v = 6.71 \text{ kg/cm}^2 - 11.7 \text{ psi} = 100 - 11.7 = 88.3 \text{ psi.}$$

Hence, $q_{max} = 4\sqrt{(\Delta p_v/1)} = 4\sqrt{(88.3/1)}$ = 37.58 gpm

so

$$y = q/q_{max} = 20/37.5 = 0.53 = x \text{ (for linear valve)}$$

Similarly, assuming other flow rates, y and x can be determined as summarized below:

gpm (q)	m/sec (u)	Reynold's No. (NRe)	Friction Factor (f)	Line pr. Drop psi (Δp_T)	Valve pr. Drop psi (Δp_v)	gpm (q_{max})	q/q_{max} (y)	L/L_{max} (x)
10	1.26	21000	0.0055	3.09	96.9	39.37	0.25	0.25
20	2.52	42000	0.0052	11.7	88.3	37.58	0.53	0.53
30	3.76	62666	0.005	25.08	74.9	34.6	0.866	0.866

Hence, flow rate at 50% open valve is about 19 cc/min from the plot of q and x.

EXERCISE 1.11

Repeat the problem for other control valves as listed below:

(1) Equal %, $\text{Log}_e\,(y/0.01) = 10x$
(2) Hyperbolic, $y = 1/(10^{-9}x)$
(3) Square root, $y = \sqrt{x}$

Solution:

q	v	y	Equal%, x	Hyperbolic, x	Square Root, x
10	1.26	0.25	0.32	0.67	0.06
20	2.52	0.53	0.39	0.90	0.28
30	3.76	0.866	0.45	0.98	0.75

Different values for the parameters of the above control valves will generate different characteristics of q and x.

EXERCISE 1.12

In a plant, a product is obtained from a gaseous raw material that is preheated by a heat exchanger followed by a reaction in a catalytic fixed-bed reactor. The reaction is exothermic, and the hot product has to be cooled by counter-currently preheating the cold feed gas followed by product cooling. Draw a pragmatic process-instrumentation-diagram of the plant.

Answer:

The feed rate is maintained at the controlled flow rate, and the temperature of the reactor is controlled by manipulating the coolant flow rate through the jacket surrounding the reactor. The temperature of the product stream exiting the reactor is measured and transduced to the temperature controller, which actuates the control valve to manipulate the coolant flow rate (Figure 1.16).

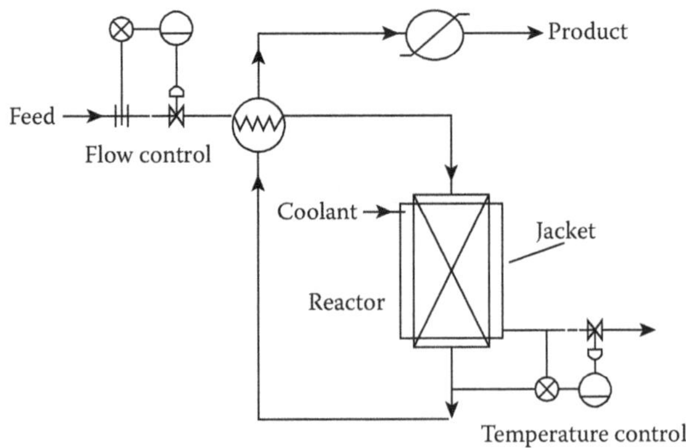

FIGURE 1.16 Instrumentation diagram of Exercise 1.13.

EXERCISE 1.13

The EMF of a thermocouple with its hot junction at different temperatures as tabulated below and cold junction at 0°C are observed as

Hot junction temperature °C	0	20	50	70	90	110
Thermo EMF (mV)	0.00	0.50	1.8	2.1	3.5	5.0

If the cold junction of the thermocouple is placed in a bath at 20°C, what will be the EMF of the thermocouple for the above hot junctions?

Answer:

According to the law of intermediate temperature of a thermocouple, i.e.,

$$E(T_h, T_{c1}) = E(T_h, T_{c2}) + E(T_{c2}, T_{c1})$$

so

$$E(T_h, 0) = E(T_h, 20) + E(20, 0).$$

Hence, $E(T_h, 20) = E(T_h, 0) - E(20, 0) = E(T_h, 0) - 0.50$

These values are tabulated next.

Hot junction temperature °C	0	20	50	70	90	110
Thermo EMF $E(T_h, 20)$, mV	–0.5	0.0	1.3	1.6	3.0	4.5

EXERCISE 1.14

If two thermocouples, one a chromel-constantan and the other an iron-constantan, both having the same hot and cold junction temperatures develop 2.8 mV and 1.5 mV, respectively, what will be the value of EMF generated in a thermocouple made of chromel-iron having the same hot and cold junction temperatures?

Answer:

According to the law of intermediate metal wires,

$$E(A,B) = E(A,C) - E(B,C)$$

i.e.,

$$E(\text{chromel-iron}) = E(\text{chromel-constantan}) - E(\text{iron-constantan})$$

$$= 2.8 + 1.5 = 4.3 \text{ mV}$$

EXERCISE 1.15

A thermometer having a scale from 0 to 100°C with ±1% accuracy reads 50°C of a liquid in a bath. What should the true bath temperature be?

Answer:

Accuracy = (True measurement – indicated value)/span × 100%
For the problem, span = 100 – 0 = 100°C
Therefore, true temperature = 50 ± 0.01 × 100 = 51 or 49°C.

EXERCISE 1.16

A float-tape level gauge is used to indicate the level of liquid in a tank as shown in Figure 1.17. The rotation of the pulley during the up and down displacement of the float drives the shaft of a small resistance regulator with a maximum resistance of R kΩ and a constant supply voltage of 24 volt DC. Determine the resistance required to indicate the lowest level as 4 mA and the maximum level as 20 mA current signals.

Answer:

At the minimum level $h = h_{min}$, the resistance is R_0 ohm, then

$$I = 4 \text{ mA} = 4 \times 10^{-3} \text{ amp} = V/R_0$$

or

$$R_0 = 24/(4 \times 10^{-3}) = 6 \text{ k}\Omega$$

and at the maximum level, h_{max}, the resistance is R_m ohm, then

$$I = 20 \text{ mA} = 20 \times 10^{-3} \text{ amp} = V/R_m$$

or

$$R_m = 24/(20 \times 10^{-3}) = 1.2 \text{ k}\Omega.$$

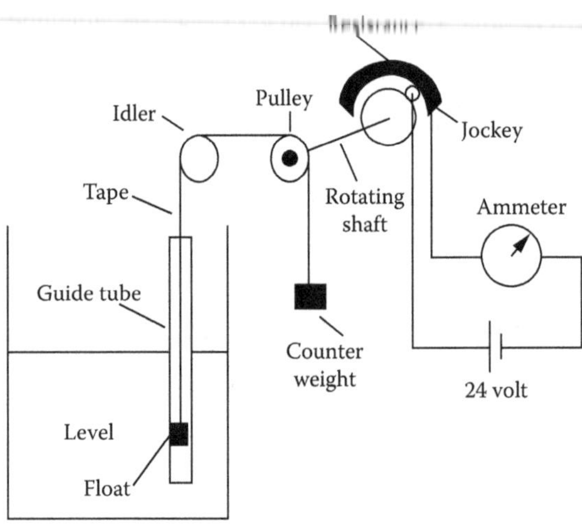

FIGURE 1.17 Float-tape level transducer in Exercise 1.16.

EXERCISE 1.17

For the above float-tape transducer problem, if the maximum and minimum levels are 0.4 m and 10 m, respectively, determine the current transduced for every meter of level until the maximum is achieved. Assume that the change of resistance is proportional to the change in level.

Answer:

Take the proportional relationship between the level (h meter) and the resistance R in kΩ as

$$R = a + bh$$

As $R = 6$ kΩ at $h = 0.4$ m so that the current will be 4 mA and $R = 1.2$ kΩ at $h = 10$ m so that current will be 20 mA.

So

$$6 = a + b0.4$$

and

$$1.2 = a + b10.$$

Solving, $a = 6.2$ and $b = -0.5$

So

$$R = 6.20 - 0.5h.$$

The values of h, R, and I are determined for each meter of depth and presented in Table 1.2 and Figure 1.18.

TABLE 1.2
Calibration of Level Transducer

Level, m	Volt	Resistance, kΩ	Current, mA
0.0	24	6	4
0.4	24	6	4
1.0	24	5.7	4.21
2.0	24	5.2	4.61
3.0	24	4.7	5.10
4.0	24	4.2	5.71
5.0	24	3.7	6.48
6.0	24	3.2	7.50
7.0	24	2.7	8.9
8.0	24	2.2	10.91
9.0	24	1.7	14.11
10.0	24	1.2	20.00

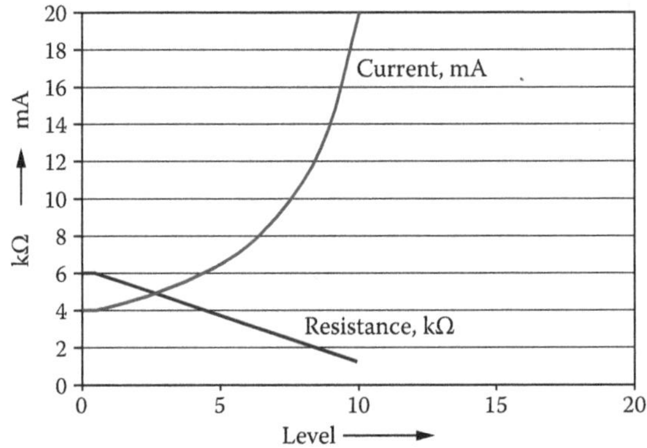

FIGURE 1.18 Variation of resistance and current with change in level.

EXERCISE 1.18

A strain gauge is to be constructed for pressure measurement as shown in Figure 1.19. The strain element has a resistance R_1, and the Wheatstone bridge resistances are R_2, R_3, and R_4 and are known and assumed to be fixed at 1 kΩ each. The current in the unbalanced bridge is measured by an ammeter in a circuit with a resistance of 0.5 kΩ. Determine the relationship between the current and resistance of the strain element.

Answer:

The out of balance potential is determined as

$$E = 24\ \{R_2/(R_2 + R_1) - R_4/(R_3 + R_4)\}$$

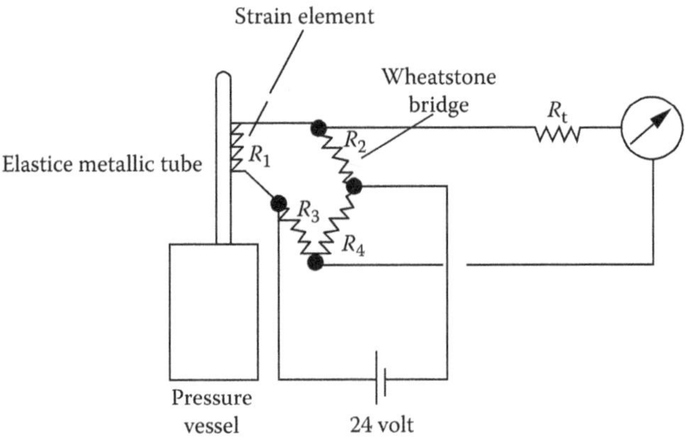

FIGURE 1.19 Strain gauge measurement for Exercise 1.18.

And the current

$$I = E/R_t = E/0.5 = 2E = 48\ \{1/(R_1 + 1) - 0.5\}$$

i.e.,

$$I = \{48/(R_1 + 1) - 24\}\ \text{mA}$$

where R_1 is the resistance of the strain element mounted on the elastic element strained by the pressure in the vessel. The other resistances of the Wheatstone bridge are each 1 kΩ, and the resistance in the measuring arm is 0.5 kΩ.

EXERCISE 1.19

For the previous exercise, if the minimum and maximum pressure in the vessel to be measured are 100 and 10 psi, respectively, determine the pressure, resistance, and current for pressures of 10, 20, 40, 60, 80, and 100 psi.

Answer:

For $P = 100$ psi, $I = 20$ mA, and for $P = 10$ psi, $I = 4$ mA.

Taking the linear relationship of pressure and strain relationships, it can be assumed that the pressure in the vessel will be linearly related with the resistance R_1.

Hence,

$$R_1 = a + bP.$$

From the current and resistance of the strain element, we get

$$\text{I} = \{48/(\text{R1} + 1) - 24\}\ \text{mA}.$$

So $20 = 48/(R_1 + 1) - 24$ for $P = 100$ psi or $R_1 = 1/11 = 0.0909$ kΩ and $4 = \{48/(R_1 + 1) - 24\}$ for $P = 10$ psi

So, $R_1 = 5/7 = 0.7141$ kΩ

Hence,

$$0.0909 = a + b100$$

and

$$0.7141 = a + b10.$$

Solving for a and b

$$a = 0.7834 \text{ and } b = -0.006925.$$

So $R_1 = 0.7834 - 0.006925P$

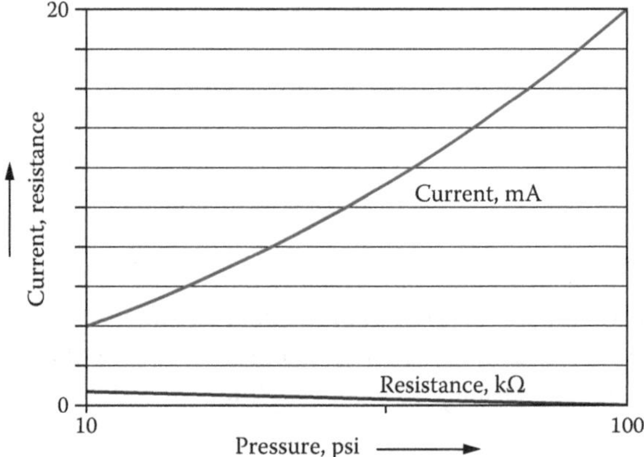

FIGURE 1.20 Variation of resistance and current for changes in pressure.

The values of R_1 and current are evaluated for different pressures and presented in the following table and also presented in Figure 1.20.

Pressure, psi	Strain Element Resistance, kΩ	Current, mA
10	0.71415	4.00
20	0.6449	5.18
40	0.5064	7.864
60	0.3679	11.09
80	0.2294	15.043
100	0.0909	20.00

EXERCISE 1.20

A current-to-pressure converter as shown in Figure 1.21 uses a solenoid and a globe valve where a supply pressure of air at 20 psig is continuously available. The solenoid lifts the valve stem proportionally to the current applied to it such that when the current signal to the solenoid is 20 mA, the pressure delivered is 15 psi, and when the

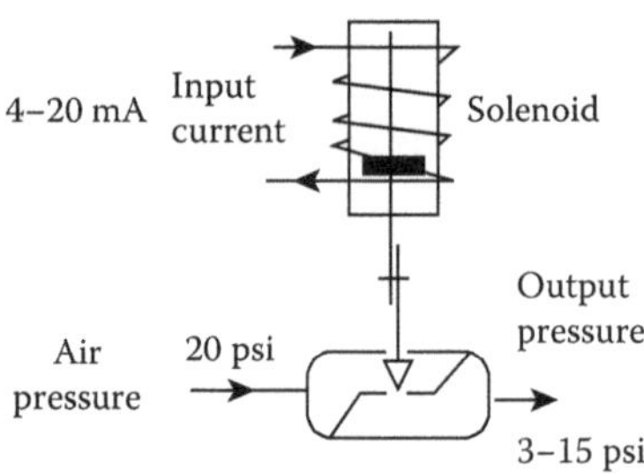

FIGURE 1.21 A current-to-pressure converter.

current signal becomes 4 mA, the pressure delivered is 3 psi. Determine the current and delivered pressure of air.

Answer:

As $I = 20$ mA, $P = 15$ psi, and when $I = 4$ mA, $P = 3$ psi.

Taking a linear relation as

$$P = a + bI$$

$$a + b20 = 15$$

and

$$a + 4b = 3.$$

Solving

$$P = 3/4I.$$

Current and pressure is evaluated as

Current, mA, *I*	**Pressure Signal Delivered, psi, *P***
4	3
6	4.5
8	6.0
10	7.5
12	9.0
16	12.0
18	13.5
20	15.0

2 Open-Loop Process Dynamics and Transfer Functions

2.1 LAPLACIAN MATHEMATICS

Mathematical analysis of a control system is desirable for the predefined design of a control system. Because a control variable has to be tracked with time, representation of this variable in relation to other variables, such as load and set point, results in differential equations that can be handled comfortably by Laplace's method of solution. Therefore, some of the basic Laplacian mathematical tools that we will frequently use in control system analysis are given next.

If $f(t)$ is a function of time (t), the Laplace transformation of this function is given as

$$Lf(t) = \int_0^\infty f(t)e^{-st}\,dt \tag{2.1}$$

where s is the Laplacian operator. Laplace transformations of some of the time functions are given below.

However, the Laplace transformation method is applicable only for the solution of linear differential equations. A linear differential equation is given as

$$a_n \frac{d^n y}{dt} + a_{n-1} \frac{d^{n-1} y}{dt} + a_{n-2} \frac{d^{n-2} y}{dt} + a_1 \frac{dy}{dt} + a_0 y = x \tag{2.2}$$

where y and x are the output and input functions, respectively, of time t and a_0, a_1, a_n are constant coefficients. Such a linear differential equation can be solved by the Laplace transformation method. The Laplace transformation of a function $f(t)$ is defined as listed below.

Some of the Common Laplace Transformations:

Function	Laplace Transform
$f(t) = A$	A/s
$f(t) = At$	A/s^2
$f(t) = At^n$	$An!/s^{n+1}$
$f(t) = e^{-at}$	$1/(s + a)$

$f(t) = Ae^{-at}$	$A/(s + a)$
$f(t) = At^n e^{-at}$	$An!/(s + a)^{n+1}$
$f(t) = A \sin kt$	$Ak/(s^2 + k^2)$
$f(t) = A \cos kt$	$As/(s^2 + k^2)$
$f(t) = e^{-at} A \sin kt$	$Ak/\{(s + a)^2 + k^2\}$
$f(t) = e^{-at} A \cos kt$	$A(s + a)/\{(s + a)^2 + k^2\}$

Some important properties of Laplace transformations:

1. Law of addition:

$$L\,\{a(t) + bg(t)\} = aL\,f(t) + bL\,g(t) \tag{2.3}$$

2. Laplace transformation of nth derivative of $f(t)$:

$$L\frac{d^n f(t)}{dt^n} = s^n f(s) - s^{n-1} f(0) - s^{n-2} f^1(0) - sf^{n-2}(0) - sf^{n-1}(0) \tag{2.4}$$

where $f^1(0), f^2(0), f^n(0)$ are the first, second, and nth derivatives of $f(t)$ at $t = 0$.

3. Initial value theorem:

$$\lim f(t) = \lim sf(s) \tag{2.5}$$
$$\text{as } t \to 0 \quad \text{as } s \to \infty$$

4. Final value theorem:

$$\lim f(t) = \lim sf(s) \tag{2.6}$$
$$\text{as } t \to \infty \quad \text{as } s \to 0$$

[illegible] FIRST ORDER SYSTEMS

Consider a mercury-in-glass thermometer that is used to measure the temperature of a bath varying with time as shown in Figure 2.1. Let the temperature of the bath be x and the corresponding indication by the thermometer be y as the column of mercury in the capillary at any instant of time t. The thermal expansion of mercury within the thermometer takes place as a result of the absorption of heat through the thermometer bulb dipped in the liquid of the bath. Therefore, the enthalpy balance at any instant of time is given by the following relationship:

rate of heat input to the bulb – rate of heat exit from the bulb
= rate of accumulation of heat in the mercury in the bulb.

So,

$$UA_b(x - y) - 0 = \frac{d(m\,C_p y)}{dt}$$

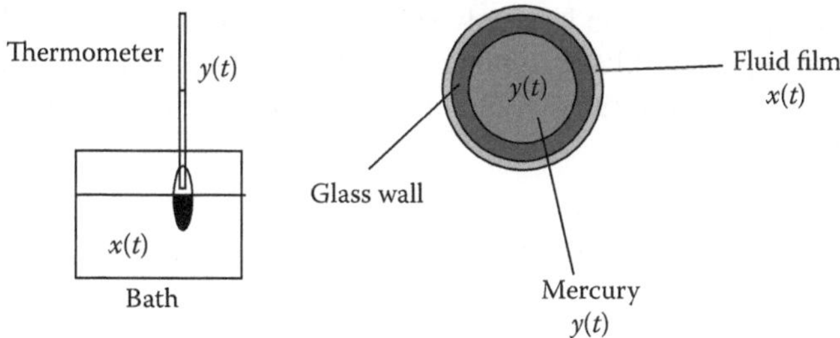

FIGURE 2.1 Thermometer in a bath and bulb sections.

or

$$UA_b(x-y)-0=m\,C_p\frac{dy}{dt} \tag{2.7}$$

where

U = overall heat transfer coefficient of the bulb–mercury surface
A_b = surface area of the bulb in contact with the liquid in the bath
m = combined mass of the glass and mercury in the bulb
C_p = combined specific heat of the glass and mercury in the bulb

At steady state, when the thermometer indication equals the bath temperature, i.e.,

$$\frac{dy}{dt}=0,\ \text{then}\ x=xs,\ y=ys,\ \text{and}\ xs=ys \tag{2.8}$$

where the subscript s indicates steady-state values.

The above equation at steady state becomes

$$UA_b\,(xs-ys)-0=0. \tag{2.9}$$

Subtracting Equation 2.9 from Equation 2.7 and taking $X = x - xs$ and $Y = y - ys$, the relationship becomes

$$\tau\frac{dY(t)}{dt}+Y(t)=X(t) \tag{2.10}$$

where

$$\tau=\frac{mC_p}{UA_b}.$$

The parameter τ is called the time constant of the thermometer and is measured in time units. Equation 2.10 is a first-order differential equation; hence, the system is called a first-order system. Now, if the bath temperature suddenly shoots up to

"A" units of temperature from its steady value xs, then $X(t) = x(t) - xs = A$ and $Y(t)$ is obtained from Equation 2.10 by solving using the Laplace transformation method. Taking the Laplace transformation of both the left- and right-hand sides of the differential Equation 2.10 as

$$L\left\{\tau\frac{dY(t)}{dt}+Y(t)\right\}=L\{X(t)\} \tag{2.11}$$

or

$$s\tau Y(s) - Y(0) + Y(s) = X(s) \tag{2.12}$$

because

$$Y(0) = Y(0)_{t=0} = y(0) - y(0) = 0$$

or

$$\frac{Y(s)}{X(s)}=\frac{1}{(\tau s+1)}. \tag{2.13}$$

As the bath temperature is changed as a step function by a magnitude of A, then

$$\begin{cases} X(t)=0 \text{ at time } =0, \text{ and earlier, } t<0 \\ X(t)=A \text{ at time } =0, \text{ and more, } t>0 \end{cases}$$
$$X(s)=\frac{A}{s}. \tag{2.14}$$

Therefore,

$$Y(s)=\frac{A}{s(\tau s+1)} \tag{2.15}$$

or

$$Y(s)=A\left\{\frac{1}{s}-\frac{1}{s+\frac{1}{\tau}}\right\}. \tag{2.16}$$

Inverting the Laplace transformations of both sides,

$$Y(t) = A\ \{1 - e^{-t/\tau}\} \tag{2.17}$$

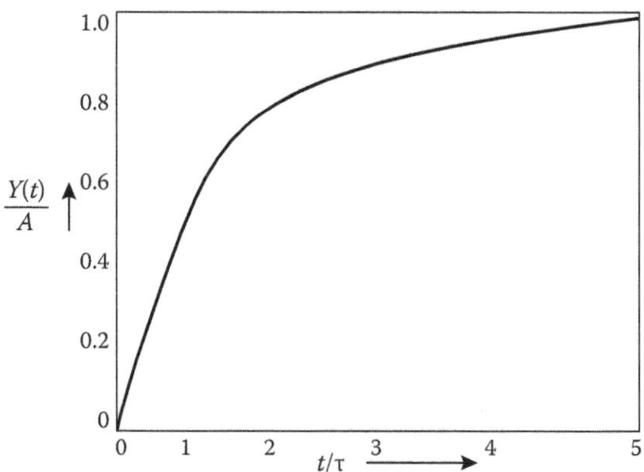

FIGURE 2.2 Thermometer reading resulting from a step change in bath temperature.

so that,

$$y(t) = ys + A\ \{1 - e^{-t/\tau}\} \tag{2.18}$$

At $t = \tau$,

$$Y(t) = A\ \{1 - e^{-1}\} = 0.632A \tag{2.19}$$

Thus, at one time constant period, the indication of the thermometer will be 63.2% of the sudden change of bath temperature A above the initial indicated value (ys or xs). Thus, it is theoretically proved that a thermometer cannot read without lag or delay. This dynamic lag is inevitable for all instruments, and this fact can be proved based on fundamental measuring phenomena, which may be a first-, second-, third-, or higher-order differential equation or any other model. The response relationship (Equation 2.17) is graphically explained in Figure 2.2.

2.2.1 First-Order Lag

Consider a process that consists of a steam-heated tank as shown in Figure 2.3, where a liquid enters and leaves at a constant rate, w mass/time. The purpose of the process is to heat the incoming cold liquid at a temperature T_i to a desired elevated temperature T_0.

Here, we can prove mathematically that the desired T_0 cannot be achieved instantaneously even if the steaming rate is quickly raised. Let the volume of the tank be V, density of the liquid ρ and the specific heat C_p. Then, the enthalpy balance relationship is given as

$$wC_pT_i - wC_pT_0 + q = \rho VC_p\frac{dT_0}{dt} \tag{2.20}$$

where q is the rate of heat supplied by the steam coil.

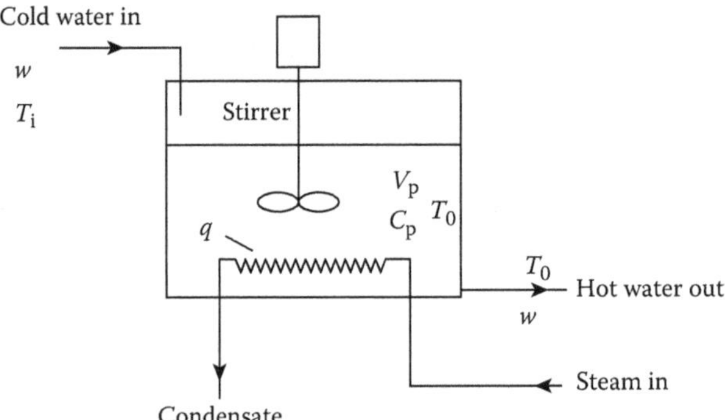

FIGURE 2.3 Steam-heated tank.

Now, at steady state,

$$T_0 = T_s = T_{is},\ q = q_s$$

and

$$dT_0/dt = 0.$$

Then the heat balance equation becomes

$$wC_pT_{is} - wC_pT_{0s} + q_s = 0 \tag{2.21}$$

Subtracting Equation 2.21 from Equation 2.20, the relation becomes

$$wC_p(T_i - T_{is}) - wC_p(T_0 - T_{0s}) + (q - q_s) = \rho VC_p \frac{d(T_0 - T_{0s})}{dt}. \tag{2.22}$$

Designating $X(t) = T_i - T_{is}$, $Y(t) = T_0 - T_{0s}$ and $Q(t) = q - q_s$, Equation 2.22 becomes

$$wC_p\, X(t) - wC_pY(t) + Q(t) = \rho VC_p \frac{dY(t)}{dt}. \tag{2.23}$$

Taking the Laplace transformation of both the left and right sides of Equation 2.23,

$$wC_p\, X(s) - wC_pY(s) + Q(s) = \rho VC_p\, \{sY(s) - Y(0)\} \tag{2.24}$$

Because $Y(0) = 0$, hence,

$$X(s) + \frac{Q(s)}{w\,C_p} = \left\{\frac{\rho V}{w}s + 1\right\}Y(s) \tag{2.25}$$

or

$$X(s)+\frac{Q(s)}{w\,C_p}=\{\tau s+1\}Y(s) \tag{2.26}$$

where $\tau = \rho V/w$ is the time constant of the process, or

$$Y(s)=\frac{X(s)+\dfrac{Q(s)}{wC_{\mathrm{p}}}}{\{\tau s+1\}}. \tag{2.27}$$

If T_i is unchanged at T_{is}, then $X(t) = 0$, and then

$$Y(s)=\frac{\dfrac{Q(s)}{wC_{\mathrm{p}}}}{\{\tau s+1\}}. \tag{2.28}$$

Now, if the steam rate is suddenly increased from its steady initial value of q_s by A units, i.e., $Q(t) = A$, then solving Equation 2.28, $Y(t)$ is obtained as

$$Y(t)=\frac{A(1-e^{(-t/\tau)})}{wC_{\mathrm{p}}}. \tag{2.29}$$

Thus, in order to raise the temperature of the outgoing liquid to T_0 from its initial temperature T_{0s}, a certain amount of time will be spent. It reaches 63.2% of A/wC_p of temperature from T_{0s} after one time constant. This is shown in Figure 2.4.

If the steam rate is held unchanged while the inlet temperature T_i is suddenly disturbed by a similar step change, the response of the tank temperature will be obtained. Students can verify this following the same procedure.

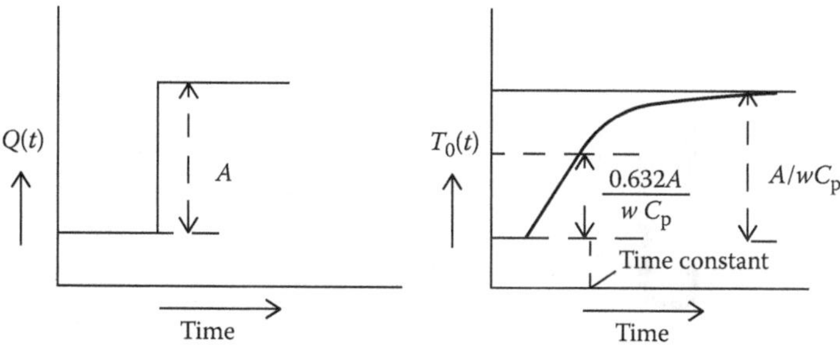

FIGURE 2.4 First-order behavior of the tank temperature $T_0(t)$ of the steam-heated tank resulting from a step change in heating rate $Q(t)$.

2.3 SECOND-ORDER SYSTEMS

Consider a manometer measuring fluid pressure $x(t)$, which is varying with time and indicated by the manometer as $y(t)$. One end of the manometer is open to the atmosphere as shown in Figure 2.5.

Let the length of the manometric fluid in the manometer be L, density ρ_m, and mass m. By dynamic force balance, the following relationship is obtained:

$$m\frac{d^2y(t)}{dt^2} = x(t)a - aC_d\frac{dy(t)}{dt} - a\rho_m g y(t) \tag{2.30}$$

where a is the column cross section and C_d is the drag coefficient of the manometric fluid and the tube wall. Considering $X(t) = x(t) - xs$ and $Y(t) = y(t) - ys$, then Equation 2.20 is modified as

$$m\frac{d^2Y(t)}{dt^2} + aC_d\frac{dY(t)}{dt} + a\rho_m gY(t) = X(t)a. \tag{2.31}$$

This is a second-order differential equation and is expressed as

$$\tau^2\frac{d^2Y(t)}{dt^2} + 2\tau\xi\frac{dY(t)}{dt} + Y(t) = \frac{X(t)}{\rho_m g}.$$

Taking the Laplace transformation,

$$\frac{Y(s)}{X(s)} = \frac{\frac{1}{\rho_m g}}{\tau^2 s^2 + 2\tau\xi s + 1} \tag{2.32}$$

where

$$\tau = \sqrt{\left(\frac{m}{\rho_m a g}\right)}.$$

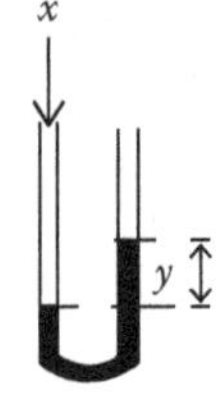

U tube manometer

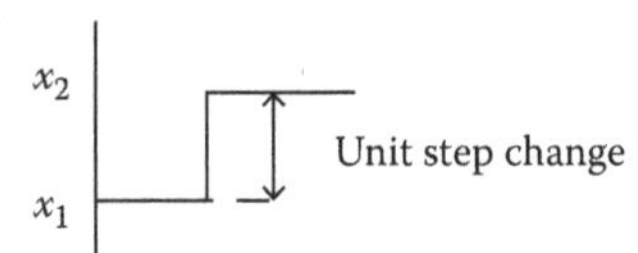

Unit change in pressure instantaneously or unit step change

FIGURE 2.5 Pressure $x(t)$ is indicated as reading y in a manometer.

is the second-order time constant of the manometer and the damping coefficient (ξ) is given as

$$\xi = \frac{Cd\sqrt{a}}{2\sqrt{(\rho_m mg)}}.$$

For simplicity, we assume $\rho_m g = 1$ such that the right-hand side of Equation 2.32 is $X(t)$. If the fluid pressure $x(t)$ is suddenly changed from its steady value x_s to another by A units, then $X(t) = A$ and the corresponding $Y(t)$ can be found by solving the differential Equation 2.32, depending upon the value of ζ; there are three types of solutions as given below:

For $\xi < 1$,

$$Y(t) = A\left[1 - \frac{e^{(-t\xi/\tau)}}{\sqrt{(1-\xi^2)}}\sin\left(wt + \tan^{-1}(w\tau/\xi)\right)\right] \quad (2.33)$$

where, $w = \sqrt{(1 - \xi^2)}/\tau$.
For $\xi > 1$,

$$Y(t) = A\,[1 - e^{(-t\xi/\tau)}\,\{\cosh wt + \xi/\sqrt{(\xi^2 - 1)}\sinh wt\}] \quad (2.34)$$

where, $w = \sqrt{(\xi^2 - 1)}/\tau$.
For $\xi = 1$,

$$Y(t) = A\,[1 - e^{(-t\xi/\tau)}\,\{1 + t/\tau\}]. \quad (2.35)$$

These different types of dynamic indications or responses to change in true value of pressure are presented graphically in Figure 2.6. Equations 2.33, 2.34, and 2.35 are underdamped, overdamped, and critically damped behaviors of the instrument. Most of the control systems behave like an underdamped system, and control systems are characterized by certain parameters as indicated in Figure 2.7.

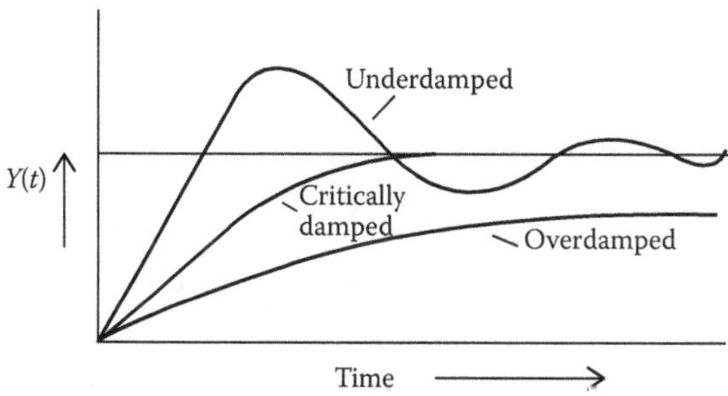

FIGURE 2.6 Second-order behavior of a manometer.

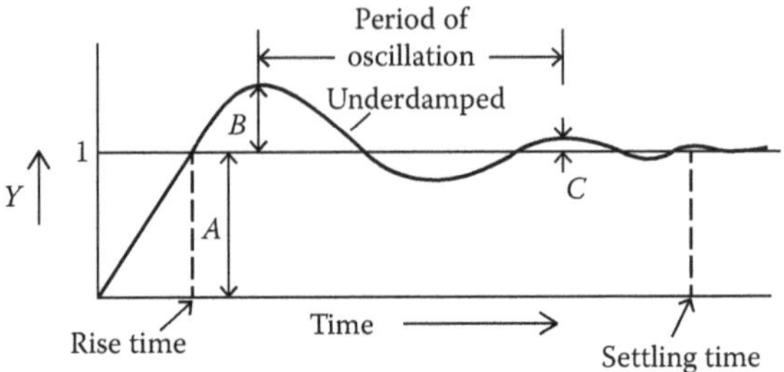

FIGURE 2.7 Parameters of an underdamped system.

where A is the unity gain or the ultimate value of Y, B is the overshoot, and C is the decay. The overshoot ratio is B/A, and the decay ratio is C/B. The rise time and settling time are also important parameters used for the response analysis of a controlled system. For instance, control performance is considered to be good when the overshoot ratio is low and the decay ratio is below 0.25 (quarter decay) with shorter rise and settling times.

2.3.1 Second-Order Lags of Tank in Series

2.3.1.1 Noninteracting System

Consider two tanks connected in series as shown in Figure 2.8, where liquid enters in the first tank at a volumetric rate f_1, then exits to the next tank by gravity at a volumetric rate f_2, and leaves the second tank at a volumetric rate f_3. If the tanks have different areas at cross sections A_1 and A_2, respectively, and the liquid has a constant density ρ, then material balances of the liquid are given by the following relationships:

$$\text{Tank 1} \quad f_1 - f_2 = A_1 \frac{dh_1}{dt} \tag{2.36}$$

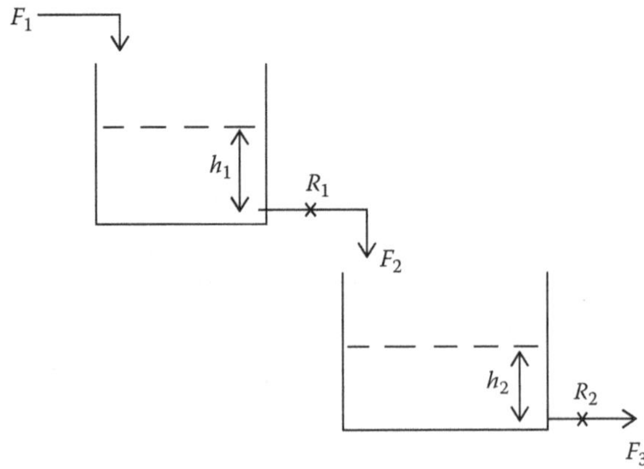

FIGURE 2.8 Two tanks connected in series—a noninteracting system.

$$\text{Tank 2} \quad f_2 - f_3 = A_2 \frac{dh_2}{dt} \tag{2.37}$$

where h_1 and h_2 are the levels of the liquid in the first and second tanks, respectively. Because f_2 and f_3 are the flow rates resulting from gravity, these are dependent on the levels h_1, h_2, and the resistance to flow through the exit pipes. Thus, additional approximate relationships of flow are obtained as

$$\text{Tank 1} \quad f_2 \cong h_1/R_1 \tag{2.38}$$

$$\text{Tank 2} \quad f_3 \cong h_2/R_2 \tag{2.39}$$

where R_1 and R_2 are the resistances of flow in the exit pipes. Thus, replacing f_2 and f_3 by Equations 2.38 and 2.39, Equations 2.36 and 2.37 become

$$f_1 - \frac{h_1}{R_1} = A_1 \frac{dh_1}{dt} \tag{2.40}$$

$$f_2 - \frac{h_2}{R_2} = A_2 \frac{dh_2}{dt}. \tag{2.41}$$

At steady state, $f_1 = f_{1s}, f_2 = f_{2s}$, and the derivatives of h_1 and h_2 are zero. Then

$$f_{1s} - \frac{h_{1s}}{R_1} = 0 \tag{2.42}$$

$$f_{2s} - \frac{h_{2s}}{R_2} = 0. \tag{2.43}$$

Subtracting Equations 2.42 and 2.43 from Equations 2.40 and 2.41, respectively, the equations become

$$\left(f_1 - f_{1s}\right) - \frac{(h_1 - h_{1s})}{R_1} = A_1 \frac{d(h_1 - h_{1s})}{dt} \tag{2.44}$$

$$\left(f_2 - f_{2s}\right) - \frac{(h_2 - h_{2s})}{R_2} = A_2 \frac{d(h_2 - h_{2s})}{dt}. \tag{2.45}$$

Designating $f_1 - f_{1s} = F_1(t), f_2 - f_{2s} = F_2(t)$, $h_1 - h_{1s} = H_1(t)$, and $h_2 - h_{2s} = H_2(t)$,

$$F_1(t) - \frac{H_1(t)}{R_1} = A_1 \frac{dH_1(t)}{dt} \tag{2.46}$$

$$F_2(t) - \frac{H_2(t)}{R_2} = A_2 \frac{dH_2(t)}{dt}. \tag{2.47}$$

If the flow rate $F_1(t)$ is suddenly increased over its steady-state value by an amount B units, then $H_1(t)$ will be obtained by solving a differential equation and is given as

$$H_1(t) = B\,\{1 - \exp(-t/\tau_1)\} \tag{2.48}$$

where $\tau_1 = R_1A_1$, the first-order time constant for tank 1.

In order to get a relationship of $H_2(t)$ with F_1, Equation 2.47 is rewritten as

$$\frac{H_1(t)}{R_1} - \frac{H_2(t)}{R_2} = A_2 \frac{dH_2(t)}{dt}. \tag{2.49}$$

Eliminating $H_1(t)$ from Equation 2.46 by using Equation 2.49,

$$F_1(t) = \frac{H_2(t)}{R_2} + A_2 \frac{dH_2(t)}{dt} + A_1R_1 \frac{H_2(t)}{R_2} + A_1R_1A_2 \frac{d^2H_2(t)}{dt^2} \tag{2.50}$$

or

$$\tau_1\tau_2 \frac{d^2H_2(t)}{dt^2} + (\tau_1 + \tau_2)\frac{dH_2(t)}{dt} + H_2(t) = F_1R_2 \tag{2.51}$$

which is a second-order differential equation for the level in the second tank where $\tau_2 = A_2R_2$. It is clear from Equation 2.51 that $H_2(t)$ is independent of $H_1(t)$ but is affected by a change in F_1, the forcing function. Such systems are called noninteracting systems. The change of $H_2(t)$ with time can be found by solving Equation 2.51 for any sudden change in $F_1(t)$. The solution can be performed rapidly if Equation [illegible] compared with Equation 2.32 and we find the equivalent τ and ζ. The solution will be given by the similar equations to those already given in Equations 2.33 through 2.35.

2.3.2.2 Interacting System

Consider the two tanks in series in such a way that the outflow from the first tank affects the level in the second tank as shown in Figure 2.9. The material balance equations for this system are:

$$F_1(t) - \frac{H_1(t) - H_2(t)}{R_1} = A_1 \frac{dH_1(t)}{dt} \tag{2.52}$$

$$F_2(t) - \frac{H_2(t)}{R_2} = A_2 \frac{dH_2(t)}{dt} \tag{2.53}$$

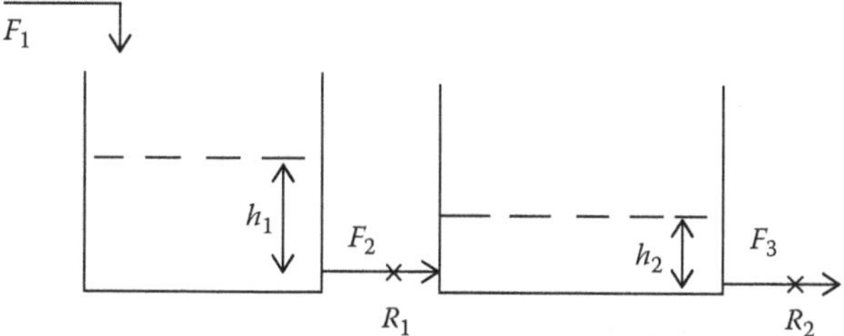

FIGURE 2.9 Two tanks connected in series—an interacting system.

where

$$F_2(t) = \frac{H_1(t) - H_2(t)}{R_1}. \tag{2.54}$$

Combining the above equations also yields a second-order differential equation:

$$\begin{aligned}
F_1(t) &= F_2(t) + A_1 \frac{dH_1(t)}{dt} \\
&= \frac{H_2(t)}{R_2} + A_2 \frac{dH_2(t)}{dt} + A_1 \frac{d}{dt}\left[R_1 \left\{ A_2 \frac{dH_2(t)}{dt} + \frac{H_2(t)}{R_2} \right\} + H_2(t) \right] \\
&= \frac{H_2(t)}{R_2} + A_2 \frac{dH_2(t)}{dt} + A_1 R_1 A_2 \frac{d^2 H_2(t)}{dt^2} + \left\{ \frac{A_1 R_1}{R_2} + A_1 \right\} \frac{dH_2(t)}{dt} \\
&= \frac{H_2(t)}{R_2} + \left\{ \frac{A_1 R_1}{R_2} + A_1 + A_2 \right\} \frac{dH_2(t)}{dt} + A_1 R_1 A_2 \frac{d^2 H_2(t)}{dt^2}
\end{aligned}$$

or

$$\tau_1 \tau_2 \frac{d^2 H_2(t)}{dt^2} + \left(\tau_1 + \tau_2 + A_1 + R_2\right) \frac{dH_2(t)}{dt} + H_2(t) = F_1 R_2. \tag{2.55}$$

Taking the Laplace transformation of both sides of Equations 2.51 and 2.55, the ratio of $H_2(s)/F_1(s)$, the transfer function is obtained as follows:

for the noninteracting tanks

$$\frac{H_2(s)}{F_1(s)} = \frac{R_2}{\tau_1 \tau_2 s^2 + \left(\tau_1 + \tau_2\right) s + 1} \tag{2.56}$$

and for the interacting tanks

$$\frac{H_2(s)}{F_1(s)} = \frac{R_2}{\tau_1\tau_2 s^2 + \left(\tau_1 + \tau_2 + A_1R_2\right)s + 1} \tag{2.57}$$

2.4 HIGHER-ORDER SYSTEMS

If the number of tanks in series is increased from two to three, a third-order differential equation will be obtained for the level in the third tank. Thus, the order of the differential equation for the last tank level will be an nth-degree differential relation if n number of tanks in series are taken. Solution of the differential equations (linear) are conveniently obtained by the Laplace transformation method.

For n tanks in series for the noninteracting tanks,

$$\text{1st tank,}\quad F_1(t) - \frac{H_1(t)}{R_1} = A_1\frac{dH_1(t)}{dt} \tag{2.58}$$

$$\text{2nd tank,}\quad F_2(t) - \frac{H_2(t)}{R_2} = A_2\frac{dH_2(t)}{dt} \tag{2.59}$$

$$\text{3rd tank,}\quad F_3(t) - \frac{H_3(t)}{R_3} = A_3\frac{dH_3(t)}{dt} \tag{2.60}$$

....

$$n\text{th tank,}\quad F_n(t) - \frac{H_n(t)}{R_n} = A_n\frac{dH_n(t)}{dt}. \tag{2.61}$$

Taking Laplace transfers for each tank as above,

$$\frac{H_1(s)}{F_1(s)} = \frac{R_1}{\tau_1 s + 1} \tag{2.62}$$

$$\frac{H_2(s)}{F_2(s)} = \frac{R_1H_2(s)}{H_1(s)} = \frac{R_2}{\tau_2 s + 1} \tag{2.63}$$

$$\frac{H_3(s)}{F_3(s)} = \frac{R_2H_3(s)}{H_2(s)} = \frac{R_3}{\tau_3 s + 1} \tag{2.64}$$

....

$$\frac{H_n(s)}{F_n(s)} = \frac{R_{n-1}H_n(s)}{H_{n-1}(s)} = \frac{R_n}{\tau_n s + 1} \tag{2.65}$$

The overall transfer function relating the level of the *n*th tank and the inlet flow to the first tank is obtained by multiplying all the above equations as

$$\frac{H_n(s)}{F_1(s)} = \frac{R_n}{(\tau_1 s+1)(\tau_2 s+1)(\tau_3 s+1)\ldots(\tau_n s+1)}. \tag{2.66}$$

If $\tau_1 = \tau_2 = \tau_3 = \tau_n = \tau$, then

$$\frac{H_n(s)}{F_1(s)} = \frac{R_n}{(\tau s+1)^n}. \tag{2.67}$$

Such tanks in a series process always behave like an overdamped system. Consider the problem of two tanks in series. The transfer function is given by Equation 2.56, and the denominator is

$$\tau_1\tau_2 s^2 + (\tau_1 + \tau_2)s + 1. \tag{2.68}$$

Comparing this with the denominator of Equation 2.32,

$$\tau^2 s^2 + 2\tau\xi s + 1 \tag{2.69}$$

$$\tau^2 = \tau_1\tau_2 \tag{2.70}$$

and

$$2\tau\xi = \tau_1 + \tau_2 \tag{2.71}$$

Thus,

$$\xi = \frac{\tau_1 + \tau_2}{2\tau} = \frac{\tau_1 + \tau_2}{2\sqrt{\tau_1\tau_2}} \tag{2.72}$$

Hence,

$$\xi - 1 = \frac{(\sqrt{\tau_1} - \sqrt{\tau_2})^2}{2\sqrt{\tau_1\tau_2}} \tag{2.73}$$

which indicates that the right side of Equation 2.73 is a positive quantity; hence, $\xi > 1$, i.e., the response of the system resulting from a step disturbance will be an overdamped one as shown in Figure 2.6. Similarly, the same is true for interacting tanks in series; verification is left for the students. In the case of *n* number of tanks in series, $H_n(t)$ will increase as a result of a step increase in $F_1(t)$, will behave as an

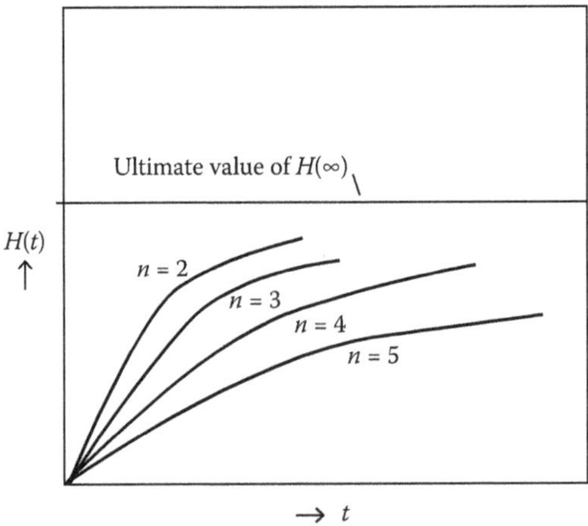

FIGURE 2.10 Overdamped response levels of tanks in series.

overdamped system, and will become more sluggish (slower) as the number of tanks (n) increases, as shown in Figure 2.10, where the ultimate value of $H(t)$ is achieved at infinite time, i.e., $H(\infty)$.

2.5 TRANSPORTATION LAG

If two functions $X(t)$ and $Y(t)$ are such that $Y(t)$ trails behind $X(t)$ by a time lag of τ_d sec, then $Y(t)$ has a transportation lag. The relationship between these functions is given mathematically as

$$Y(t) = X(t - \tau_d). \tag{2.74}$$

This implies that $Y(t)$ achieves the value of X at $t = 0$ after τ_d sec, or it achieves a value of X at $t = 1$ sec after $(\tau_d + 1)$ sec, and so on. This is explained in Figure 2.11, where $X(t)$ is a step function.

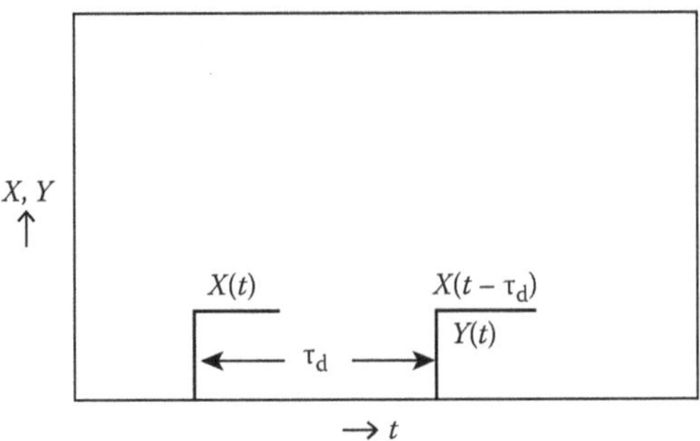

FIGURE 2.11 Transportation lag between two step functions.

Equation 2.74 can be expressed in terms of the Laplacian parameter s as follows. Taking the Laplace transformation of both sides of Equation 2.74,

$$LY(t) = LX(t - \tau_d) \tag{2.75}$$

or

$$Y(s) = \int_0^\infty X(t - \tau_d)e^{-st}\,dt. \tag{2.76}$$

Taking $p = t - \tau_d$, in terms of p the equation becomes,

$$Y(s) = \int_0^\infty X(p)e^{-s(p+\tau_d)}\,dp \tag{2.77}$$

or

$$Y(s) = e^{-s\tau_d}\int_0^\infty X(p)e^{-sp}\,dp = e^{-s\tau_d}X(s) \tag{2.78}$$

$$\frac{Y(s)}{X(s)} = e^{-s\tau_d} \tag{2.79}$$

2.6 TRANSFER FUNCTION

The ratio, $Y(s)/X(s)$ is called the transfer function and is defined as the ratio of the Laplace transformations of the output variable to the input variable, both in deviation form, where "deviation" indicates the difference of the variable from its steady value at $t = 0$. So far, we have determined the transfer functions of a few instruments and processes. As we have seen in the thermometer, the input to the thermometer is the temperature of a bath sensed by the bulb, and the output is the indicated temperature. Similarly, in a manometer, the input is the differential pressure sensed by the U-tube ends, and the output is the height of the column of manometric fluid indicated. In the case of processes, the tank temperature of a bath is the input to the processes (or heating rate), and the output is the exit temperature. In tank level processes, the input is the inlet flow, and the output is the liquid level in the tank. A list of transfer functions can be found in Table 2.1.

If the transfer function of any instrument or process is known (or evaluated), the output (as a function of Laplacian operator s) is determined by multiplying the input (also as a function of s) with the transfer function as explained in Figure 2.12.

TABLE 2.1
Transfer Functions of a Few Processes and Instruments

Process	Input Variable	Output Variable	Transfer Function
Steam-heated tank	Inlet temperature	Exit temperature	$\frac{1/wc_p}{(\tau s+1)}$
Level of a tank	Inlet flow	Liquid level	$\frac{R_1}{\tau_1 s+1}$
Level of second tank in series (noninteracting)	Inlet flow to first tank	Liquid level in second tank	$\frac{R_2}{\tau_1\tau_2 s^2+(\tau_1+\tau_2)s+1}$
Level of second tank in series (interacting)	Inlet flow to first tank	Liquid level in second tank	$\frac{R_2}{\tau_1\tau_2 s^2+(\tau_1+\tau_2+A_1R_2)s+1}$
Thermometer	Bath temperature	Indicated temperature	$\frac{1}{(\tau s+1)}$
Manometer	Differential pressure	Column height	$\frac{\frac{1}{\rho_m g}}{\tau^2 s^2+2\tau\xi s+1}$
Transportation	$X(t)$	$Y(t)$	$e^{-s\tau}d$

Note: a and k are constants and t is time in the above relationships.

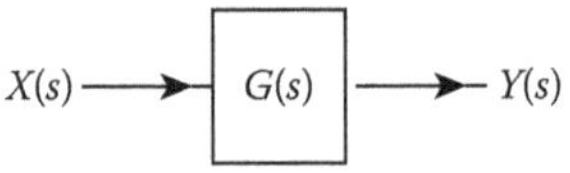

$$X(s) \longrightarrow \boxed{G(s)} \longrightarrow Y(s)$$

FIGURE 2.12 Output from input and transfer function.

where the output $Y(s)$ is obtained by multiplying the transfer function $G(s)$ by the input $X(s)$ as

$$Y(s) = G(s)\,X(s). \tag{2.80}$$

The value of $Y(t)$ can be obtained by inverting the right-hand side, i.e.,

$$Y(t) = L^{-1}\,Y(s) = L^{-1}\,[G(s)\,X(s)]. \tag{2.81}$$

2.7 QUESTIONS AND ANSWERS

EXERCISE 2.1

A mixing tank has a volume of V m^3 and is fed with a liquid stream at a rate of F m^3/sec containing an inorganic component having a concentration of c_i kg/m^3. The flow rate and volume of liquid in the tank are unchanged with time, but the concentration of the feed component varies. If the instantaneous exit concentration of the

component is c_0, determine the process transfer function of a mixing tank relating the outlet and inlet concentrations in deviation forms. Assume the mixture in the tank is well stirred, so concentration of the component is uniform throughout the tank.
Answer:

Instantaneous (at any instant of time) material balance of the component under unsteady-state conditions is

$$FC_i - FC_0 = V\frac{dC_0}{dt}. \tag{2.82}$$

Taking the Laplace transformation

$$\frac{C_0(s)}{C_i(s)} = \frac{1}{(\tau s+1)} \tag{2.83}$$

where $\tau = V/F$ and C_i and C_0 are the concentration in deviation form from steady values.

EXERCISE 2.2

In a continuously stirred tank reactor, an elementary first-order reaction A → B with a rate constant k is carried out at a constant temperature. A feed containing a reactant A enters with a composition of C_{A0} kmoles/m^3 and at a rate of F m^3/sec. If the volume of the reactor V is constant, determine the transfer function relating the concentration of A in the product and feed streams. Assume the concentration in the reactor is the same as the effluent concentration.
Hints:

A material balance of the component A is

$$FC_{A_0} - FC_A - kC_AV = V\frac{dC_A}{dt} \tag{2.84}$$

where C_A is the concentration of A in the product stream, and the rate of disappearance of A is $k\,C_AV$ moles/time.

EXERCISE 2.3

A thermocouple thermometer is used to measure the temperature of a bath, which is at a temperature $T(t)$ at any instant of time. The bulb of the thermometer is placed in a protective tube in order to avoid contact with the liquid in the bath (Figure 2.13). The surface area of the bulb and the tube for heat transfer are taken as A_b and A_t, respectively, and the film heat transfer coefficients of the bulb–air in the tube and tube–bath fluid are, respectively, h_b and h_t. Determine the transfer function between the bath temperature and the indicated temperature. Take mass and specific heats of the bulb and tube as (m_b, m_t) and (C_{pb}, C_{pt}), respectively.

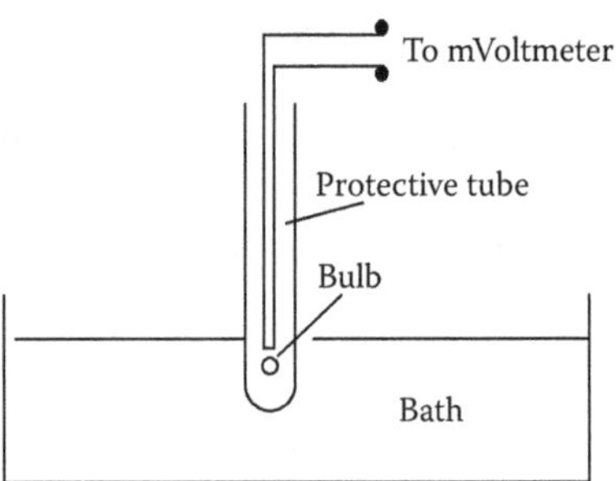

FIGURE 2.13 Thermometer in Exercise 2.3.

Answer:

The heat balance between the bath and the outer surface of the tube contact is

$$h_t A_t (T - T_t) = m_t C_{pt} \frac{d}{dt} T_t \tag{2.85}$$

or

$$T = T_t + \frac{m_t C_{pt}}{h_t A_t} \frac{dT_t}{dt} \tag{2.86}$$

and between the inner tube surface and the bulb it is

$$h_b A_b (T_t - T_b) = m_b C_{p_b} \frac{d}{dt} T_b \tag{2.87}$$

or

$$T_t = T_b + \frac{m_b C_{P_b}}{h_b A_b} \frac{d}{dt} T_b \tag{2.88}$$

$$T = T_t + \frac{m_t C_{pt}}{h_t A_t} \frac{dT_t}{dt} = T_b + \frac{m_b C_{P_b}}{h_b A_b} \frac{d}{dt} T_b + \frac{m_t C_{pt}}{h_t A_t} \left(\frac{dt_b}{d_t} + \frac{m_b C_{P_b}}{h_b A_b} \frac{d^2}{dt^2} T_b \right)$$

$$= T_b + \left(\frac{m_b C_{P_b}}{h_b A_b} + \frac{m_t C_{pt}}{h_t A_t} \right) \frac{d}{dt} T_b + \frac{m_b C_{P_b}}{h_b A_b} \frac{d^2}{dt^2} T_b$$

i.e.,

$$T = T_b + (\tau_t + \tau_b) \frac{dT_t}{dt} + \tau_t \tau_b \frac{d^2}{dt^2} T_b \tag{2.89}$$

where τ_t and τ_b are the time constants of the outer wall and the inner wall of the thermometer bulb. Taking the Laplace transformations of both sides of the equation, the transfer function relating T_b and T can be obtained as

$$\frac{T_b(s)}{T(s)} = \frac{1}{\tau_t \tau_b s^2 + (\tau_t + \tau_b)s + 1}$$

where T_b and T are expressed in deviation form with respect to the steady-state values.

EXERCISE 2.4

A mercury-in-glass thermometer indicates an ambient temperature of 30°C when the same is suddenly dipped in an ice bath at 0°C. Temperature is recorded as shown below:

Time, sec	Temperature, °C	Time, sec	Temperature, °C
0	30	1	6
0.1	25	1.5	2.46
0.2	22	2.0	0.0
0.5	13		

Determine the first-order time constant of the thermometer.

Solution:

Plot a temperature vs. time curve and locate temperature = 30 – 0.632 × (30 – 0) = 11.04°C from the graph and find the corresponding time. This is found as 0.6 sec. Hence, the time constant is 0.6 sec. Ans.

EXERCISE 2.5

Invert the following Laplace transformation of Y in time domain:

$$Y(s) = \frac{1}{s(s^2 + 5.5s + 7.5)}.$$

Solution:

By partial fraction,

$$Y(s) = \frac{A}{s} + \frac{B}{(s+3)} + \frac{C}{(s+2.5)}$$

where A, B, and C are obtained as $A = 0.133$, $B = 0.7$, and $C = -0.8$ and inverting to time domain

$$Y(t) = 0.133 + 0.7e^{-3t} - 0.8e^{-2.5t}$$

EXERCISE 2.6

A second-order transfer function is given as

$$\frac{Y(s)}{X(s)} = \frac{20}{(s^2 + 3.2s + 9)}$$

If a unit step change occurs in $X(t)$, determine the overshoot, decay ratio, rise time, period of oscillation, and maximum and ultimate values of $Y(t)$.

Answer:

Rearranging the given transfer function,

$$\frac{Y(s)}{X(s)} = \frac{20/9}{(1/9s^2 + 3.2/9s + 1)}$$

and comparing with the standard one as

$$\frac{Y(s)}{X(s)} = \frac{K_p}{(\tau^2 s^2 + 2\tau\xi s + 1)}.$$

Then, $\tau^2 = 1/9$ and $2\tau\xi = 3.2/9$, or $\tau = 1/3$ and $\xi = 3.2/6 = 0.533 < 1$

Hence, the $Y(t)$ will be an underdamped function for the unit step disturbance in $X(t)$. Therefore,

$$Y(t) = K_p\,[1 - \exp(-t\,\xi/\tau)/\sqrt{(1-\xi^2)}\,\sin(wt + \tan^{-1}(w\tau)]$$

where

$$w = \sqrt{(1-\xi^2)}/\tau = 2.54$$

Overshoot ratio $= \exp\{(-\Pi\,\xi/\sqrt{(1-\xi^2)}\}$

$$= \exp(-3.14 \times 0.533)/\sqrt{(1 - 0.533^2)} = 0.138 = 13.8\% \text{ of } K_p$$

where

$K_p = 20/9 = 2.22 =$ ultimate $Y(t)$

Decay ratio $= \exp\{(-2\Pi\,\xi/\sqrt{(1-\xi^2)}\} = (0.138)^2 = 0.019 = 1.9\%$ of K_p

Maximum value $Y(t) =$ ultimate $Y(t)$ + overshoot $= (1 + 0.138) \times 2.22 = 2.53$

Rise time is the minimum time when $Y(t)$ reaches its ultimate value, i.e., 2.22. This occurs when $\sin\{wt + \tan^{-1}(w\tau)\} = 0 = \sin(180°)$. Or rise time $t = \{180 - \tan^{-1}(2.54 \times 0.533)\}/2.54 = 48$ time unit. The period of oscillation $= 2\Pi/w = 2.47$ time unit.

EXERCISE 2.7

A proportional–integral (PI) temperature controller has an output signal of 13.33 mA when the set point (SP) and process value (PV) are equal. If a deviation between the SP and PV is suddenly displaced by 0.5°C, the output of the controller is obtained at different times, as listed below.

Time, sec	mA
0–	13.33
0+	10.33
20	18.66
60	6.66
90	4.66

Determine the parameters (K_c and τ_i) of the controller.

Answer:

The equation for a PI controller is

$$I = I_s + K_c\varepsilon + \frac{K_c}{\tau_i}\int \varepsilon dt$$

where ε = SP – PV = 0.5 as the step change in error, and I_s = 13.33 mA.

A plot of I vs. t will then be a straight line with intercept (13.33 + 0.5 K_c) and slope $\frac{K_c}{\tau_i}$.

Thus, values K_c and τ_i will be obtained from the plot of output of controller vs. time (Figure 2.14).

From the plot,

$$13.33 + 0.5\,K_c = 10.66.$$

So

$$K_c = -2.67/0.5 = -5.34 \text{ mA/c}$$

$$\text{Slope} = 0.5\,K_c/\tau_i = -\,(10.66 - 4.66)/(90 - 0) = -6/90 = -1/15$$

or

$$\tau_i = 0.5 \times 5.34 \times 15 = 40.05 \text{ sec}$$

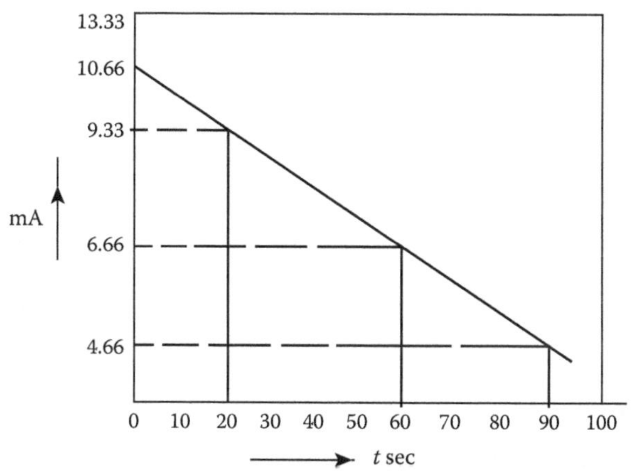

FIGURE 2.14 Plot for Exercise 2.7.

3 Disturbances and Responses

3.1 STEP DISTURBANCE

A step indicates a jump up or down like a step on a ladder. It jumps from one value to another higher or lower value, i.e., a step up or step down. Practically, when a switch is off, there is no power supply, and when the switch is on, full power exists. Practical examples of step disturbance events are listed in Table 3.1.

Thus, mathematical representation of such disturbances can be used to predict the effects on the output.

3.2 STEP DISTURBANCE TO A FIRST-ORDER SYSTEM

If the transfer function is represented as a first-order transfer function, then

$$G(s) = \frac{Y(s)}{X(s)} = \frac{1}{\tau s + 1}. \tag{3.1}$$

Such a first-order system has a "unity gain" as the value of the numerator in Equation 3.1. Also, the ratio of output to input in the Laplace transformations is the transfer function.

Now, a unit step disturbance in the input $X(t)$ occurs as $X(t) = 0$ at $t < = 0$, which means the previous value jumped to unity instantaneously ($t = 0$) and continued for the rest of the time ($t > 0$) (Figure 3.1).

The transfer function of $X(t) = 1$ is determined as

$$X(s) = \int_0^\infty X(t)e^{-st}\,dt = \int_0^\infty e^{-st}\,dt = \left[-\frac{e^{-st}}{s}\right]_0^\infty = \frac{1}{s} \tag{3.2}$$

Hence,

$$Y(s) = \frac{1}{s(\tau s + 1)}. \tag{3.3}$$

By partial fraction method,

$$Y(s) = \frac{1}{s} - \frac{1}{s + \frac{1}{\tau}}. \tag{3.4}$$

TABLE 3.1
Different Types of Mathematical Disturbances

Event	Step	Result	Mathematical Equivalence	Graphical Presentation
Switch to electric bulb of 100 watts	On	Bulb glows with full 100 watt power	$X(t) = 0$ at $t <= 0$ $X(t) = 100$ at $t => 0$	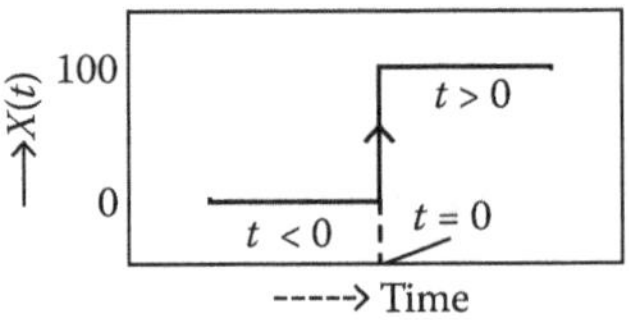 $\rightarrow X(t)$; 100; 0; $t > 0$; $t < 0$; $t = 0$; ----> Time
	Off	Bulb stops glowing with zero watts	$X(t) = 100$ at $t <= 0$ $X(t) = 0$ at $t >= 0$	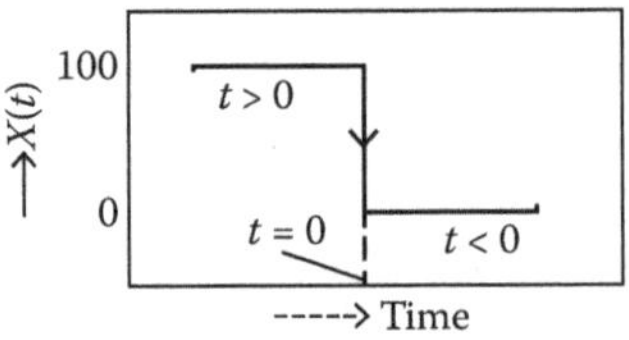 $\rightarrow X(t)$; 100; 0; $t > 0$; $t = 0$; $t < 0$; ----> Time
A valve in the pipeline	Fully open	Flow takes place to maximum, say 10 m^3/hr	$X(t) = 0$ at $t <= 0$ $X(t) = 10$ at $t >= 0$	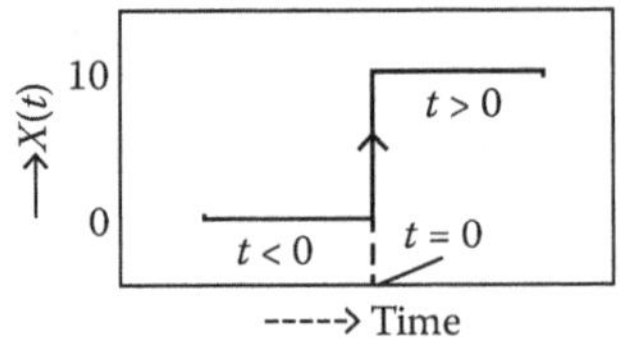 $\rightarrow X(t)$; 10; 0; $t > 0$; $t < 0$; $t = 0$; ----> Time
	Fully closed	Flow becomes zero	$X(t) = 10$ at $t >= 0$ $X(t) = 0$ at $t <= 0$	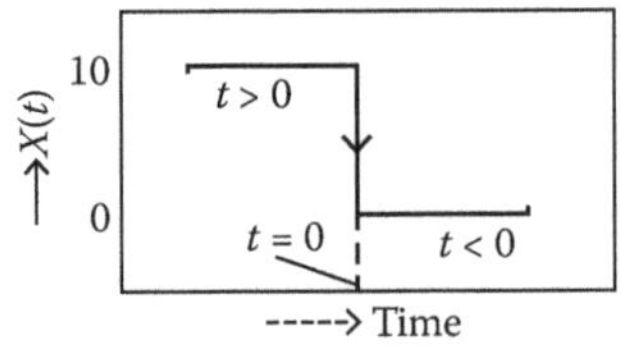 $\rightarrow X(t)$; 10; 0; $t > 0$; $t = 0$; $t < 0$; ----> Time
An alarm	On	Sounds to maximum, say 100 decibles	$X(t) = 0$ at $t <= 0$ $X(t) = 100$ at $t >= 0$	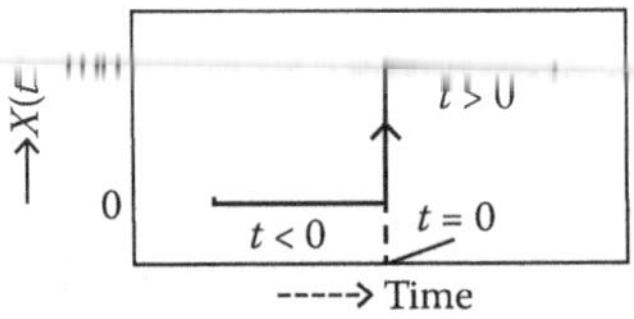 $\rightarrow X(t)$; 0; $t > 0$; $t < 0$; $t = 0$; ----> Time
	Off	Sound stops	$X(t) = 100$ at $t <= 0$ $X(t) = 0$ at $t >= 0$	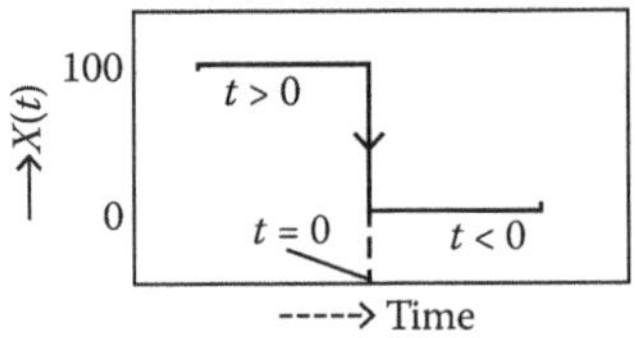 $\rightarrow X(t)$; 100; 0; $t > 0$; $t = 0$; $t < 0$; ----> Time

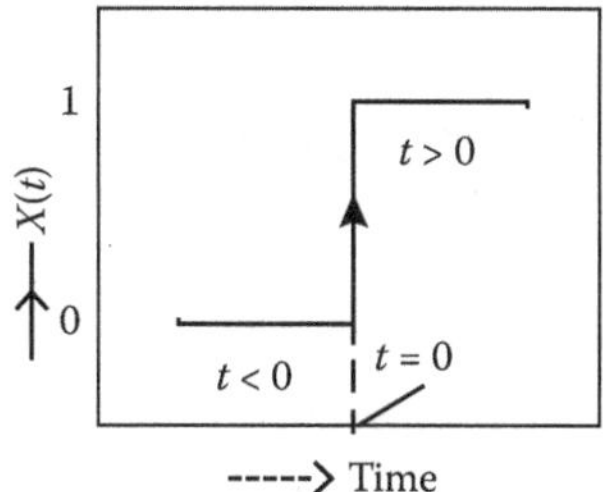

FIGURE 3.1 Unit step disturbance of the input variable $X(t)$.

Inverting the transforms,

$$Y(t) = (1 - e^{-t/\tau}). \tag{3.5}$$

Thus, we have seen that if the transfer function is known and the mathematical representation of the disturbance is known, the output can be determined. In this case, the output is an exponential increase in the value until it reaches unity as $t \rightarrow$ infinity. It is evident from Equation 3.5, if $t = \tau$, then $Y(t) = 0.632$. Thus, the value of τ is obtained at the time when $Y(t)$ reaches 0.632 as shown in Figure 3.2.

Alternatively, it can be said that a first-order time constant is defined as the time at which the output reaches 63.2% of its ultimate value (which is the value $Y(t)$ reaches at infinite time).

If the gain of the first-order system is A and the step disturbance of magnitude B is applied, the value of $Y(t)$ is given as

$$Y(t) = AB(1 - e^{-t/\tau}). \tag{3.6}$$

The ultimate value of $Y(t)$ as t approaches infinity is AB.

An increase in time constant decreases the response, i.e., the value of $Y(t)$ increases slowly for higher values of the time constant as shown in Figure 3.3.

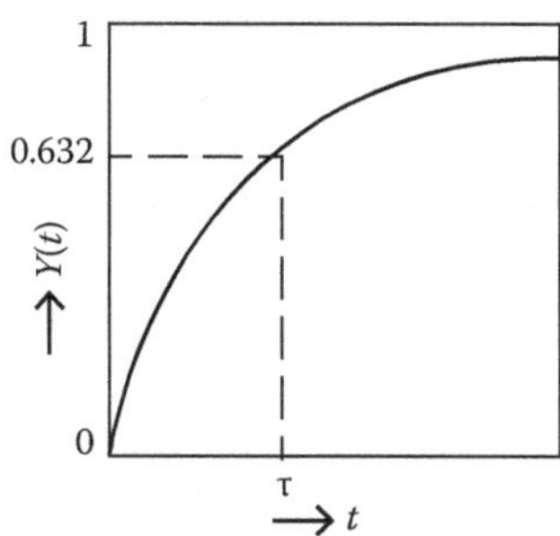

FIGURE 3.2 Output variation with time resulting from step disturbance.

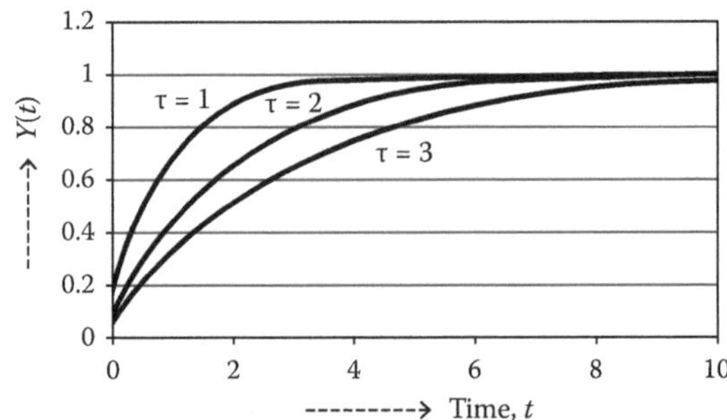

FIGURE 3.3 Response of $Y(t)$ to increasing time constants.

3.3 STEP DISTURBANCE TO A SECOND-ORDER SYSTEM

A unity gain second-order transfer function is given as

$$G(s) = \frac{1}{\tau^2 s^2 + 2\tau\xi s + 1} \tag{3.7}$$

and $X(s) = 1/s$.

So that

$$Y(s) = \frac{1}{s(\tau^2 s^2 + 2\tau\xi s + 1)}. \tag{3.8}$$

In order to make a partial fraction, it is necessary to factorize the denominator as

$$Y(s) = \frac{1}{\tau^2 s(s-\alpha)(s-\beta)} \tag{3.9}$$

where

$$\alpha, \beta = \frac{-\xi \pm \sqrt{(\xi^2 - 1)}}{\tau}. \tag{3.10}$$

Depending upon the values of $\xi < 1$, >1, or $=1$, the values of α and β will be real, complex, or equal.

Considering, $\xi < 1$, the discriminant becomes negative, so the values of α and β will be given as

$$\alpha, \beta = \frac{-\xi \pm i\sqrt{(1 - \xi^2)}}{\tau} \tag{3.11}$$

where $i = \sqrt{-1}$.

Using the partial fraction method,

$$Y(s)=\frac{1}{\tau^2 s(s-\alpha)(s-\beta)}=\frac{A}{s}+\frac{B}{(s-\alpha)}+\frac{C}{(s-\beta)}. \tag{3.12}$$

Solving for the constants A, B, and C by using identity properties, for instance to get A, both sides of Equation 3.12 are multiplied by s and setting $s = 0$; thus,

$$A=\frac{1}{\tau^2(0-\alpha)(0-\beta)}=\frac{1}{\tau^2\alpha\beta}=\frac{1}{\tau^2\frac{1}{\tau^2}}=1 \tag{3.13}$$

(as $\alpha\beta = 1/\tau^2$).

Similarly, B and C are evaluated by multiplying both sides of Equation 3.12 by $(s - a)$ and setting $s = a$; thus,

$$\begin{aligned}B&=\frac{1}{\tau^2\alpha(\alpha-\beta)}=\frac{\beta}{\tau^2\alpha\beta(\alpha-\beta)}\\&=\frac{\beta}{\tau^2\frac{1}{\tau^2}(\alpha-\beta)}=\frac{\beta}{(\alpha-\beta)}\end{aligned} \tag{3.14}$$

and

$$C=\frac{1}{\tau^2\beta(\beta-\alpha)}=\frac{\alpha}{\tau^2\alpha\beta(\alpha-\beta)}=-\frac{\alpha}{\tau^2\frac{1}{\tau^2}(\alpha-\beta)}=-\frac{\alpha}{(\alpha-\beta)}. \tag{3.15}$$

Thus,

$$\begin{aligned}Y(t)&=1+\frac{\beta}{(\alpha-\beta)}e^{\alpha t}-\frac{\alpha}{(\alpha-\beta)}e^{\beta t}\\&=1+\frac{1}{(\alpha-\beta)}(\beta e^{\alpha t}-\alpha e^{\beta t})\end{aligned}$$

Substituting the values of α and β, we obtain

$$\begin{aligned}Y(t)=1+\frac{\tau}{2i\surd(1-\xi^2)}&\left[\left(-\frac{\xi}{\tau}-i\frac{\surd(1-\xi^2)}{\tau}\right)e^{\left\{-\frac{\xi}{\tau}+i\frac{\surd(1-\xi^2)}{\tau}\right\}t}\right.\\&\left.-\left(-\frac{\xi}{\tau}+i\frac{\surd(1-\xi^2)}{\tau}\right)e^{\left\{-\frac{\xi}{\tau}-i\frac{\surd(1-\xi^2)}{\tau}\right\}t}\right].\end{aligned}$$

Applying De Moivre's rule of conversion from exponential complex numbers,

$$Y(t) = 1 + \frac{e^{-\frac{\xi}{\tau}t}}{2i\sqrt{(1-\xi^2)}}\left[-2i\xi\sin\left(\frac{\sqrt{(1-\xi^2)}}{\tau}t\right) - 2i\sqrt{(1-\xi^2)}\cos\left(\frac{\sqrt{(1-\xi^2)}}{\tau}t\right)\right]$$

$$= 1 - \frac{e^{-\frac{\xi}{\tau}t}}{\sqrt{(1-\xi^2)}}\left[\xi\sin\left(\frac{\sqrt{(1-\xi^2)}}{\tau}t\right) + \sqrt{(1-\xi^2)}\cos\left(\frac{\sqrt{(1-\xi^2)}}{\tau}t\right)\right].$$

Taking $r\cos\Phi = \xi$ and $r\sin\Phi = \sqrt{(1-\xi^2)}$.

Thus, $r = 1$ and $\tan\Phi = \dfrac{\sqrt{(1-\xi^2)}}{\xi}$

$$Y(t) = 1 - \frac{e^{-\frac{\xi}{\tau}t}}{\sqrt{(1-\xi^2)}}\left[r\cos\Phi\sin\left(\frac{\sqrt{(1-\xi^2)}}{\tau}t\right) + r\sin\Phi\cos\left(\frac{\sqrt{(1-\xi^2)}}{\tau}t\right)\right]$$

$$= 1 - \frac{e^{-\frac{\xi}{\tau}t}}{\sqrt{(1-\xi^2)}}\sin\left(\frac{\sqrt{(1-\xi^2)}}{\tau}t + \tan^{-1}\frac{\sqrt{(1-\xi^2)}}{\xi}\right). \tag{3.16}$$

For $\xi > 1$, α and β are given by Equation 3.10.

$$Y(t) = 1 + \frac{1}{(\alpha-\beta)}(\beta e^{\alpha t} - \alpha e^{\beta t})$$

$$= 1 + \frac{e^{-\frac{\xi}{\tau}t}}{\frac{2\sqrt{(\xi^2-1)}}{\tau}}\left[-\frac{\xi}{\tau}\left(e^{\frac{\sqrt{(\xi^2-1)}}{\tau}t} - e^{-\frac{\sqrt{(\xi^2-1)}}{\tau}t}\right) - \frac{\sqrt{(\xi^2-1)}}{\tau}\left(e^{\frac{\sqrt{(\xi^2-1)}}{\tau}t} + e^{-\frac{\sqrt{(\xi^2-1)}}{\tau}t}\right)\right]$$

$$= 1 - \frac{e^{-\frac{\xi}{\tau}t}}{2\sqrt{(\xi^2-1)}}\left[\xi\left(e^{\frac{\sqrt{(\xi^2-1)}}{\tau}t} - e^{-\frac{\sqrt{(\xi^2-1)}}{\tau}t}\right) + \sqrt{(\xi^2-1)}\left(e^{\frac{\sqrt{(\xi^2-1)}}{\tau}t} + e^{-\frac{\sqrt{(\xi^2-1)}}{\tau}t}\right)\right]$$

$$= 1 - \frac{e^{-\frac{\xi}{\tau}t}}{2\sqrt{(\xi^2-1)}}\left[\xi 2\sinh\frac{\sqrt{(\xi^2-1)}}{\tau}t + \sqrt{(\xi^2-1)}\,2\cosh\frac{\sqrt{(\xi^2-1)}}{\tau}t\right]$$

$$= 1 - e^{-\frac{\xi}{\tau}t}\left[\frac{\xi}{\sqrt{(\xi^2-1)}}\sinh\frac{\sqrt{(\xi^2-1)}}{\tau}t + \cosh\frac{\sqrt{(\xi^2-1)}}{\tau}t\right]. \tag{3.17}$$

For $\xi = 1$,

$$Y(s) = \frac{1}{\tau^2 s^2 + 2\tau s + 1} = \frac{1}{\tau^2\left(s + \frac{1}{\tau}\right)^2}$$

$$= \frac{1}{s} - \frac{1}{\left(s + \frac{1}{\tau}\right)} - \frac{1}{\tau\left(s + \frac{1}{\tau}\right)^2} \tag{3.18}$$

$$Y(t) = 1 - e^{-\frac{t}{\tau}} - \frac{t}{\tau}e^{-\frac{t}{\tau}}$$

$$= 1 - e^{-\frac{t}{\tau}}\left(1 + \frac{t}{\tau}\right).$$

For $\xi = 0$, α and β are given as

$$\alpha, \beta = \pm\frac{i}{\tau}. \tag{3.19}$$

Substituting these values in the equation,

$$Y(t) = 1 - \frac{1}{2}\left(e^{t/\tau} + e^{-t/\tau}\right).$$

$$Y(t) = 1 - \cosh(t/\tau). \tag{3.20}$$

Thus the output $Y(t)$ for $\xi < 1$, >1, $= 1$ and $= 0$, are given by Equations 3.16, 3.17, 3.18, and 3.20, respectively. The nature of the outputs for these are presented in Figures 3.4 through 3.7. For $\xi < 1$, the output rises to a high value and then is

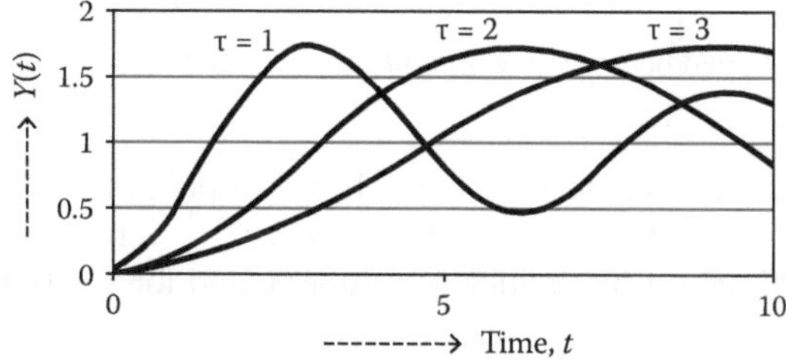

FIGURE 3.4 Underdamped behavior at different time constants at $\xi = 0.1$.

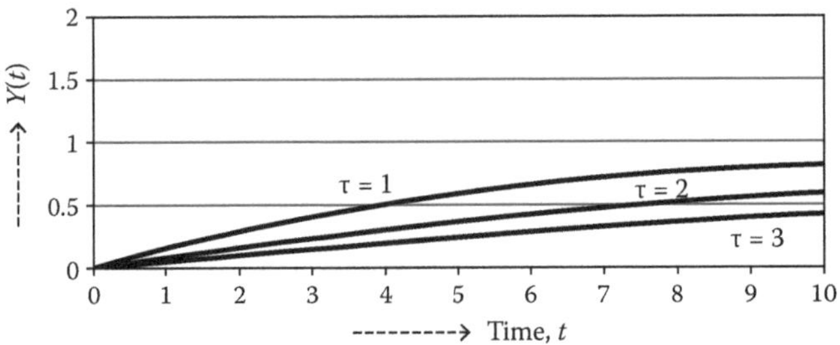

FIGURE 3.5 Overdamped behavior at different time constants at $\xi = 3.0$.

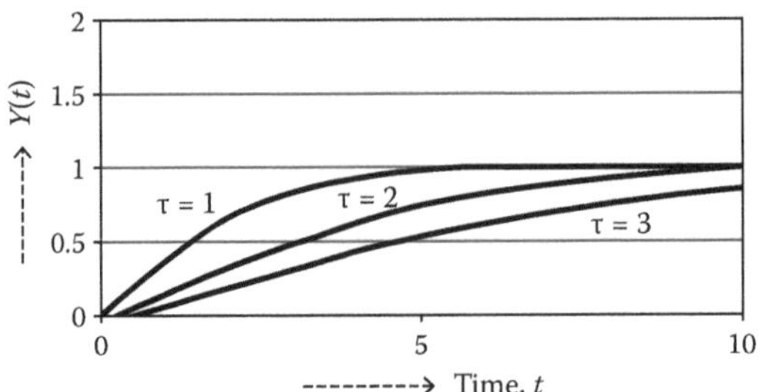

FIGURE 3.6 Critically damped behavior at different time constants for $\xi = 1$.

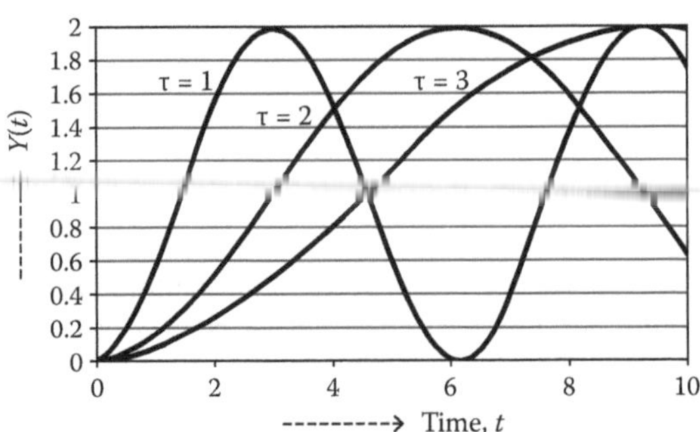

FIGURE 3.7 Undamped behavior at different time constants for $\xi = 0$.

gradually damped (underdamped); for $\xi > 1$, the output gradually rises but always remains below the ultimate value (overdamped); for $\xi = 1$, the output rises gradually and equals the ultimate value (critically damped); and for $\xi = 0$, the fluctuation of output never stops (undamped).

As shown in Figure 3.4, for an underdamped response when $\xi < 1$, the value of Y is above and below the ultimate value of unity, but finally, it is nearing unity after some time. The greater the value of the time constant, the slower or more sluggish it is in reaching near unity, the ultimate value. For the overdamped response, the value of Y never reaches unity. The effect of the increasing time constant is to make the response slower as shown in Figure 3.5. For the critically damped response, the value of Y reaches unity and never exceeds it as shown in Figure 3.6. In Figure 3.7, the value of response is continuously fluctuating as a sine curve up and down around unity.

3.4 RAMP DISTURBANCE

The ramp is the ladder. Mathematically, it is the equation of a line passing through a point and having a definite slope (Figure 3.8). In reality, this is represented by a constant rate of increase (or decrease) of an entity with time. It is represented by a function:

$$X(t) = kt \tag{3.21}$$

where k is the slope or the rate of increase (or decrease) of $X(t)$ with time t.

The Laplace transformation of a unit ramp (i.e., $k = 1$) function is obtained as

$$\begin{aligned} LX(t) = X(s) &= \int_0^\infty X(t)e^{-st}\,dt = \int_0^\infty te^{-st}\,dt \\ &= \left[t\int e^{-st}\,dt - \int \frac{d(t)}{dt}\int e^{-st}\,dt \right]_0^\infty \\ &= \left[-t\frac{e^{-st}}{s} - \frac{e^{-st}}{s^2} \right]_0^\infty = \frac{1}{s^2}. \end{aligned} \tag{3.22}$$

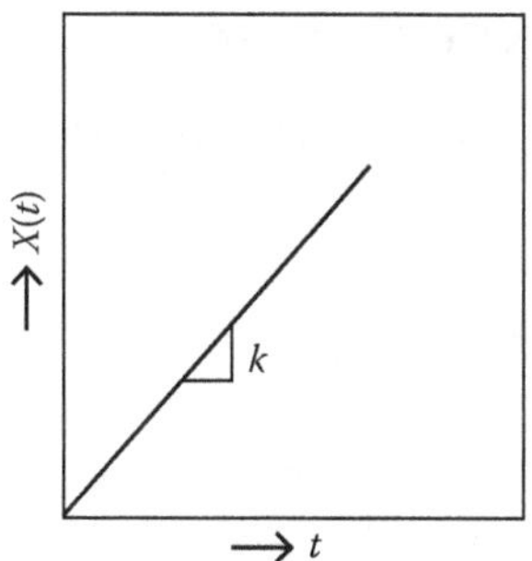

FIGURE 3.8 Ramp function $X(t)$ with a slope k.

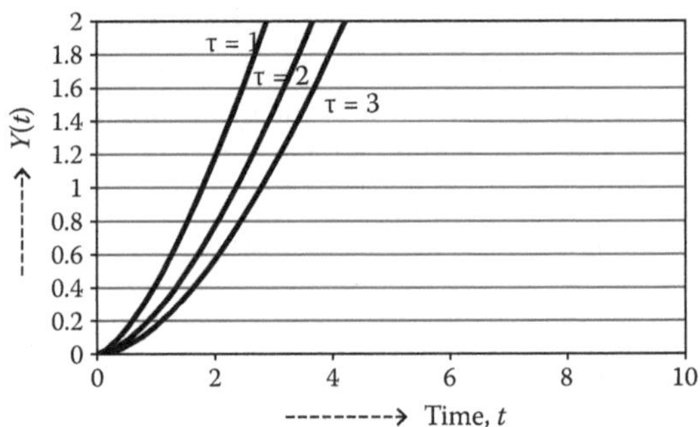

FIGURE 3.9 Behavior of output $Y(t)$ for a ramp disturbance on a unity first-order system at three different time constants.

3.5 RAMP DISTURBANCE TO A FIRST-ORDER SYSTEM

If the transfer function is a first-order with a unity gain,

$$G(s) = Y(s)/X(s) = 1/(\tau s + 1). \tag{3.23}$$

Then $Y(t)$ can be evaluated as

$$\begin{aligned} Y(s) &= \frac{1}{s^2(\tau s+1)} = \frac{1}{s^2} - \tau\left(\frac{1}{s} - \frac{1}{s+\frac{1}{\tau}}\right) \\ Y(t) &= t - \tau\left(1 - e^{-t/\tau}\right). \end{aligned} \tag{3.24}$$

This behavior is shown in Figure 3.9.

3.6 RAMP DISTURBANCE TO A SECOND-ORDER SYSTEM

Ramp disturbance on a unity gain second-order system is obtained as

$$Y(s) = \frac{1}{s^2\left(\tau^2 s^2 + 2\tau\xi s + 1\right)}. \tag{3.25}$$

The analytical solution is complex, and there will be four types of solutions depending upon the values of ξ (0, <1, >1, and =1). The nature of the responses of $Y(t)$ are shown in Figures 3.10, 3.11, 3.12, and 3.13 for different values of ξ and the effect of increasing values of τ.

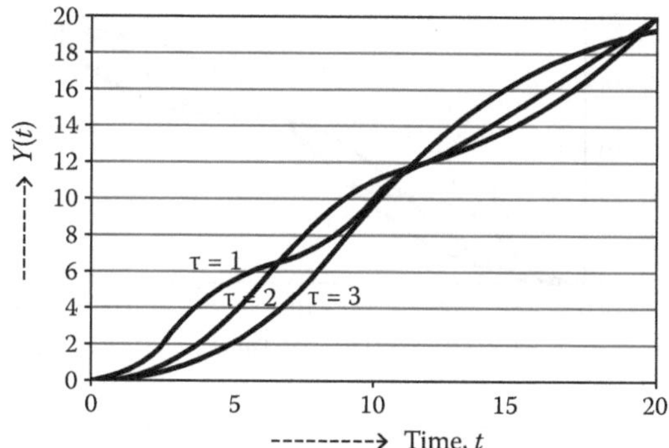

FIGURE 3.10 Responses resulting from a unit ramp for $\xi = 0.1$.

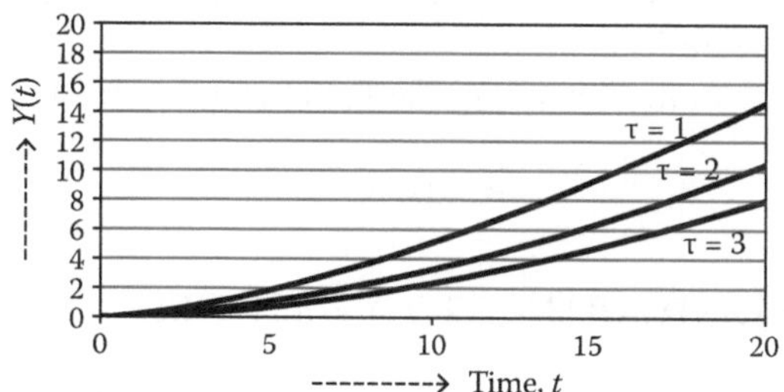

FIGURE 3.11 Responses resulting from a unit ramp for $\xi = 3$.

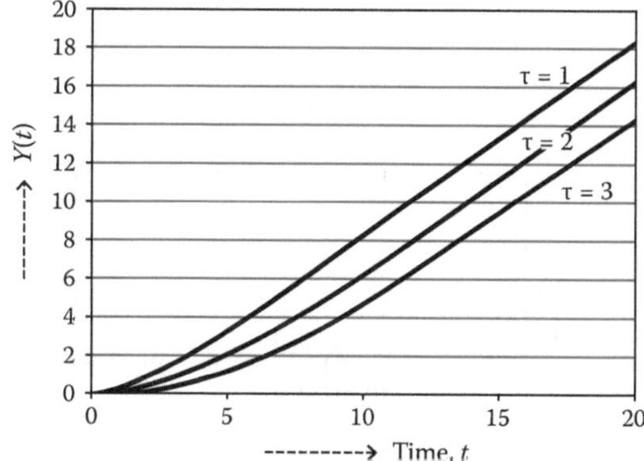

FIGURE 3.12 Responses resulting from a unit ramp for $\xi = 1$.

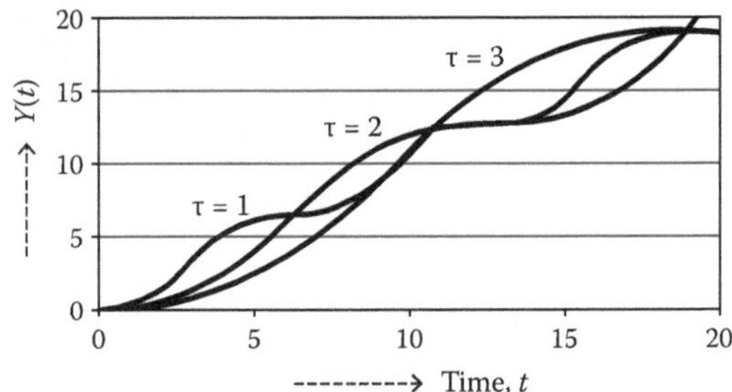

FIGURE 3.13 Responses resulting from a unit ramp for $\xi = 0$.

3.7 IMPULSE DISTURBANCE

If a function $X(t)$ is suddenly increased by a step and then decreased immediately to its original state, the function is an impulse. For example, when a switch is turned on and off very quickly, then an impulse of electricity is given. Similarly, if a valve is suddenly opened and very quickly closed down, an impulse of flow will take place. This is shown in Figure 3.14.

The Laplace transformation of impulse $X(t)$ of unit magnitude is unity, i.e.,

$$X(s) = 1.$$

3.8 IMPULSE DISTURBANCE TO A FIRST-ORDER SYSTEM

Thus, if an impulse disturbance of unit magnitude is applied to a first-order system, then output $Y(t)$ is obtained as given by Equation 3.26.

$$\frac{Y(s)}{X(s)} = \frac{1}{\tau s + 1} \tag{3.26}$$

or

$$Y(s) = \frac{1}{\tau s + 1} \tag{3.27}$$

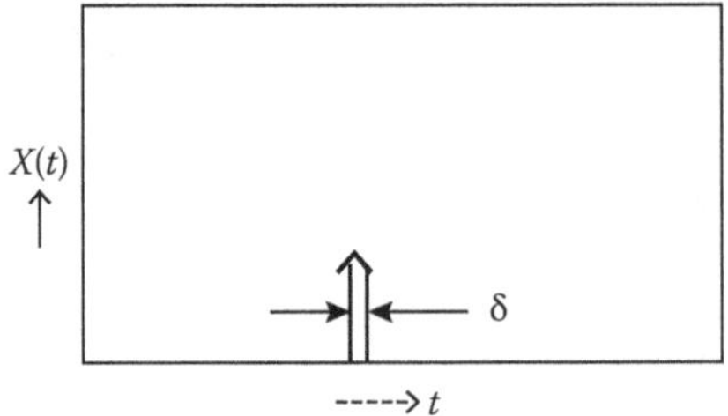

FIGURE 3.14 An impulse disturbance where δ is the delay time approaching zero.

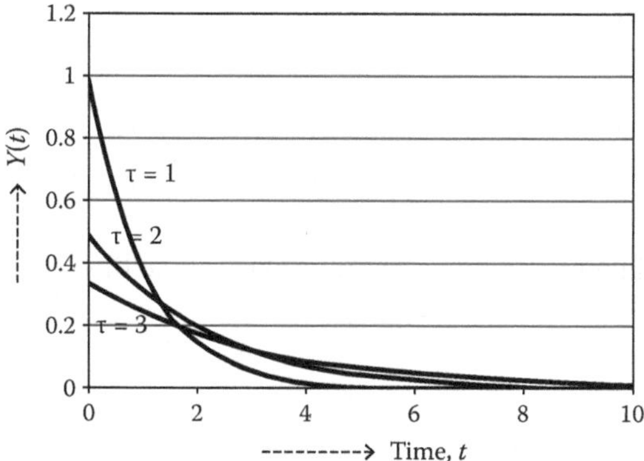

FIGURE 3.15 Output response resulting from the unity impulse disturbance of a first-order system with different time constants.

as the Laplace transformation of a unit impulse is unity, $X(s) = 1$; hence,

$$Y(s) = \frac{1}{\tau\left(s + \frac{1}{\tau}\right)}$$

inverting,

$$Y(t) = \frac{e^{-t/\tau}}{\tau}. \tag{3.28}$$

This is graphically shown in Figure 3.15.

3.9 IMPULSE DISTURBANCE TO A SECOND-ORDER SYSTEM

Considering an impulse disturbance to a second-order system,

$$Y(s) = \frac{1}{(\tau^2 s^2 + 2\tau\xi s + 1)} \tag{3.29}$$

$$Y(s) = \frac{1}{\tau^2 (s-\alpha)(s-\beta)} = \frac{B}{(s-\alpha)} + \frac{C}{(s-\beta)} \tag{3.30}$$

$$Y(t) = \frac{1}{\tau^2(\alpha-\beta)}(e^{\alpha t} - e^{\beta t})$$

For $\xi < 1$, proceeding as before,

$$Y(t) = \frac{e^{-\xi\frac{t}{\tau}}}{\tau\sqrt{(1-\xi^2)}}\left[\sin\sqrt{(1-\xi^2)}\frac{t}{\tau}\right] \tag{3.31}$$

This is shown in Figure 3.16 for increasing values of the time constant.

For $\xi > 1$,

$$Y(t) = \frac{e^{-\xi\frac{t}{\tau}}}{\tau\sqrt{(\xi^2-1)}}\left[\sinh\sqrt{(\xi^2-1)}\frac{t}{\tau}\right] \tag{3.32}$$

This is shown in Figure 3.17 for increasing values of the time constant.

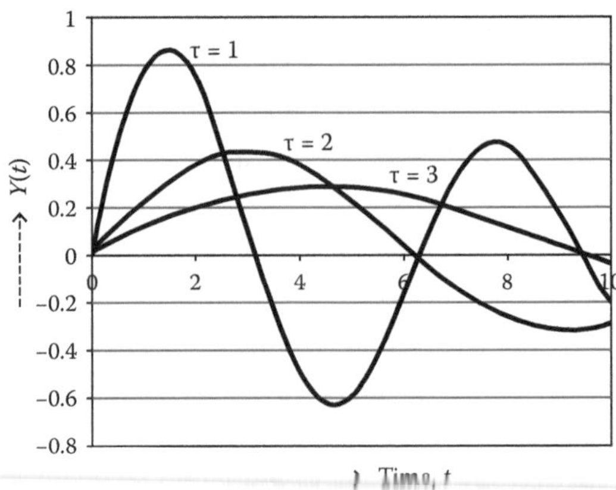

FIGURE 3.16 Responses of $Y(t)$ for a unit impulse disturbance for $\xi < 1$.

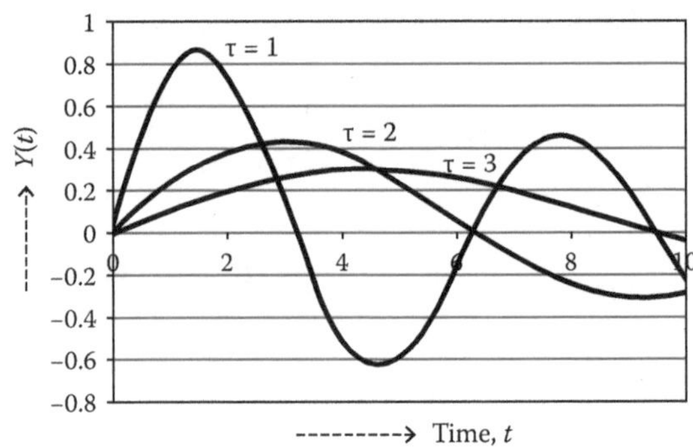

FIGURE 3.17 Responses of $Y(t)$ for a unit impulse disturbance for $\xi > 1$.

For $\xi = 1$,

$$Y(t) = \frac{te^{-\frac{t}{\tau}}}{\tau^2} \tag{3.33}$$

This response is shown in Figure 3.18.

For $\xi = 0$,

$$\begin{aligned} Y(s) &= \frac{1}{\tau^2 s^2 + 1} \\ Y(t) &= \frac{1}{\tau^2} \sin\left(\frac{t}{\tau^2}\right) \end{aligned} \tag{3.34}$$

The responses of $Y(t)$ are shown in Figure 3.19.

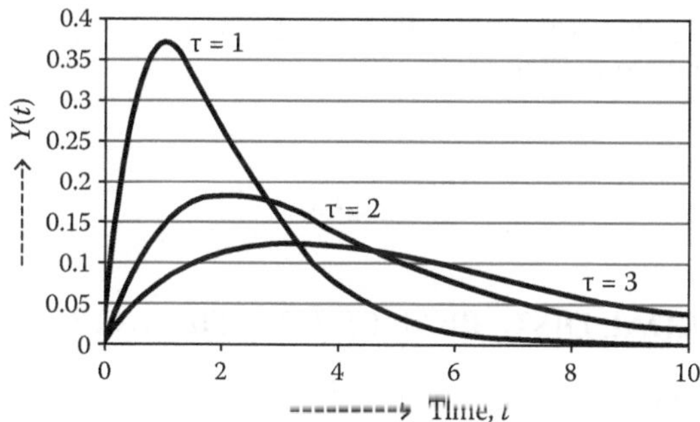

FIGURE 3.18 Responses of $Y(t)$ at increasing time constants at $\xi = 1$.

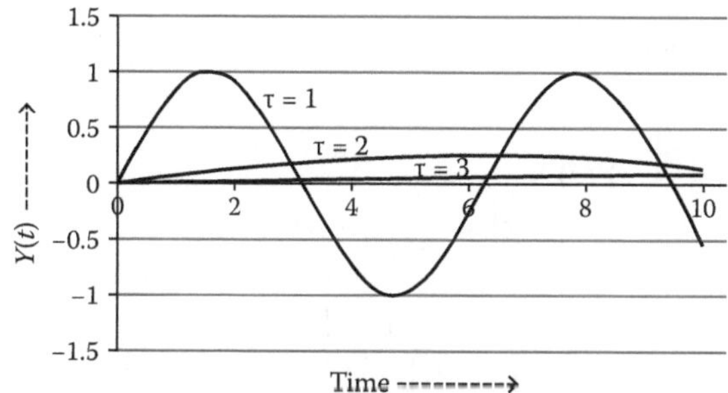

FIGURE 3.19 Responses of $Y(t)$ at increasing time constants at $\xi = 0$.

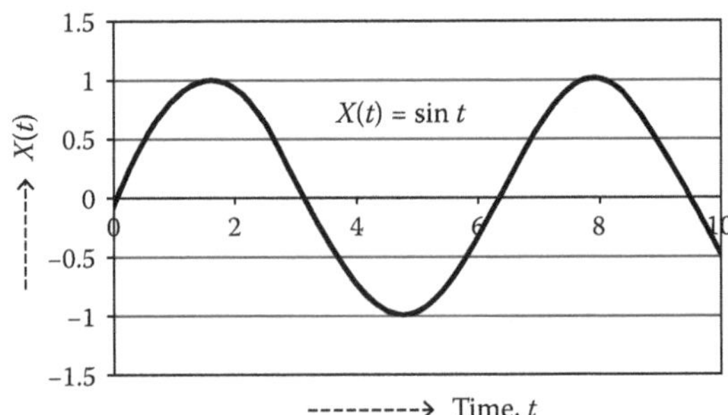

FIGURE 3.20 Sinusoidal disturbance for $X(t) = \sin t$.

3.10 SINUSOIDAL DISTURBANCE

Consider $X(t) = A \sin wt$ where w is the radian frequency and t is the time. The nature of the sine function is given in Figure 3.20 where it fluctuates between +1 and −1 when $A = 1$.

Practically, this type of behavior is manifested by alternating current or voltage, a point on a rotating circular ring, etc. It is also approximated by a flow rate through a valve, which fully opens and fully shuts periodically.

The Laplace transformation of $X(t)$,

$$X(s) = \frac{Aw}{s^2 + w^2}. \tag{3.35}$$

3.11 SINUSOIDAL DISTURBANCE TO A FIRST-ORDER SYSTEM

If such a disturbance is applied to a first-order system, the response of the output $Y(s)$ is given as

$$Y(s) = \frac{Aw}{(s^2 + w^2)(\tau s + 1)}. \tag{3.36}$$

By partial fraction,

$$\begin{aligned} Y(s) &= \frac{Aw}{\tau\left(s + \frac{1}{\tau}\right)(s + iw)(s - iw)} \\ &= \frac{B}{\left(s + \frac{1}{\tau}\right)} + \frac{C}{(s + iw)} + \frac{D}{(s - iw)}. \end{aligned} \tag{3.37}$$

where

$$B = \frac{Aw\tau}{1+w^2\tau^2}$$

$$C = \frac{Aw}{-2jw(jw\tau - 1)}$$

$$D = \frac{Aw}{2jw(jw\tau + 1)}.$$

Hence,

$$Y(t) = \frac{Aw\tau}{1+w^2\tau^2}e^{-t/\tau} - \frac{Aw}{2jw(jw\tau - 1)}e^{-jwt} + \frac{Aw}{2jw(jw\tau + 1)}e^{jwt}.$$

Using De Moivre's rule,

$$e^{-jwt} = \cos(wt) - j\sin(wt)$$
$$e^{jwt} = \cos(wt) + j\sin(wt)$$

or

$$\begin{aligned} Y(t) &= \frac{Aw\tau}{1+w^2\tau^2}e^{-t/\tau} - \frac{Aw\tau}{1+w^2\tau^2}\cos(wt) + \frac{A}{1+w^2\tau^2}\sin(wt) \\ &= \frac{Aw\tau}{1+w^2\tau^2}e^{-t/\tau} + \frac{A}{\sqrt{(1+w^2\tau^2)}}\sin(wt+\Phi) \end{aligned} \tag{3.38}$$

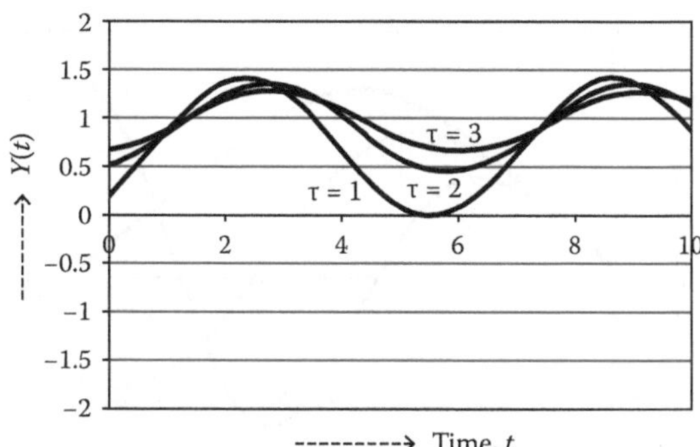

FIGURE 3.21 Responses of $Y(t)$ by sinusoidal disturbance for $X(t) = \sin t$ for different time constants.

where

$$\Phi = \tan^{-1}(-w\tau). \tag{3.39}$$

The responses of the output $Y(t)$ for $A = 1$ and $w = 1$ at different time constants are shown in Figure 3.21.

3.12 SINUSOIDAL DISTURBANCE TO A SECOND-ORDER SYSTEM

If the same sinusoidal disturbance is applied to a second-order system, the output is evaluated in the same way and is given next.

$X(s)$ is given by Equation 3.35, so $Y(s)$ is given as

$$Y(s) = \frac{Aw}{(s^2 + w^2)(2\tau^2 s^2 + 2\tau\xi s + 1)}.$$

By partial fractions,

$$Y(s) = \frac{1}{\tau^2 (s-\alpha)(s-\beta)(s^2 + w^2)}$$

$$Y(s) = \frac{Aw}{\tau^2 (s-\alpha)(s-\beta)(s+jw)(s-jw)}$$
$$= \frac{a_1}{(s+jw)} + \frac{a_2}{(s-jw)} + \frac{a_3}{(s-\alpha)} + \frac{a_4}{(s-\beta)}.$$

Responses for $\xi < 1$, >1, $=1$ and 0 are shown in Figures 3.22, 3.23, 3.24, and 3.25, respectively.

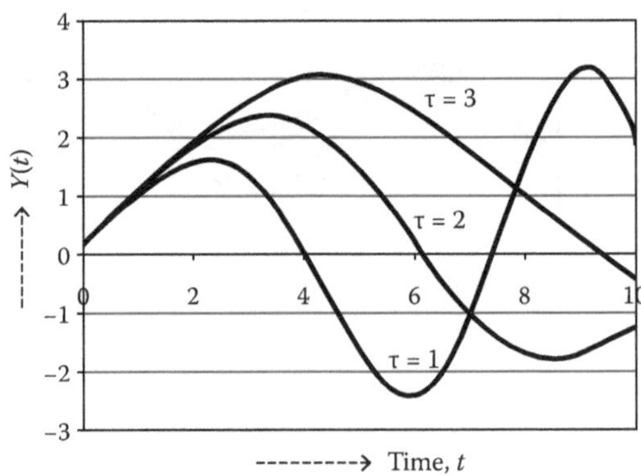

FIGURE 3.22 Response of output for $\xi = 0.1$ at three different time constants.

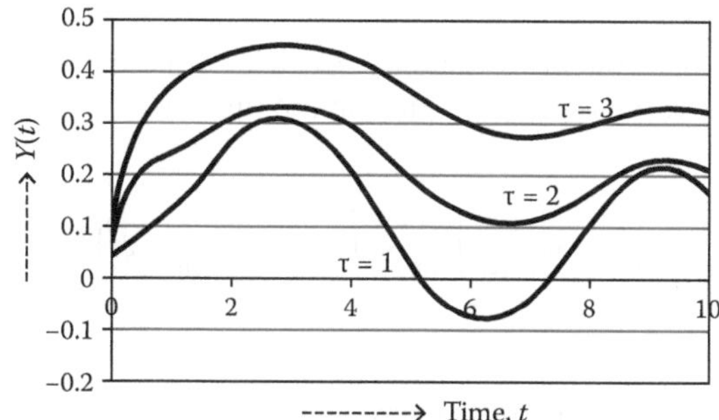

FIGURE 3.23 Response of output for $\xi = 3$ at three different time constants.

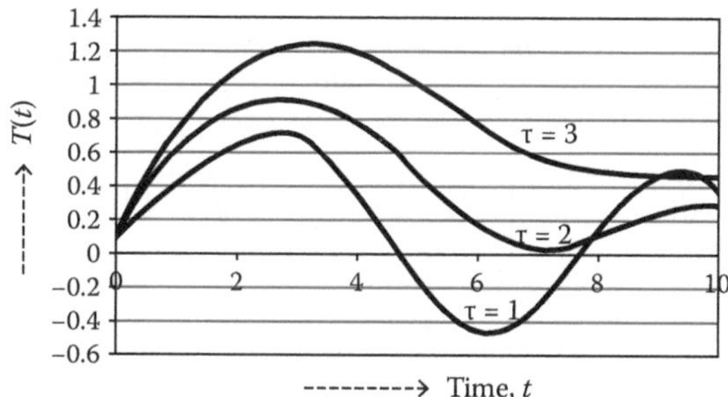

FIGURE 3.24 Response of output for $\xi = 1$ at three different time constants.

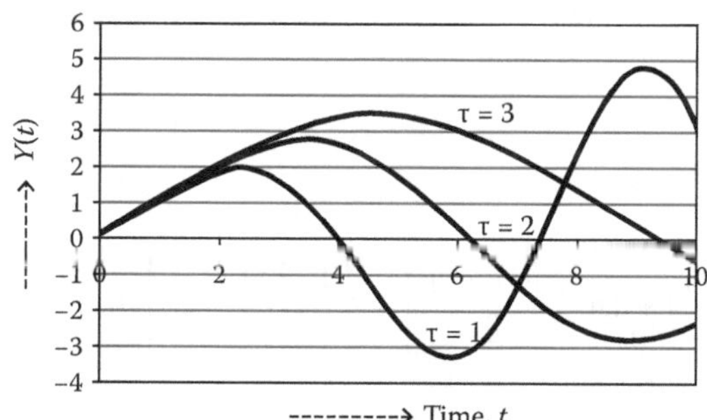

FIGURE 3.25 Response of output for $\xi = 0$ at three different time constants.

3.13 QUESTIONS AND ANSWERS

EXERCISE 3.1

For a second-order system having the transfer function as

$$\frac{Y(s)}{X(s)} = \frac{1}{(\tau^2 s^2 + 2\tau\xi s + 1)}$$

determine the response owing to a unit step disturbance in $X(t)$.

Answer:

See Section 3.3

EXERCISE 3.2

For the second-order system in Exercise 3.1, determine the overshoot and decay ratios. Because the response *Y*(*t*) as a function of time *t* is

$$Y(t) = 1 - \frac{e^{-\frac{\xi}{\tau}t}}{\sqrt{(1-\xi^2)}}\left[\sin\frac{\sqrt{(1-\xi^2)}}{\tau}t + \tan^{-1}\frac{\sqrt{(1-\xi^2)}}{\xi}\right]$$

in the event of overshoot, the *Y*(*t*) reaches the maximum value at which $dY(t)/dt = 0$, i.e.,

$$\frac{\xi e^{-\frac{\xi}{\tau}t}}{\tau\sqrt{(1-\xi^2)}}\sin\left[\frac{\sqrt{(1-\xi^2)}}{\tau}t + \tan^{-1}\frac{\sqrt{(1-\xi^2)}}{\xi}\right] - \frac{e^{-\frac{\xi}{\tau}t}}{\sqrt{(1-\xi^2)}}$$

$$-\frac{\sqrt{(1-\xi^2)}}{\tau}\cos\left[\frac{\sqrt{(1-\xi^2)}}{\tau}t + \tan^{-1}\frac{\sqrt{(1-\xi^2)}}{\xi}\right] = 0$$

i.e.,

$$\tan\left[\frac{\sqrt{(1-\xi^2)}}{\tau}t + \tan^{-1}\frac{\sqrt{(1-\xi^2)}}{\xi}\right] = \frac{\sqrt{(1-\xi^2)}}{\xi} = \tan\left[\tan^{-1}\frac{\sqrt{(1-\xi^2)}}{\xi}\right]$$

or

$$\tan\left[\frac{\sqrt{(1-\xi^2)}}{\tau}t + \tan^{-1}\frac{\sqrt{(1-\xi^2)}}{\xi}\right] = \frac{\sqrt{(1-\xi^2)}}{\xi} = \tan\left[n\pi + \tan^{-1}\frac{\sqrt{(1-\xi^2)}}{\xi}\right]$$

or

$$\frac{\sqrt{(1-\xi^2)}}{\tau}t = n\pi$$

or

$$t = \frac{n\pi\tau}{\sqrt{(1-\xi^2)}}$$

so,

$$Y(t) = 1 + \frac{e^{-\frac{\xi n\pi\tau}{\tau\sqrt{(1-\xi^2)}}}}{\sqrt{(1-\xi^2)}} \sin\left[\frac{\sqrt{(1-\xi^2)}}{\tau}\frac{n\pi\tau}{\sqrt{(1-\xi^2)}} + \tan^{-1}\frac{\sqrt{(1-\xi^2)}}{\xi}\right]$$

$$= 1 + \frac{e^{-\frac{\xi n\pi}{\sqrt{(1-\xi^2)}}}}{\sqrt{(1-\xi^2)}} \sin\left[n\pi + \tan^{-1}\frac{\sqrt{(1-\xi^2)}}{\xi}\right]$$

$$= 1 + \frac{e^{-\frac{\xi n\pi}{\sqrt{(1-\xi^2)}}}}{\sqrt{(1-\xi^2)}} \sqrt{(1-\xi^2)} = 1 + e^{-\frac{\xi n\pi}{\sqrt{(1-\xi^2)}}}.$$

Hence, the overshoot occurs at the first time when $n = 1$; therefore,

$$\text{Overshoot} = Y(t) - 1$$
$$= e^{-\frac{\xi\pi}{\sqrt{(1-\xi^2)}}}.$$

The decay ratio will be the next value of n, i.e., $n = 2$, and the

$$\text{Decay ratio} = Y(t) - 1$$
$$= e^{-\frac{2\pi\xi}{\sqrt{(1-\xi^2)}}}.$$

4 Process Control Loop System and Analysis

4.1 FEEDBACK CONTROL

In the usual process control, a feedback control system is widely used. In such a system, information about the variable to be controlled is obtained from the transducer, and the controller then calculates the difference between the desired set point (SP) and the current process value (PV) as measured. It then calculates the output signal, which drives the actuator of the control valve for manipulating the appropriate streams for necessary correction. Such a control system is presented in Figure 4.1.

In this control scheme, the difference between the SP and PV is first calculated by the controller. This difference is called the deviation or error ε, i.e.,

$$\varepsilon = SP - PV. \tag{4.1}$$

Because the measured value PV is deducted from the SP, the feedback is, in fact, a negative feedback system. The correction signal or the output O_c from the controller is a function of the error as

$$O_c = \text{funct}\ (\varepsilon). \tag{4.2}$$

4.2 TEMPERATURE CONTROL

The feedback control loop of a temperature-control system in a steam-heated tank as shown in Figure 1.2 in Chapter 1 is explained schematically in Figure 4.2 where a temperature transducer will feed back to the controller.

In this control scheme, the difference between the SP temperature T_{set} and measured process temperature T is first calculated by the controller. This difference is called the deviation or error ε, i.e.,

$$\varepsilon = T_{set} - T. \tag{4.3}$$

The correction signal or the output O_c from the controller is a function of the error as

$$O_c = A + K_c\,\varepsilon + \frac{Kc}{\tau_i}\int \varepsilon\,dt + K_o\tau_d\frac{d\varepsilon}{dt}. \tag{4.4}$$

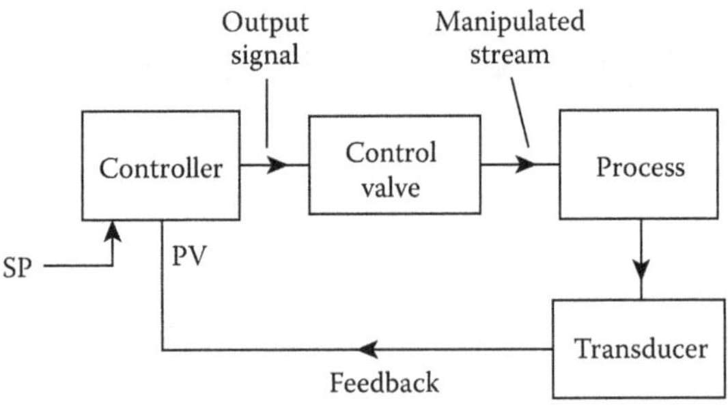

FIGURE 4.1 A feedback control loop.

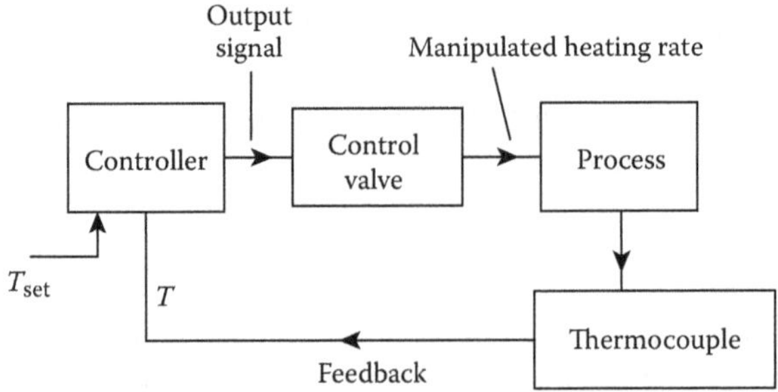

FIGURE 4.2 A feedback temperature-control loop.

The output function is known as the proportional–integral–derivative (PID) control equation, where A is the bias value, which is equal to the output signal, and error ε is zero, i.e., when the SP equals the PV. K_c, τ_i, and τ_d are three constants that will be discussed in detail in subsequent chapters.

4.3 LEVEL CONTROL

In a storage tank as shown in Figure 4.3, the level-control loop will use a controller to manipulate either the influent or effluent flow rates. The level transducer will feed back to the controller.

In this control scheme, the difference between the SP of level L_{set} and measured level L is first calculated by the controller as the error

$$\varepsilon = L_{set} - L. \tag{4.5}$$

The correction signal or the output O_c from the controller is a function of the error as given in Equation 4.4.

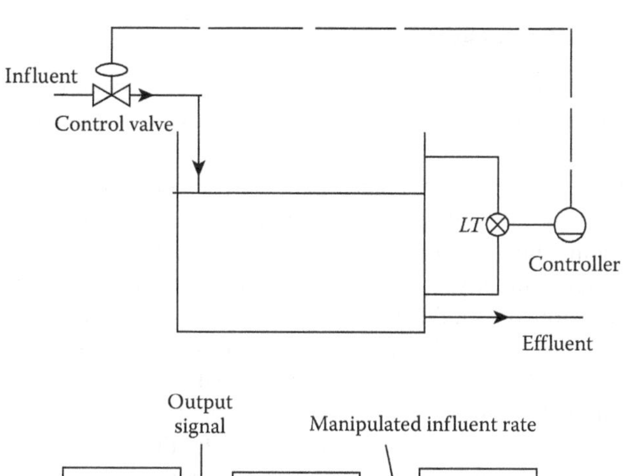

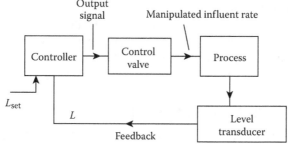

FIGURE 4.3 A feedback level-control loop.

4.4 FLOW CONTROL

In a flow control system as shown in Figure 1.1 in Chapter 1, the error is defined with respect to the set value of flow rate F_{set} and the measured value of flow rate F and is given as

$$\varepsilon = F_{set} - F. \tag{4.6}$$

The correction signal or the output O_c from the controller is a function of the error as given in Equation 4.4.

4.5 PRESSURE CONTROL

In a gas pressure control system as shown in Figure 1.3 in Chapter 1, the error is defined in terms of set pressure P_{set} and measured pressure P as

$$\varepsilon = P_{set} - P \tag{4.7}$$

and the output of the controller is given by Equation 4.4.

4.6 QUALITY CONTROL

The quality of a chemical product is measured by various properties, such as physical, chemical, optical, mechanical, electrical properties, etc. Very often, it is specified by a chemical composition, which may depend on the properties of influent, temperature, pressure, flow rate, and many other factors. For example, in a vapor-liquid separator, the composition of separated vapor and liquid streams will be dependent on the temperature and pressure in the separator drum. Hence, in order to maintain the desired composition of vapor and liquid streams, both the temperature and pressure must be controlled with the help of two controllers as explained in Figure 4.4.

The pressure transducer (PT) and pressure controller (PC) control the pressure in the separator drum. The error is given in Equation 4.7 with the controller output as given by Equation 4.4. Similarly, the temperature control is carried out with a temperature transducer (TT) and a temperature controller (TC) manipulating the steam rate in the separator drum. Remember also that a change in composition of a feed stream will change the quality of the vapor and liquid products. Hence, the SPs of temperature and pressure must be changed to keep the quality unaltered.

4.7 DIRECT AND REVERSE ACTING

Consider the level control system in a tank as shown in Figure 4.3. If the liquid level in the tank goes above the SP, it is necessary to reduce the inflow rate. Similarly, if the level falls, the inflow rate has to be increased. These types of actions are known as reverse acting, which is defined, in short, as far as the cause and its action are concerned, as increase-decrease and decrease-increase actions. Alternatively, a level-control action could be made by manipulating the outflow rate. For instance, while the level increases, the outflow has to be increased and vice versa. This type of action is called direct acting. Thus, the direct acting mode is expressed in terms of the

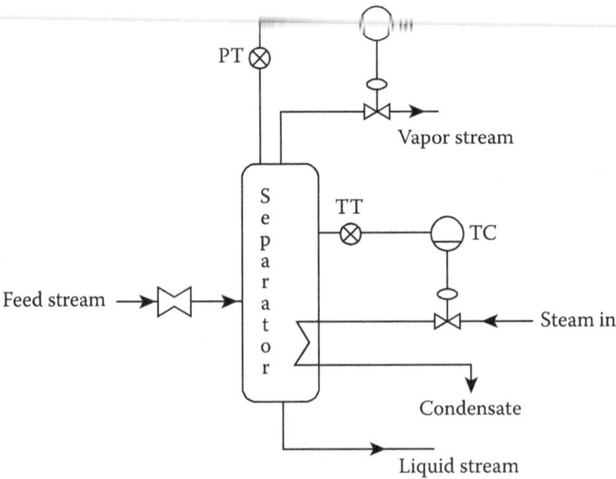

FIGURE 4.4 Composition control of a vapor–liquid separator.

cause and action as increase-increase and decrease-decrease actions. In any modern controller, these actions either direct- or reverse-acting modes, must be chosen by the operator. If direct-acting mode is selected wrongly for a control system requiring the reverse-acting mode instead, the control action will be impossible and may be devastating. For the level-control system, while the inflow rate is manipulated for control if direct-acting mode is selected, i.e., as the level increases, the inflow rate will also increase, making the level increase, and finally, liquid will overflow the tank. Similarly, if the level falls, the inflow rate will also fall causing the level to decrease and make the tank empty.

4.8 PROPORTIONAL CONTROL

As discussed above, in case the inflow rate is manipulated for level control in the reverse-acting mode, the output of the controller will decrease proportionally with negative error (i.e., when level goes above SP) and vice versa. The equation of this is given as

$$O_c = A + K_c \varepsilon. \tag{4.8}$$

Thus O_c will decrease while ε is negative, and O_c will increase while ε is positive. The overall reverse action will be effective while the control valve increases the flow rate while O_c increases and decreases the flow while O_c decreases, i.e., the valve has to be directly acting with O_c.

In case outflow is manipulated, a direct-acting mode has to be implemented by a proportional controller with a direct-acting control valve. The controller output equation will be

$$O_c = A - K_c \varepsilon. \tag{4.9}$$

such that as ε is negative, O_c will increase and vice versa, and the action will be direct acting.

In case the control valve is a reverse-acting one (normally open type), i.e., flow through the valve decreases as O_c increases, and flow increases while the O_c decreases, if the controllers chosen are reverse acting, the overall control action will be a direct-acting one, and if direct-acting mode is chosen for the controller, the overall control action will be reverse acting. Proportional control mode is found to be successful for flow-control systems. It can also be noted that the offset (which is defined as the error at infinite time, i.e., when sufficient time is allowed for the control system to settle the difference between the SP and the control variable) will not be zero.

4.9 PROPORTIONAL–INTEGRAL (PI) CONTROL

Consider a situation where ε becomes unchanged whether negative or positive. The value of O_c will be unchanged, and as a result, the flow rate through the valve will be fixed. Hence, the level in the tank will be unchanged and may be different than the

SP value. For such a situation, a proportional equation is added with integration of the error as given in Equation 4.10.

$$O_c = A + K_c \varepsilon + \frac{K_c}{\tau_i} \int \varepsilon \, dt \tag{4.10}$$

It is understood that O_c will be changing even if the error becomes fixed as a cumulative error under the integration function will go on changing.

The value of K_c will be negative for direct-acting mode, and K_c will be positive if reverse-acting mode is selected. PI-control mode is found to be suitable for level-control systems. However, PI-control mode generates oscillation of the control variable around the SP and ultimately settles to the SP value without offset.

4.10 PROPORTIONAL–INTEGRAL–DERIVATIVE CONTROL

In some process control systems, PI control may not be successful. In this case, a derivative of the error term is added with a PI equation. This is given in Equation 4.4. This will be discussed in more detail in the next chapter. The PID mode of control is found to be successful for temperature-control systems. In addition to the benefit of zeroing the offset, the amplitude of oscillation of the control variable is reduced as compared to PI-control mode.

4.11 ON–OFF CONTROL

This control action either fully opens a valve (or a switch) or shuts a valve (or a switch). In fact, an on/off action is not a control action; rather, it is a security or safety operation in case the process value exceeds the highest value or falls below the minimum value. This is the case when the sensor breaks, power fails, or operating parameters exceed the limits, which can lead to accidents. For example, a solenoid-operated valve trips the power supply in case the level of a tank crosses a certain safe limit. A pressure safety valve pops in case pressure exceeds a certain limit; this is also an example of an on/off situation.

However, a control valve may behave like an on/off valve if the K_c is so selected that the output of the controller O_c reaches a maximum or minimum limit (typically 4 ma or 20 ma current), and the valve is either fully open or fully closed. Alarms are usually fitted to indicate the events when the limits are exceeded, and corrective action must be made by the operator or engineer.

4.12 QUESTIONS AND ANSWERS

EXERCISE 4.1

A flow control (feedback) system uses a proportional controller with the following equation delivering the signal O_c in mA DC every 0.1 second (100 millisecond) interval:

$$O_c = 12 + K_c\, e(t) \tag{4.11}$$

where K_c is to be selected by the user for which a good control performance will be achieved and instantaneous deviation or error signal $e(t)$ is

$$e(t) = \mathrm{SP} - \mathrm{PV}. \tag{4.12}$$

The input and output of the controller is in the range of 4–20 mA DC. A control valve having the following relationship with the flow F is connected with the output signal line of the controller.

$$F = 6.25\,(O_c - 4) \tag{4.13}$$

where F is the flow rate in cu.m/hr in the pipeline.

The flow transmitter produces a signal (PV) in the range of 4–20 mA DC corresponding to a flow rate from 0 to 100 cu.m/hr.

$$\mathrm{PV} = 4 + 0.16F. \tag{4.14}$$

If the flow rate is at a steady value of 50 cu.m/hr initially, determine the flow fluctuations as a function of time (at every 100 milliseconds until 1.0 second) while the SP is changed from 50 cu.m/hr to 60 cu.m/hr in no time.

Solution:

Let $K_c = 0.9$

Time, sec	SP, cu.m/hr	SP, mA	*F*, cu.m/hr	PV, mA	e(*t*), mA	O_c, mA
$t < 0$	50	12	50	12	0	12
0	60	13.6	50	12	1.6	13.44
0.1	60	13.6	59	13.44	0.16	12.144
0.2	60	13.6	50.9	12.144	1.456	13.31
0.3	60	13.6	58.18	13.31	0.29	12.26
0.4	60	13.6	51.63	12.26	1.339	13.205
0.5	60	13.6	57.53	13.20	0.3949	12.35
0.6	60	13.6	52.18	12.35	1.25	13.125
0.7	60	13.6	57.83	13.125	0.475	12.42
0.8	60	13.6	52.62	12.42	1.172	13.05
0.9	60	13.6	56.56	13.05	0.55	12.5
1.0	60	13.6	53.09	12.5	1.10	12.99

The above results indicate that F has not reached the new SP value of 60 cum/hr after 1.0 second. Rather, it has reached 58.18 cu.m/hr with an offset of (60 – 58.18) = 1.82 cu.m/hr after 0.3 second but fluctuated below the SP. This is clear from Figure 4.5.

Try with other values of K_c as an exercise left to the students to find the behavior of F as a function of time.

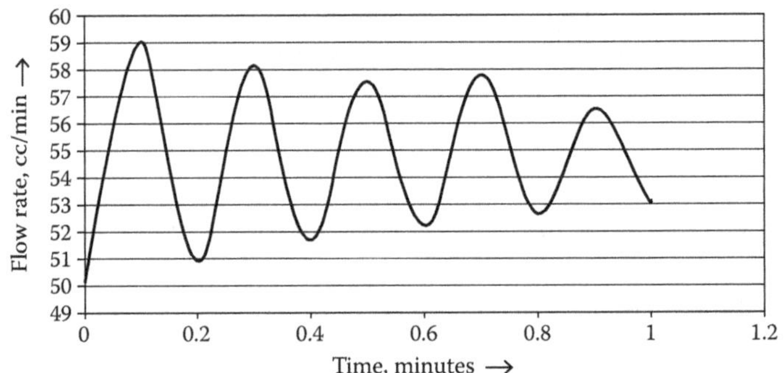

FIGURE 4.5 Flow rate fluctuation for a proportional controller.

EXERCISE 4.2

Repeat Exercise 4.1 with a PI controller and determine the response for any set of values of K_c and integration time (T_i).

The equation of the controller is

$$O_c = 12 + K_c\, e(t) + (K_c/T_i) \int e(t)\, dt \tag{4.15}$$

Solution:

The controller equation can be approximately written as

$$O_c = 12 + K_c\, e(t) + (K_c/T_i)\, \Sigma e(t)\, \Delta t \tag{4.16}$$

where $\Delta t = 0.1$ sec.

Take $T_i = 1$ sec and $K_c = 0.9$. Calculations are shown in the following table.

Time, sec	SP, cu.m/hr	SP, mA	F, cu.m/hr	PV, mA	e(t), mA	O_c, mA
$t < 0$	50	12	50	12	0	12
0	60	13.6	59	12	1.6	13.44
0.1	60	13.6	59.9	13.44	0.16	13.58
0.2	60	13.6	59.99	13.58	0.02	13.60
0.3	60	13.6	60	13.6	0.00	13.60
0.4	60	13.6	60	13.6	0.00	13.60
0.5	60	13.6	60	13.6	0.00	13.60
0.6	60	13.6	60	13.6	0.00	13.60
0.7	60	13.6	60	13.6	0.00	13.60
0.8	60	13.6	60	13.6	0.00	13.60
0.9	60	13.6	60	13.6	0.00	13.60
1.0	60	13.6	60	13.6	0.00	13.60

Here, the response is faster as the SP is reached in about 0.2 seconds.

This is evident from Figure 4.6.

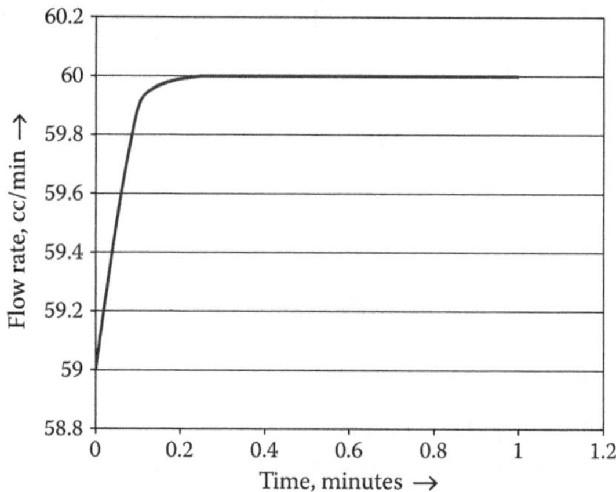

FIGURE 4.6 Flow rate variation resulting from PI controller.

EXERCISE 4.3

A proportional controller generates 4 mA when the error signal is –16 mA and 20 mA when the error is 16 mA. The control valve is normally a closed-type valve. It is fully open when the input current to it is 20 mA. Determine the proportional gain, proportional band, and the bias signal of the controller.

$$Y = A + K_c\varepsilon \tag{4.17}$$

where

Y = the output of the proportional controller (mA)
K_c = is the proportional gain of the controller
ε = is the error or deviation of the PV from the SP

$$\varepsilon = \text{SP} - \text{PV} \tag{4.18}$$

A = the bias value of the controller when $\varepsilon = 0$, mA
Y = 4 mA, and $\varepsilon = -16$ mA, so $4 = A - 16\ K_c$

and

Y = 20 mA, and ε = 16 mA, so $20 = A + 16K_c$
So $A = 24/2 = 12$ mA and $K_c = 16/32 = 0.5$ mA/mA
The proportional gain $K_c = 0.5$

Proportional band (PB) is defined as the change in error as a percentage of the span of the scale of the controller that causes the control valve to be fully closed or fully open. In this case, because the output of the controller is 4 to 20 mA, which causes the valve to be fully closed or open

$$\text{PB} = \Delta\varepsilon/\text{span} \times 100 = (16 + 16)/(20 - 4) \times 100 = 32/16 \times 100 = 200\%$$

Because all the input and output signals are in the same units, i.e., in mA, K_c and PB are related as

$$K_c = 100/\text{PB}.$$

Hence,

$$K_c = 100/200 = 0.5.$$

EXERCISE 4.4

A differential pressure cell (DPC) transmitter is used to transmit an electrical signal equivalent to the differential pressure across the orifice. The signal from the DPC varies from 4 to 20 mA DC (milliampere direct current) for a flow rate from 0 to 100 cc/min. Because flow rate is proportional to the square root of the pressure drop, it is necessary to install another instrument called a square root extractor (SRE) in the line, which linearize the current signal from the DPC, so the current signal is proportional to the flow rate. The output signal from the SRE is within the range of 4 to 20 mA for the input signal coming from the DPC in the same range. The relationship of the signals to the flow rate is mathematically expressed as

$$\text{DPC: } I_{\text{dpc}} = 4 + 16\,(q/100)^2 \tag{4.19}$$

$$\text{SRE: } I_{\text{sre}} = 4 + 4\sqrt{(I_{\text{dpc}} - 4)} \tag{4.20}$$

where I_{dpc} and I_{sre} are the current signals in mA DC, and q is the flow rate in cc/min. The following table shows the signal values as they are available in a calibration.

q in cc/min	I_{dpc} in mA DC	I_{sre} in mA DC
0	4	4
10	4.16	5.6
20	4.64	7.2
30	5.44	8.8
40	6.56	10.4
50	8	12
60	9.76	13.6
70	11.84	15.2
80	14.24	16.8
90	16.96	18.4
100	20	20

Note: SRE output is linear with the flow rate.

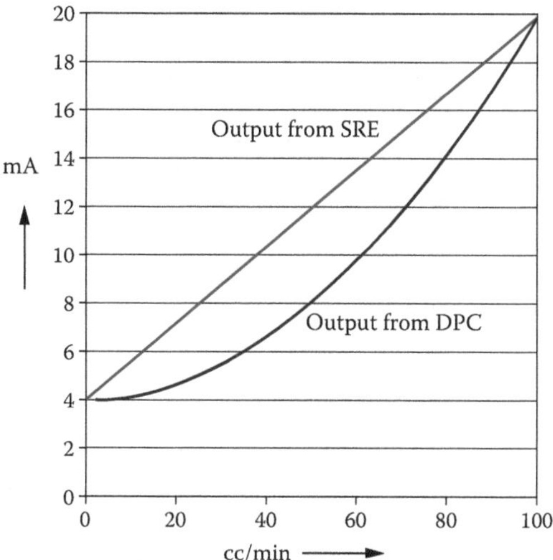

FIGURE 4.7 Output signals from DPC and SRE.

A flow controller having the following control logic is connected with the SRE as described above

$$I_{\text{controller}} = 12 + 0.6e + 0.6 \int e\, dt \tag{4.21}$$

where e is the difference between the SP (desired flow rate in current equivalent) and the measured flow signal from SRE, i.e.,

$$e = (I_{\text{sp}} - I_{\text{sre}}) \tag{4.22}$$

The output signal from this flow controller varies from 4 to 20 mA DC for the variation in e.

A motor-operated valve is a final control element that is driven by the current signal such that it fully closes the valve when there is no current or when the current applied is less than and equal to 4 mA DC. The valve starts opening when the current applied is greater than 4 mA DC and fully opens when the current is 20 mA DC or more. Such a control valve is connected with the controller described above. A flow control system is shown in Figure 4.8. As the desired SP is different from the measured flow rate, the controller sends a signal to the control valve either to reduce the flow rate when the SP is more than the measured flow and vice versa. Determine the flow rate as a function of time (take the interval of $\Delta t = 0.1$ sec) for a change in flow rate from 20 to 25 cc/min.

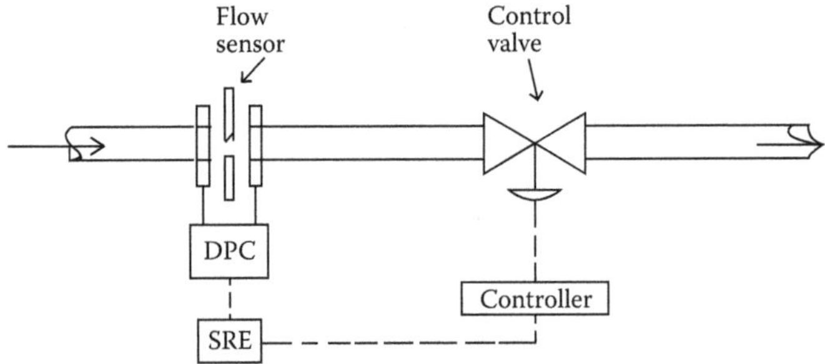

FIGURE 4.8 Flow rate control through a pipe.

Computer solution:

Time (sec)	I_{dpc} mA	I_{sre} mA	I_{sp} mA	q cc/min
Δt	4.64	7.2	8	20
$2\Delta t$	8.5	12.5	8	36.2
$3\Delta t$	6	9.8	8	29.48
$4\Delta t$	5.4	8.7	8	26.79
$5\Delta t$	5.14	8.3	8	25.71
$6\Delta t$	5.05	8.11	8	25.11
$7\Delta t$	5	8	8	25
$8\Delta t$	5	8	8	25

Note that a new SP of 25 cc/min is achieved after the sixth interval, i.e., after 0.6 second.

This is depicted in Figure 4.9.

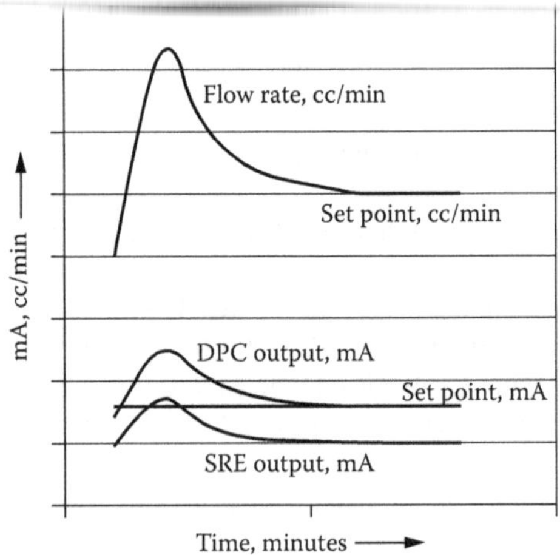

FIGURE 4.9 Output signals from the DPC and SRE and flow rate.

EXERCISE 4.5

A controller circuit is constructed using resistors, a capacitor, and an inductor of the coil with a voltage amplifier as shown in Figure 4.10. Prove that the output voltage of the circuit is given by the following relationship

$$-e_0 = \varepsilon R_3 + 1/C \int \varepsilon dt + L\, d\varepsilon/dt \tag{4.23}$$

where $\varepsilon = (e_1/R_1 - e_2/R_2)$, C is the capacitance, and L is the inductance of the coil.
Answer:

Taking the high gain of the amplifier, $e_0 >> e_g$, i.e., e_g is neglected.
Hence,

$$i = i_1 + i_2 = \frac{e_1}{R_1} - \frac{e_2}{R_2} \tag{4.24}$$

and

$$e_g - e_0 \approx -e_0 = iR_3 + \frac{\int i\,dt}{c} + L\frac{di}{dt} \tag{4.25}$$

or

$$e_0 = \varepsilon R_3 + \frac{\int \varepsilon\,dt}{c} + L\frac{d\varepsilon}{dt} \tag{4.26}$$

where

$$\varepsilon = \frac{e_1}{R_1} - \frac{e_2}{R_2}. \tag{4.27}$$

Equation 4.26 is equivalent to the output of a PID controller relationship.

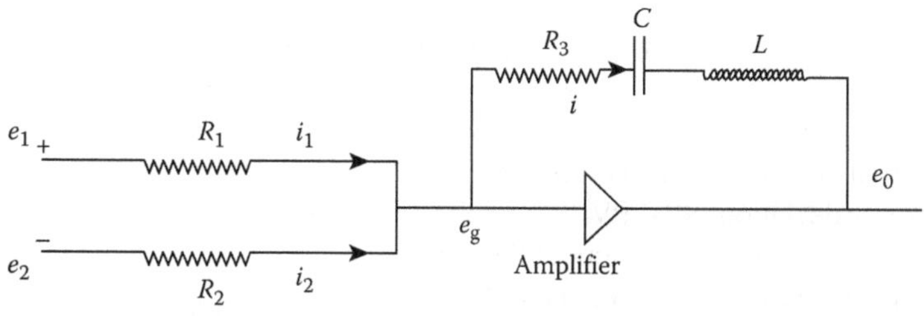

FIGURE 4.10 Controller circuit.

EXERCISE 4.6

A flow transducer measures flow rates from 0 to 100 m^3/hr and transduces current in the range from 4 to 20 mA. If the flow rate is 50 m^3/hr, the flow controller accesses this, compares it with the SP, and generates an output signal in the range of 4 to 20 mA to drive a control valve. If the control valve is normally open when the signal is 4 mA and closes completely when the signal is 20 mA, and a SP of 50 m^3/hr is set in the controller, determine the bias value of the controller.

Answer:

The relationship of the flow rate and the transducer signal is obtained as a linear one

$$\text{FT} = 4 + 0.16\ \text{Flow}$$

$$\text{SP} = 4 + 0.16\ \text{Flow}$$

where FT and SP are the current signal for the flow rate.

Because the SP and FT are the same, the error is zero; hence the output signal of the controller is the bias value. The output signal from the controller should be such that the flow through the control valve is 50 m^3/hr.

Because the control valve is fully opened at $i = 4$ mA and fully closed at $i = 20$ mA, making a linear relationship, the flow rate through the control valve is

$$\text{Flow} = 0,\ i = 20\ \text{mA}$$

$$\text{Flow} = 100,\ i = 4\ \text{mA}$$

$$\text{Flow} = 125 - 6.25i$$

Hence, to have flow = 50 m^3/hr, $i = (125 - 50)/6.25 = 12$ mA.

The bias value of the controller is 12 mA.

EXERCISE 4.7

A PID controller is used to control levels in a tank. The controller has the PID equation

$$O_c = A + K_c\varepsilon + K_c/\tau_i \int \varepsilon\, dt + K_c\varepsilon\tau_d\, d\varepsilon/dt$$

where

O_c: output signal from the controller
ε: error or deviation, SP – PV,
A, K_c, τ_i, and τ_d are parameters of the controller
A: bias value of the signal when $\varepsilon = 0$
τ_i: integration time constant
τ_d: derivative time constant

If the SP is 14 mA and the process value is 9 mA, determine the output signal of the controller in mA. Given that A = 12 mA, K_c = 0.2 mA/mA, τ_i = 0.1 sec, τ_d = 0.5 sec. Assume that the controller delivers a signal every 0.1 sec.
Answer:

$$SP = 14 \text{ mA and } PV = 9 \text{ mA, so } \varepsilon = 14 - 9 = 5 \text{ mA.}$$

So output from the controller is

$$O_c = 12 + 0.2 \times 5 + 0.2/0.1\ (5t) + 0.5 \times 0.5 \times 0 = 13 + 10t.$$

Because the controller sends a signal every 0.1 second, the output signal will vary from time to time until the error is unchanged. This is evaluated below:

Time, sec	Output, mA	Time, sec	Output, mA
0.0	13	0.1	14
0.2	15	0.3	16
0.4	17	0.5	18
0.6	19	0.7	20
0.8	20	0.9	20
1.0	20	1.1	20

Note that when the output reaches 20 mA, it will not increase further, and it will remain at 20 mA with the increasing time. The output of the controller will also not change as long as the error is unchanged.

5 Control Loop Analysis

5.1 CONTROL LOOP

A control loop is a complete control system where the transducer, controller, and control valve are connected. In such a loop, the transducer sends a signal to the controller, which then determines the appropriate output signal for manipulating a stream responsible for change in the process variable (the variable to be controlled) by actuating a final control element, usually a control valve. An example of a control loop is shown in Figure 5.1.

5.2 CONTROL STRATEGIES

A control strategy is a methodical processing of measured signals and structuring of a control loop for effective correction of the process variable to be controlled. For example, a feedback control loop involves a transducer to feed the measured variable (variable to be controlled) to the controller, which generates output based on a logic to actuate a control valve for manipulation of certain streams for necessary corrections. Such a control loop is shown in Figure 5.2. In this system, the controller accesses the feedback (measured value), and action is taken in a continuous manner. This type of feedback control is called a single-input/single-output (SISO) strategy. There may be more than one input and more than one output, i.e., multiple-input/multiple-output abbreviated as MIMO. An example of a MIMO system can be found in an inferential control. A control system may have a single input and multiple outputs (SIMO) as in the split range control strategy or also multiple inputs and single output (MISO) as in a cascade control. Various control strategies, such as cascade, feed forward, ratio, inferential, override, etc. are discussed in Chapter 7. Feedback strategy is the most widely used strategy as shown in Figure 5.2.

5.3 CONTROL LOGICS

Control logic refers to decisions made by the controller relating the existing process variable (PV) and the desired set point (SP) in order to determine the output signal to manipulate the process streams for any necessary correction of the PV. Examples of such logics are on/off, proportional–integral–derivative (PID), fuzzy, artificial neural logic, etc. On/off and PID logics with direct- or reverse-acting modes are more common. Advanced control logics, such as fuzzy logic, artificial neural logic, etc., are covered in Chapter 8. The traditional control strategy is a PID control logic. Though this has already been discussed in Chapter 4, this logic is still prevalent in

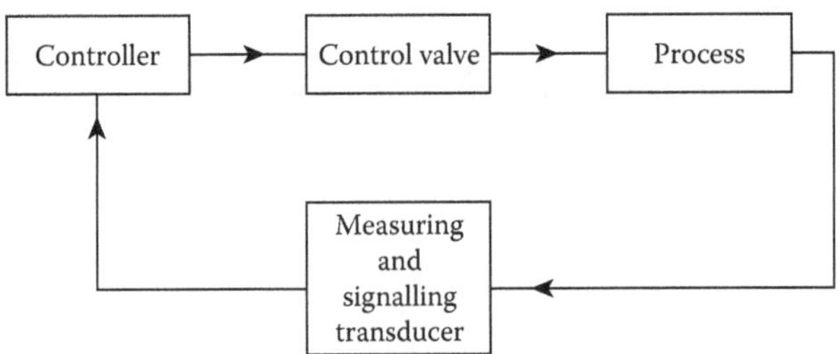

FIGURE 5.1 Control loop.

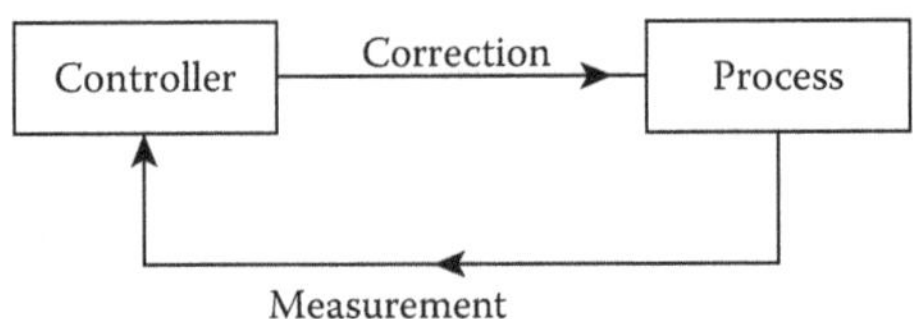

FIGURE 5.2 Feedback control strategy.

most of the modern control systems, and we will, therefore, use this logic for further studies in this book.

5.4 NEGATIVE FEEDBACK CONTROL SYSTEM

In a feedback control loop, the measured variable is compared to the desired SP. The difference obtained by subtracting the measured PV from the desired SP is taken as the deviation or error (ε). Thus, the feedback is negated (subtracted) from the SP; hence this system is known as the negative feedback system. The output of the controller is then evaluated as a function of the PID relationship. The controller output then actuates a control valve.

5.5 DEVELOPMENT OF A LAPLACIAN BLOCK DIAGRAM

In order to analyze any controlled process, the Laplace transformation method is convenient. As discussed about the Laplace transformation methods in Chapter 2, a Laplace transfer function converts (the Laplace transformation of) an input variable to (Laplace transformation of) an output variable. Schematically, the transfer function acts as the multiplier of the Laplace transformation of the input and converts it to an output. Thus, the negative feedback system can be depicted theoretically by using the transfer function of the process (G_p), controller (G_c), transducer (G_m), and control valve (G_v). The subtraction of the measured PV and the SP is represented by a comparator symbol, and the addition of the correcting signal (M) to the process is represented by an adder symbol. This is explained in Figure 5.3.

$Y_{set}(s)$, $Y_m(s)$, $M(s)$, and $U(s)$ are the Laplace transformations of SP, measured variable, manipulated variable, and load variable, respectively. The Laplacian diagram of the negative feedback control loop is presented in Figure 5.4.

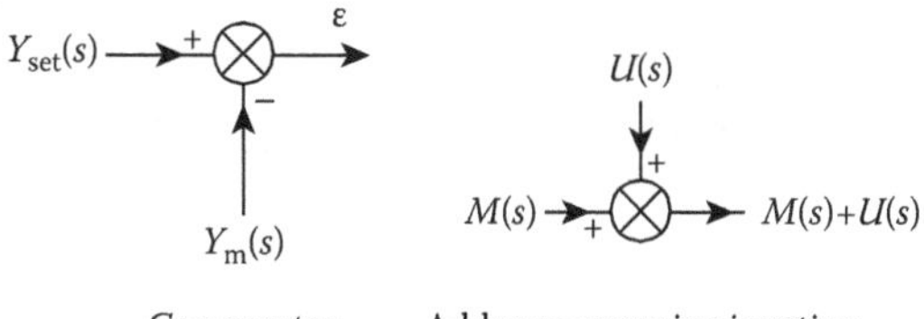

FIGURE 5.3 Symbolic representation of a comparator and an adder.

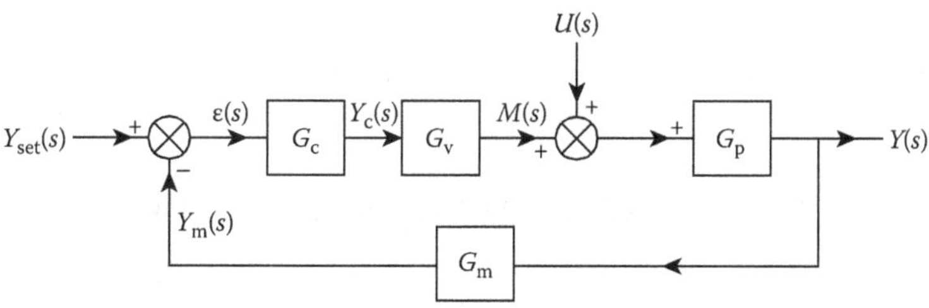

FIGURE 5.4 Laplacian block diagram of a negative feedback control loop.

The transfer functions of the outputs from the comparator, controller, control valve, adder, process, and transducer blocks are, respectively, $\varepsilon(s)$, $Y_c(s)$, $M(s)$, $M(s) + U(s)$, $Y(s)$, and $Y_m(s)$, respectively. Using the transfer function properties, the controlled variable $Y(s)$ can be related with the SP $Y_{set}(s)$ and load $U(s)$ as follows:

$$Y(s) = G_p \{M(s) + U(s)\} = G_p M(s) + G_p U(s). \tag{5.1}$$

Because

$$M(s) = Y_c(s)\, G_v \tag{5.2}$$

$$Y_c(s) = G_c \varepsilon(s) \tag{5.3}$$

$$Y_m(s) = G_m Y(s) \tag{5.4}$$

and

$$\varepsilon(s) = Y_{set}(s) - Y_m(s) \tag{5.5}$$

hence, substituting Equations 5.2, 5.3, 5.4, and 5.5 relationships in Equation 5.1 results in

$$Y(s) = G_p G_c G_v (Y_{set}(s) - Y(s) G_m) + G_p U(s) = G_p G_c G_v Y_{set}(s) - G_p G_c G_v G_m Y(s) + G_p U(s)$$

or

$$Y(s)[1 + G_pG_cG_vG_c] = G_pG_cG_vY_{set}(s) + G_pU(s)$$

or

$$Y(s) = \frac{\left[G_pG_cG_vY_{set}(s) + G_pU(s)\right]}{\left[1 + G_pG_vG_mG_c\right]}. \tag{5.6}$$

This is the closed-loop relationship for the control variable as a function of the SP and the load.

5.6 OPEN- AND CLOSED-LOOP TRANSFER FUNCTIONS

If the feedback path is disconnected from the comparator, the relationship becomes

$$Y_m(s) = G_mG_pG_cG_vY_{set}(s) + G_mG_pU(s). \tag{5.7}$$

The product of all the transfer functions, $G_mG_pG_vG_c$, is known as the open-loop transfer function or G_{ol}. Effects on the control variable for the simultaneous variation of SP and the load is difficult to analyze; rather, it is studied in the variation of either of them, keeping the variation of the other unchanged. These are separately known as regulatory and servo control analysis.

5.7 REGULATORY CONTROL ANALYSIS

In this analysis, the effect on the control variable resulting from the change of load in absence of variation of SP is analyzed and is known as the regulatory analysis. Equation 5.6 then yields the following relationship:

$$\frac{Y(s)}{U(s)} = \frac{G_p}{\left[1 + G_pG_vG_mG_c\right]}. \tag{5.8}$$

This is the closed-loop transfer function, and the block diagram is represented as

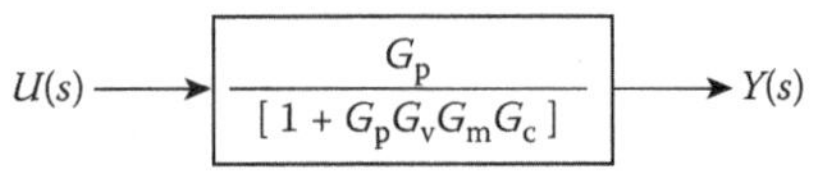

$$U(s) \longrightarrow \boxed{\frac{G_p}{[1 + G_pG_vG_mG_c]}} \longrightarrow Y(s)$$

FIGURE 5.5 Closed loop transfer function for regulatory control analysis.

5.8 SERVO CONTROL ANALYSIS

In this system, the effect of SP variation is studied alone in absence of the variation of load. Such an analysis is known as the servo mechanism or servo control analysis. Equation 5.6 then yields the following relationship:

$$\frac{Y(s)}{Y_{\mathrm{set}}(s)} = \frac{G_{\mathrm{p}}G_{\mathrm{v}}G_{\mathrm{c}}}{\left[1+G_{\mathrm{p}}G_{\mathrm{v}}G_{\mathrm{c}}G_{\mathrm{m}}\right]}. \tag{5.9}$$

This is the closed-loop transfer function, and the block diagram is represented as

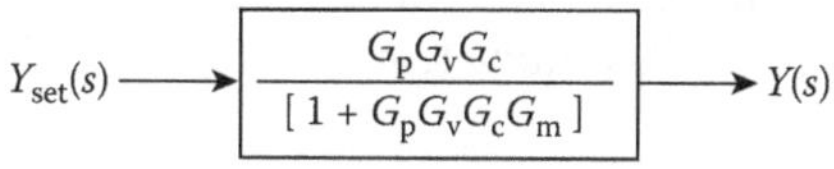

$$Y_{\mathrm{set}}(s) \longrightarrow \boxed{\frac{G_{\mathrm{p}}G_{\mathrm{v}}G_{\mathrm{c}}}{[\,1+G_{\mathrm{p}}G_{\mathrm{v}}G_{\mathrm{c}}G_{\mathrm{m}}\,]}} \longrightarrow Y(s)$$

FIGURE 5.6 Closed-loop transfer function for servo control analysis.

5.9 TEMPERATURE CONTROL LOOP ANALYSIS

A steam-heated tank with a temperature-control arrangement is shown in Figure 5.7 where a liquid enters and leaves at a constant rate w mass/time such that the level is unchanged. The flow rate of steam is manipulated to control the temperature of the liquid in the tank. A theoretical study can be made by using Laplace transfer functions.

The Laplace transfer function of the process is already determined in Chapter 2 and is given as

$$Y(s) = \frac{X(s) + \dfrac{Q(s)}{wC_{\mathrm{p}}}}{(\tau s + 1)} \tag{5.10}$$

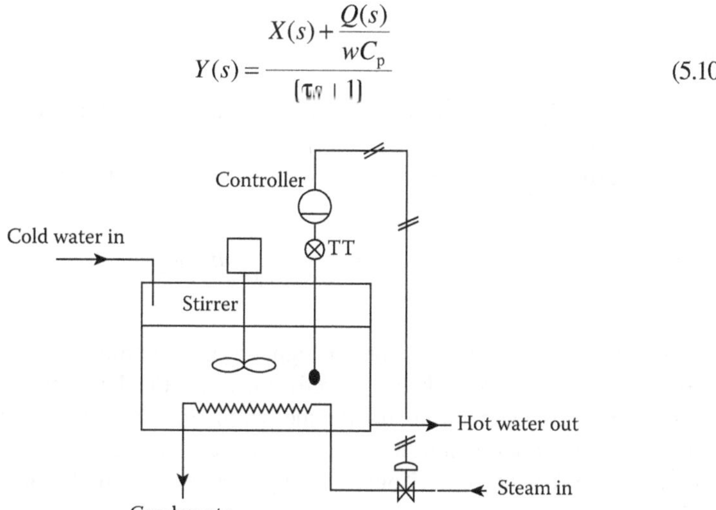

FIGURE 5.7 Temperature control arrangement in a steam-heated tank.

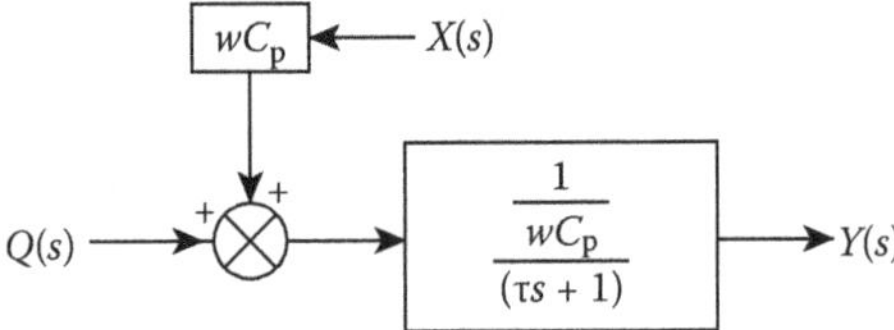

FIGURE 5.8 Transfer function diagram of the open-loop process of a steam-heated tank.

where $Y(s)$, $X(s)$, and $Q(s)$ are the Laplace transformations of the tank temperature, the inlet temperature of the liquid and heating rate, respectively, measured with respect to their steady-state values. τ and C_p are the time constant and specific heat of the liquid. The equivalent Laplacian block diagram is shown in Figure 5.8.

The transfer function of a temperature transducer has been found to be a first-order transfer function as explained in Chapter 2. If the indicated temperature by the transducer is Y_m in the deviation form with respect to its steady-state value, the transfer function is given as

$$\frac{Y_m(s)}{Y(s)} = \frac{1}{\{\tau_m s + 1\}}. \tag{5.11}$$

The Laplacian transfer function block is represented as

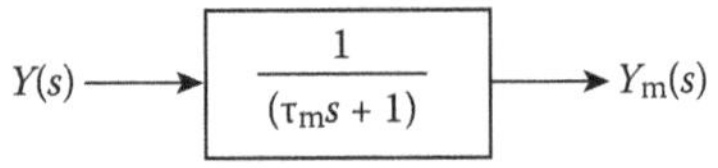

$Y(s) \longrightarrow \boxed{\dfrac{1}{(\tau_m s + 1)}} \longrightarrow Y_m(s)$

FIGURE 5.9 Transfer function block diagram for temperature transducer.

where $Y(s)$ and $Y_m(s)$ are the Laplace transformations of tank temperature and its measurement, respectively.

The controller logic of a PID controller has been given in Chapter 1 as

$$O_c(t) = A + K_c\varepsilon + \frac{K_c}{\tau_i}\int \varepsilon\, dt + K_c\tau_d\frac{d\varepsilon}{dt} \tag{5.12}$$

where $O_c(t)$ is the instantaneous output signal of the controller, A is the steady-state output signal when error (ε) becomes zero, and $\varepsilon(t)$ is the deviation or error between the SP of the tank temperature and the measured temperature $= Y_{set}(t) - Y_m(t)$. If the output signal from the controller $Y_c(t)$ is measured with respect to the steady-state value A, then $Y_c(t) = O_c(t) - A$. Thus, the controller equation in deviation form will be

$$Y_c(t) = K_c\varepsilon + \frac{K_c}{\tau_i}\int \varepsilon\, dt + K_c\tau_d\frac{d\varepsilon}{dt}. \tag{5.13}$$

Taking the Laplace transfer of Equation 5.13, the transfer function is evaluated as

$$Y_c(s)/\varepsilon(s) = K_c(1 + 1/\tau_i s + \tau_d s) \tag{5.14}$$

where K_c, τ_i, and τ_d are the proportional gain, integration time, and derivative time, respectively, of the PID controller. The Laplacian transfer function block of the controller is then represented as

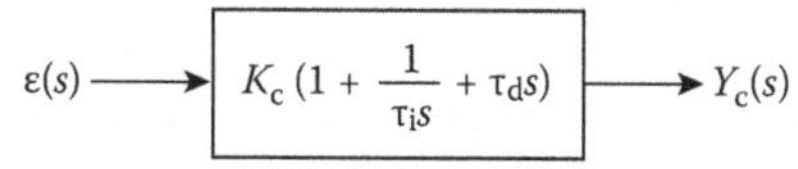

$$\varepsilon(s) \longrightarrow \boxed{K_c\left(1 + \frac{1}{\tau_i s} + \tau_d s\right)} \longrightarrow Y_c(s)$$

FIGURE 5.10 Transfer function block of controller.

The comparator is used to symbolize the determination of error as

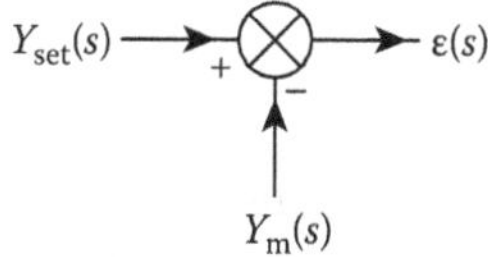

FIGURE 5.11 The comparator block in Laplacian transformations.

A pneumatic control valve is shown in Figure 5.12.

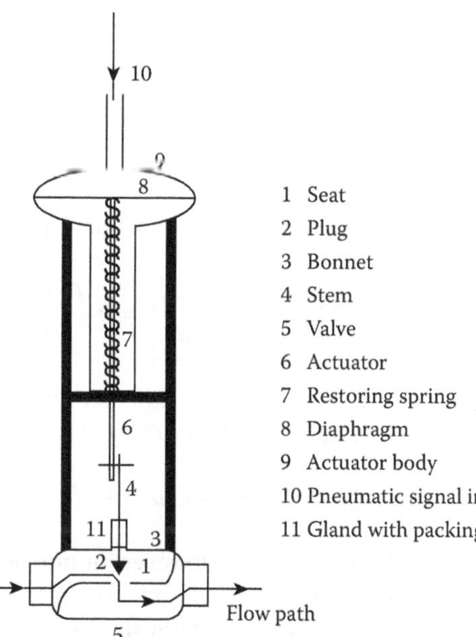

FIGURE 5.12 Pneumatic control valve.

If the output signal of the controller is a pneumatic pressure signal entering the actuator entrance position marked "10", the diaphragm ("8") will be pushed down the actuator link or shaft ("6"), which ultimately drives the stem ("4") of the valve to manipulate flow through the annular space between the plug ("2") and seat ("1"). A force balance over the diaphragm, actuator link, restoring spring ("7") and the viscous friction of the fluid over the plug can be made to evaluate the transfer function of the valve.

If the pressure signal from the controller Y_c enters the control valve, the force exerted by it on the diaphragm of the area of cross section a is aY_c. This force is then opposed by the elastic force exerted by the elastic diaphragm, restoring spring, gland packing friction, and the viscous drag on the plug. The summation of all forces is then equal to the mass and the acceleration of the actuator, consisting of the actuator diaphragm, shaft, spring, valve stem, and plug. If the combined mass of these is m, and the travel of the actuator with the stem is x, then the following relationship using the force balance is written as

$$Y_c(t)a - K_s X - C_d ap\frac{dx}{dt} = m\frac{d^2x}{dt^2} \tag{5.15}$$

where K_s is the spring constant of the combined diaphragm and the restoring spring, and ap is the effective area of viscous drag of the combined plug and the packing. The flow rate of steam R is obtained from the valve characteristic as explained in Chapter 1. For simplicity, consider a linear characteristic, so that

$$R = \beta X. \tag{5.16}$$

The rate of heat Q supplied by the steam is then obtained as the product of its mass flow rate R and the latent heat of condensation λ

$$Q(t) = R\lambda = R\lambda\beta X = CX \tag{5.17}$$

where $C = R\lambda\beta$ taken as a constant.

Substituting this value of X in Equation 5.15, the force balance yields

$$CY_c(t)a - K_s Q(t) - C_d ap\frac{dQ(t)}{dt} = m\frac{d^2Q(t)}{dt^2}. \tag{5.18}$$

Considering the variables $Y_c(t)$ and $Q(t)$ as the variation with respect to their steady-state values as usual, the Laplace transformation yields the following transfer function:

$$Q(s)/Y_c(s) = C/(ms^2 + C_d aps + ks). \tag{5.19}$$

This is, in fact, a second-order system. However, in practice, it was found that mass (m) and the acceleration term may be negligible with respect to other terms, and hence, the transfer function yields a first-order transfer function as

$$Q(s)/Y_c(s) = C/(C_d aps + ks) = C/ks/(1 + \tau_v s) \tag{5.20}$$

where τ_v is the first-order time constant of the control valve. The equivalent transfer function block is represented as

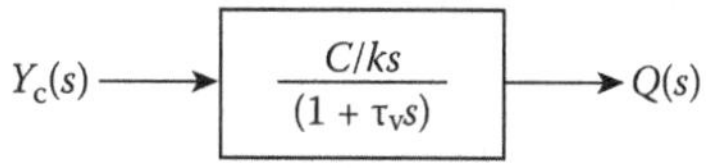

$$Y_c(s) \longrightarrow \boxed{\frac{C/ks}{(1+\tau_v s)}} \longrightarrow Q(s)$$

FIGURE 5.13 Transfer function block of control valve.

The Laplacian block diagram of the closed-loop control system can be constructed now by assembling the transfer function blocks of the comparator, controller, control valve, and transducer representing a model of an actual feedback control system as shown in Figure 5.14.

5.10 LEVEL CONTROL LOOP ANALYSIS

Consider a level control system as shown in Figure 5.15 where a control valve at the exit manipulates the effluent rate in order to control the level of liquid in the tank.

In order to constitute an equivalent Laplacian block diagram for this level–control system, the transfer functions of the open-loop process, transducer, controller, and control valve have to be determined.

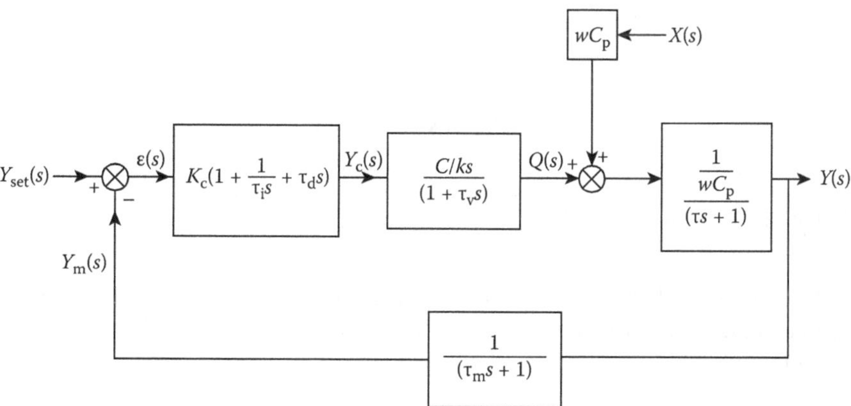

FIGURE 5.14 Laplacian block diagram of closed loop temperature control system of steam-heated tank.

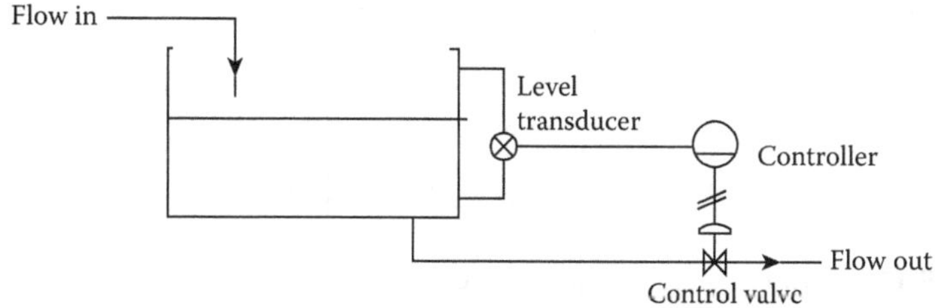

FIGURE 5.15 Level control system.

If the inlet flow and outlet flow are F_1 and F_2, respectively, h is the level in the tank, and the area of the cross section of the tank is A, then the material balance equation under unsteady-state conditions is given as

$$F_1 - F_2 = A\frac{dh}{dt} \tag{5.21}$$

expressing F_1, F_2, and h in terms of deviation forms, i.e., as a variation with respect to the steady-state values. Thus,

$$F_1(t) - F_{1s} = X(t),\ F_2(t) - F_{2s} = Y(t) \text{ and } h(t) - h_s = H(t) \tag{5.22}$$

where F_{1s}, F_{2s}, and h_s are the steady-state values of the influent rate, effluent rate, and the level in the tank, respectively. With the help of the Laplace transformation, Equation 5.21 becomes

$$X(s) - Y(s) = AsH(s). \tag{5.23}$$

The equivalent Laplacian block diagram is shown as

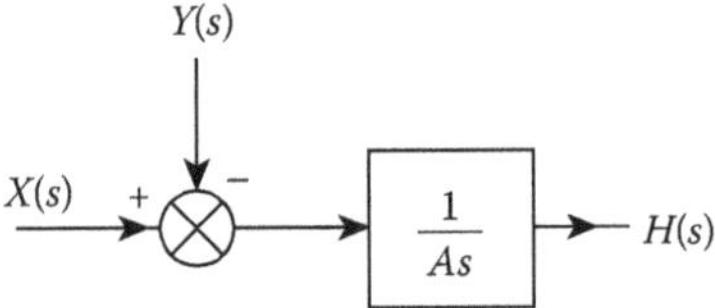

FIGURE 5.16 Laplacian block diagram of tank level.

A common level transducer is a float type gauge [illegible] the float under time-variable conditions can be made to determine the transfer function (Figure 5.17).

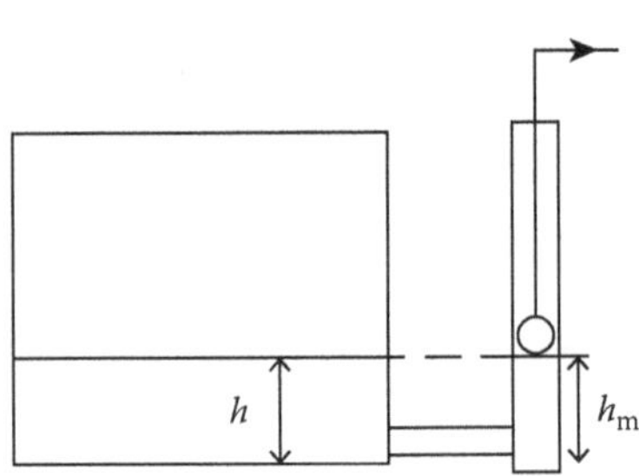

FIGURE 5.17 Float-type level transducer and the tank where h and h_m are the liquid level and the indicated level, respectively.

$$h\rho ga - h_{\rm m}\rho ga - C_{\rm d}a\frac{dh_{\rm m}}{dt} = m\frac{d^2h_{\rm m}}{dt^2} \tag{5.24}$$

where $C_{\rm d}$ is the viscous drag coefficient, a is the cross-sectional area, and g is the acceleration resulting from gravity. Expressed in terms of the deviation variables,

$$H(t) - H_{\rm m}(t) - \frac{C_{\rm d}}{\rho g}\frac{dH_{\rm m}(t)}{dt} = \frac{m}{\rho ga}\frac{d^2H_{\rm m}(t)}{dt^2}. \tag{5.25}$$

Taking the Laplace transformation, the transfer function is evaluated as

$$\frac{H_{\rm m}(s)}{H(s)} = \frac{1}{\frac{m}{\rho ga}s^2 + \frac{C_{\rm d}}{\rho g}s + 1}. \tag{5.26}$$

$$\tau_{\rm m} = \sqrt{(m/\rho g, a)} \text{ and } \xi = \frac{C_{\rm d}}{2}\sqrt{\left(\frac{a}{m\rho g}\right)} \tag{5.27}$$

where $\tau_{\rm m}$ is the time constant and ξ is the damping coefficient of the float gauge. The second-order transfer function is written as

$$\frac{H_{\rm m}(s)}{H(s)} = \frac{1}{\tau_{\rm m}^2 s^2 + 2\tau_{\rm m}\xi s + 1}. \tag{5.28}$$

The Laplacian block diagram is represented as

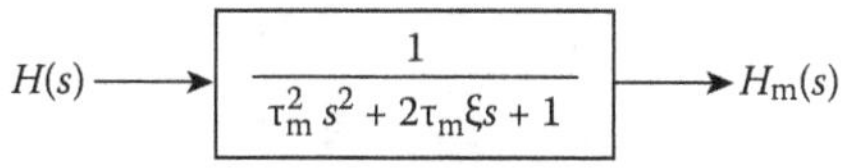

$$H(s) \longrightarrow \boxed{\frac{1}{\tau_{\rm m}^2 s^2 + 2\tau_{\rm m}\xi s + 1}} \longrightarrow H_{\rm m}(s)$$

FIGURE 5.18 Transfer function of level transducer.

The controller and the control valve transfer function will be the same as those of Equations 5.14 and 5.20. Note that, at the comparator, the $\varepsilon(s)$ is the difference between the SP level, $H_{\rm set}(s)$, and the measured level $H_{\rm m}(s)$. The Laplacian block diagram for the closed-loop level-control system can be obtained by combining the transfer functions of the process, level transducer, controller, and control valve as shown in Figure 5.19.

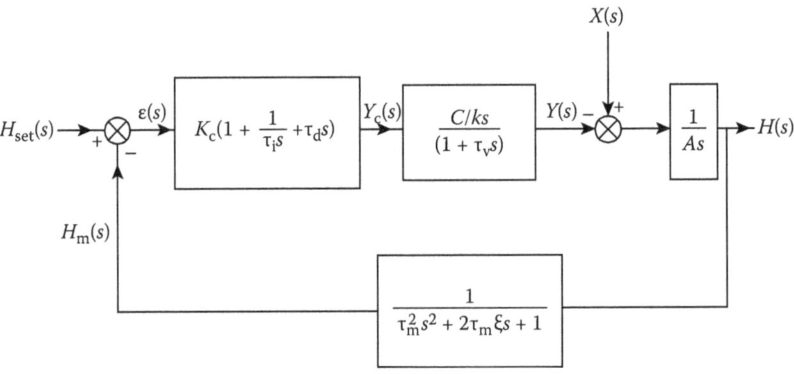

FIGURE 5.19 Laplacian block diagram of level-control system.

5.11 PRESSURE CONTROL LOOP ANALYSIS

Consider the pressure control in a vessel where a gas enters and leaves as shown in Figure 5.20.

In order to make a theoretical analysis of such a control system, it is necessary to find the Laplace transfer functions of the open-loop process, pressure transducer, controller, and control valve.

If the pressure of the influent gas, pressure in the vessel, and the pressure at which gas is expelled are p_1, p, and p_2, respectively, the following material balance equation can be set up as the process transfer function:

$$\frac{p_1 - p}{R_1} - \frac{p_2 - p}{R_2} = \frac{VM}{RT}\frac{dp}{dt} \tag{5.29}$$

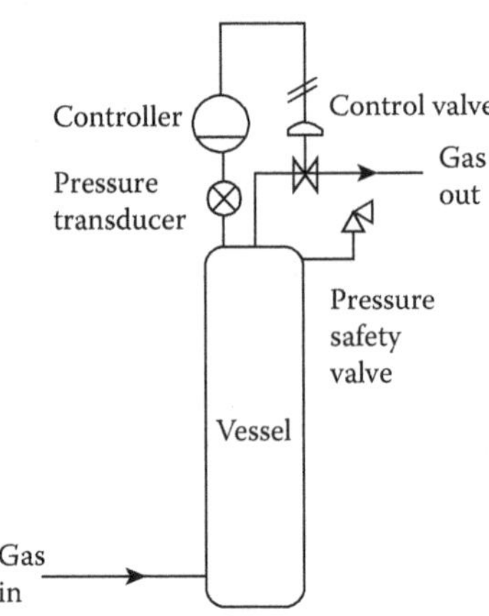

FIGURE 5.20 Pressure control system in a vessel.

where the density of the gas at a temperature T and pressure p in the vessel is taken as pM/RT assuming ideal gas law. M and V are the molecular weight of the gas and volume of the vessel, respectively. R_1 and R_2 are the resistances of the inlet and exit lines, respectively. It is assumed that the flow rate is proportional to the pressure difference. Equation 5.29 is applicable for an open-loop system when there was no control valve. But this is more appropriate in a controlled system where a control valve is present at the exit. Hence, the equation is rewritten as

$$\frac{p_1 - p}{R_1} - w = \frac{VM}{RT}\frac{dp}{dt} \tag{5.30}$$

where w is the mass flow rate of the gas leaving the vessel through the control valve and is determined by the controller. Thus,

$$p_1(t) - R_1 w(t) = p(t) + \frac{VM}{RT} R_1 \frac{dp(t)}{dt}. \tag{5.31}$$

The steady-state relationship becomes

$$p_{1s} - R_1 w_s = p_s \tag{5.32}$$

where p_{1s}, w_s, and p_s are the steady-state values of the inlet pressure, flow rate, and vessel pressure, respectively. Subtracting Equation 5.32 from Equation 5.31, the equation becomes

$$\left[p_1(t) - p_{1s}\right] - R_1\left[w(t) - w_s\right] = \left[p(t) - p_s\right] + \frac{VM}{RT} R_1 \frac{d\left[p(t) - p_s\right]}{dt}. \tag{5.33}$$

Expressing the variables in deviation form with respect to the steady-state values and taking Laplace transformations,

$$P_1(s) - R_1 w(s) = (1 + \tau s)\, P(s) \tag{5.34}$$

where $P_1(s)$ = the Laplace transformation of $(p_1(t) - p_s)$, $W(s)$ = the Laplace transformation of $(w(t) - w_s)$ and $\tau = VMR_1/RT$, the time constant of the process vessel.

The equivalent Laplacian block diagram is given as

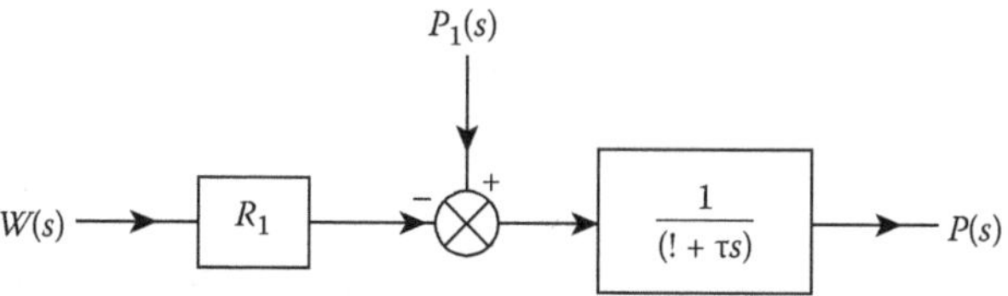

FIGURE 5.21 Laplacian block diagram of process vessel.

A pressure transducer may be taken as a manometric measurement that yields a second-order differential equation, as obtained in Chapter 2, as

$$\tau^2 \frac{d^2Y(t)}{dt^2} + 2\tau\xi \frac{dY(t)}{dt} + Y(t) = \frac{X(t)}{\rho_m g}. \tag{2.32}$$

Taking the Laplace transformation, the transfer function is given as

$$\frac{P_m(s)}{P(s)} = \frac{\frac{1}{\rho_m g}}{\tau_m^2 s^2 + 2\tau_m \xi_m s + 1} \tag{5.35}$$

where τ_m and ξ_m are the time constant and damping coefficient, respectively, of the pressure transducer. The Laplacian block diagram is given as

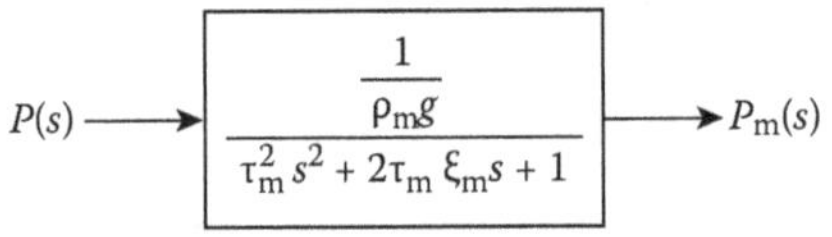

$$P(s) \longrightarrow \boxed{\frac{\frac{1}{\rho_m g}}{\tau_m^2 s^2 + 2\tau_m \xi_m s + 1}} \longrightarrow P_m(s)$$

FIGURE 5.22 Transfer function block diagram of pressure transducer.

The transfer function blocks of the controller and control valve will be same as before with the comparator, which will evaluate the difference between the P_{set} and P_m measured by the transducer. The closed-loop diagram of the pressure control system is presented as

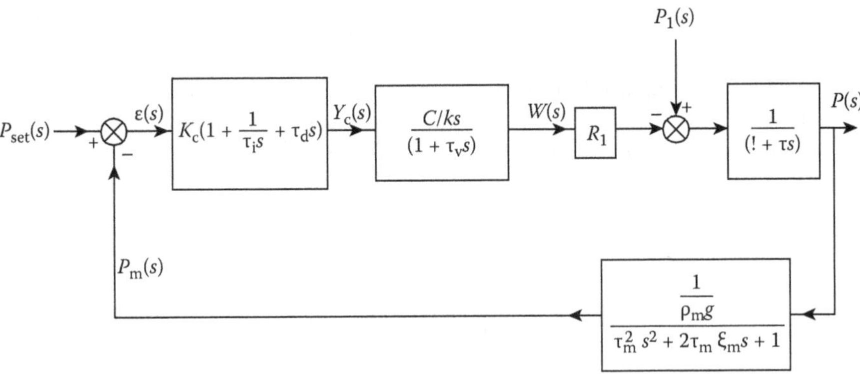

FIGURE 5.23 Laplacian block diagram of closed-loop pressure control system.

5.12 QUESTIONS AND ANSWERS

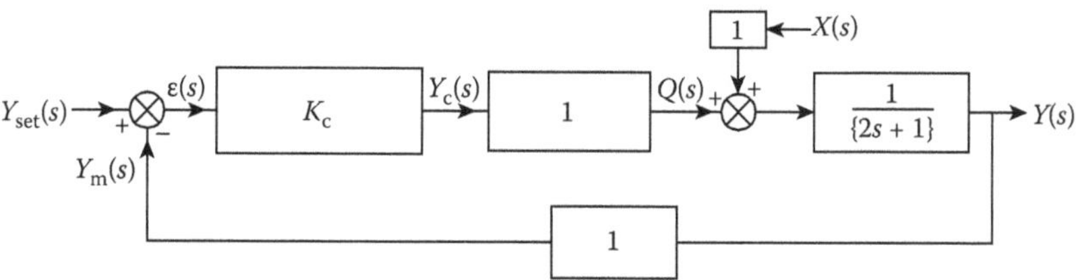

FIGURE 5.24 Block diagram of temperature-control system with proportional control and unity transfer functions of valve and temperature transducer.

EXERCISE 5.1

Considering the closed-loop control of temperature in a steam-heated tank, a proportional controller has to be used with K_c as the gain of the controller. Assume the control valve and the measuring transducer blocks have no dynamic lags, i.e., the transfer function of each of this is unity. If the product of the flow rate of liquid and its specific heat is unity, i.e., $wCp = 1$, then determine the tank-temperature variation as a function of time with any value of K_c when the SP is suddenly changed by a unit step of temperature. Also determine the offset.

Answer:

This is a event of a servo control system, i.e., while the load is unchanged and only the SP is changed. Therefore, the SP variation $Y_{set}(s) = 1/s$, and the overall transfer function is given as

$$\frac{Y(s)}{Y_{set}(s)} = \frac{\dfrac{K_c}{(2s+1)}}{1+\dfrac{K_c}{(2s+1)}} \tag{5.36}$$

or

$$\begin{aligned} Y(s) &= \frac{K_c}{s\{(2s+1)+K_c\}} = \frac{K_c}{2\{s+(K_c+1)/2\}s} \\ &= \frac{K_c}{(K_c+1)}\left\{\frac{1}{s}-\frac{1}{\{s+(K_c+1)/2\}}\right\}. \end{aligned} \tag{5.37}$$

Inverting the Laplace transformation to time parameters,

$$Y(t) = \frac{K_c}{(K_c+1)}\left\{1-e^{\frac{-(K_c+1)}{2}t}\right\}. \tag{5.38}$$

The plot of $Y(t)$ as a function of time (t) is shown in Figure 5.25.

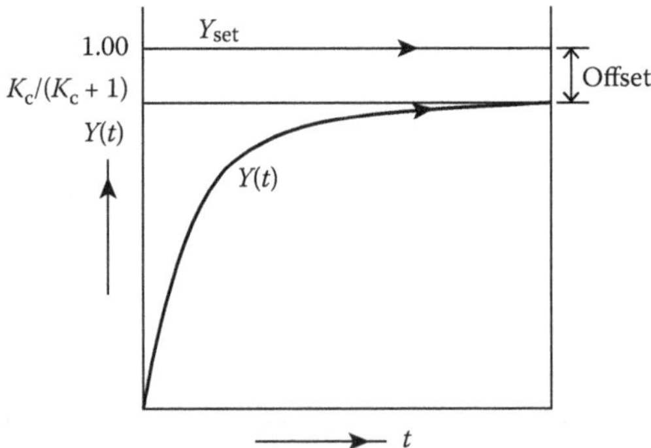

FIGURE 5.25 Variation of tank temperature for proportional-control action.

The value of $Y(t)$ ultimately will reach $K_c/(K_c + 1)$, which is less than unity. The greater the value of K_c, the greater will be the value of $Y(t)$, approaching nearest to SP. The difference of the SP and the ultimate value of $Y(t)$ is the offset, i.e.,

$$\text{Offset} = Y_{set}(\infty) - Y(\infty) = 1 - K_c/(K_c + 1) = 1/(K_c + 1). \tag{5.39}$$

Hence, this indicates the offset cannot be eliminated in a proportional-control system. The greater the value of K_c, the smaller the offset, i.e., the control temperature will be nearest to the SP. This can be evaluated as follows:

K_c	1	2	3	4	5	6	10
Offset	0.5	0.33	0.25	0.20	0.17	0.14	0.091

EXERCISE 5.2

Repeat the previous problem with a regulatory system while the SP is held unchanged but the load is changed by a step change in the unit degree [illegible]

The transfer function for such a system is given as

$$\frac{Y(s)}{wCpX(s)} = \frac{\dfrac{1}{(2s+1)}}{1 + \dfrac{K_c}{(2s+1)}}. \tag{5.40}$$

Because $wCp = 1$, then

$$\begin{aligned} Y(s) &= \frac{1}{s\{(2s+1)+K_c\}} = \frac{1}{2\{s+(K_c+1)/2\}s} \\ &= \frac{1}{(K_c+1)}\left\{\frac{1}{s} - \frac{1}{\{s+(K_c+1)/2\}}\right\}. \end{aligned} \tag{5.41}$$

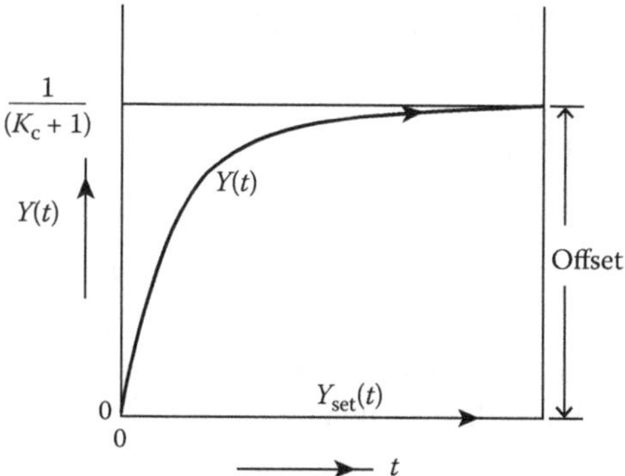

FIGURE 5.26 Response for proportional control during a load change.

Inversion of $Y(s)$ yields $Y(t)$,

$$Y(t) = \frac{1}{\left(K_c + 1\right)} \left\{1 - e^{\frac{-(K_c+1)}{2}t}\right\}. \tag{5.42}$$

So the ultimate value of $Y(t)$ as $t \to \infty$ is $1/(K_c + 1)$.

Hence, offset = $Y_{set}(\infty) - Y(\infty) = 0 - 1/(K_c + 1) = -1/(K_c + 1)$ as Y_{set} is zero as there was no change in the SP.

This is represented in Figure 5.26.

It is understood that during a load-change event, a negative offset exists. Thus, in both the cases of SP change alone in the previous exercise and during load change alone in this exercise, the offsets are always present in a proportional-control system. Of course, offsets will be reduced as the value of K_c increases. In this case, negative offsets will reduce as

K_c	1	2	3	4	5	6	10
Offset	−0.5	−0.33	−0.25	−0.20	−0.17	−0.14	−0.091

EXERCISE 5.3

Consider a control system with a PI controller having $K_c = 2$ and $\tau_i = 1$, $G_p = 1/(s + 1)$, $G_v = G_m = 1$. Determine the control variable as a function of time for a unit step change of a) SP point alone and b) load change alone.

Answer:

The relation for the control variable (Y) with the SP (Y_{set}) and load (U) from the control loop is

$$Y(s) = \frac{G_c G_p G_v Y_{set}(s) + G_p U(s)}{\left(1 + G_c G_p G_v G_m\right)}. \tag{5.6}$$

Substituting the values of $G_c = K_c(1 + 1/\tau_i s)$, $G_m = G_v = 1$, $G_p = 1/(\tau s + 1)$,

$$Y(s) = \frac{\left[K_c\left(1+1/\tau_i s\right)\frac{1}{(1+\tau s)}Y_{set}(s)\right]+U(s)\frac{1}{(1+\tau s)}}{1+K_c\left(1+1/\tau_i s\right)\frac{1}{(1+\tau s)}}. \tag{5.43}$$

(a) For SP disturbance by a unit step disturbance, keeping the load unchanged, i.e., $Y_{set}(s) = 1/s$, $Y(s)$ is then written as

$$Y(s) = \frac{K_c\left(1+\tau_i s\right)Y_{set}(s)}{\tau\tau_i s^2+\tau_i s\left(1+K_c\right)+K_c} \tag{5.44}$$

$$= \frac{2(1+s)}{s\left(s^2+3s+2\right)} \tag{5.45}$$

$$= \frac{2(s+1)}{s(s+1)(s+2)} = \frac{2}{s(s+2)} = \frac{1}{s} - \frac{1}{(s+2)}$$

Hence, inverting the Laplace transformation, $Y(t) = 1 - e^{-2t}$.

This is the response of the control variable in deviation form as a function of time.

Also, $Y(\infty) = 1$ and therefore, offset $= Y_{set}(\infty) - Y(\infty) = 1 - 1 = 0$.

(b) For load change by a unit step, keeping the SP unchanged,

$$Y(s) = \frac{K_c\tau_i s}{\tau\tau_i s^2+\tau_i s\left(1+K_c\right)+K_c} \tag{5.46}$$

Substituting the values of K_c, τ, τ_i,

$$= \frac{2s}{s\left(s^2+3s+2\right)} = \frac{2}{(s^2+3s+2)}$$
$$= \frac{2}{(s+1)(s+2)} = 2\left[\frac{1}{(s+1)} - \frac{1}{(s+2)}\right] \tag{5.47}$$

Inverting $Y(s)$, we get

$$Y(t) = 2(e^{-t} - e^{-2t})$$

This is the response of the control variable in deviation form as a function of time and also $Y(\infty) = 0$, so offset $= Y_{set}(\infty) - Y(\infty) = 0 - 0 = 0$.

Thus, it is found that the offset is eliminated when a PI controller is used in both the SP change alone or load change alone. (However, offset can be determined for any value of controller mode and parameters and a generalized process transfer function using a final value theorem.)

EXERCISE 5.4

If a PID controller is used to control a unity gain first-order process with unity time constant (τ), what will be the nature of the response of the controlled variable in deviation form with respect to time for a unit step disturbance for SP change alone? Assume that the proportional gain (K_c), integral time (τ_i), and derivative (τ_d) constants of the controller are each unity. Neglect the dynamic lags of the control valve and the measuring block. Also determine the offset.

Answer:

The transfer function of the controller in deviation form is

$$G_c = K_c(1 + 1/\tau_i s + \tau_d s) = K_c(1 + 1/s + s)$$

$$G_p = 1/(\tau s + 1) = 1/(s + 1)$$

$$G_m = G_v = 1$$

Then the closed-loop transfer function for SP change alone is

$$Y(s)/Y_{set}(s) = (s^2 + s + 1)/(2s^2 + 2s + 1)$$

for SP change alone by unit step, $Y_{set}(s) = 1/s$.

Hence,

$$Y(s) = \frac{\left(s^2+s+1\right)}{s\left(2s^2+2s+1\right)} = \frac{\left(s^2+s+\frac{1}{2}\right)+\frac{1}{2}}{s\left(2s^2+2s+1\right)} = \frac{1}{2s}+\frac{1}{2s\left(2s^2+2s+1\right)}$$

$$= \frac{1}{2s}+\frac{1}{\left(4s^2+4s+2\right)s}.$$

By partial fraction,

$$\frac{1}{\left(4s^2+4s+2\right)s} = \frac{1}{4s(s-a)(s-b)} = \frac{A}{s}+\frac{B}{(s-a)}+\frac{C}{(s-b)}$$

where a and b are the roots of the equation

$$4s^2 + 4s + 2 = 0$$

and

$$a = \frac{-1+j}{2} \text{ and } b = \frac{(-1-j)}{2}$$

Thus, the constants A, B, and C are determined as

$$A = 1/2$$

$$B = \frac{1}{4a(a-b)}$$

$$C = \frac{1}{4b(b-a)}.$$

Hence,

$$Y(s) = \frac{1}{2s} + \frac{1}{2s} + \frac{1}{4a(a-b)(s-a)} - \frac{1}{4b(a-b)(s-b)}$$

$$= \frac{1}{s} + \frac{1}{4ab(a-b)}\left(\frac{b}{(s-a)} - \frac{a}{(s-b)}\right).$$

as, $ab = 1/2$ and $a - b = j$.

Inverting $Y(s)$,

$$Y(t) = 1 + \frac{(-1-j)e^{(-1+j)t} - (-1+j)e^{(-1-j)t}}{4j}$$

$$= 1 + \frac{-e^{-t}e^{jt} + e^{-t}e^{-jt} - j\left(e^{-t}e^{-jt} + e^{-t}e^{jt}\right)}{4j}$$

$$= 1 - \frac{e^{-t}(\sin t + \cos t)}{2}.$$

This is the response of the controlled variable $Y(t)$ in deviation form.

Also the ultimate value is $Y(\infty) = 1$, so offset $= 1 - 1 = 0$.

From the previous and this exercise it is found that the offset is zero both for PI and PID controllers.

EXERCISE 5.5

Referring to Figure 5.14 for a temperature control, determine the following:

(a) The closed loop transfer function for SP change alone and for load change alone.
(b) The offset for both the SP change and load change, respectively, using the final value theorem for PID control systems.

Answer:

(a)

$$Y(s)=\frac{K_c\left(1+\dfrac{1}{\tau_i s}+\tau_d s\right)\dfrac{C/ks}{(1+\tau_v s)}\dfrac{\frac{1}{wC_p}}{\{\tau s+1\}}Y_{set}(s)+\dfrac{X(s)}{\{\tau s+1\}}}{1+K_c\left(1+\dfrac{1}{\tau_i s}+\tau_d s\right)\dfrac{C/ks}{(1+\tau_v s)}\dfrac{\frac{1}{wC_p}}{\{\tau s+1\}}\dfrac{1}{\{\tau_m s+1\}}}$$

$$=\frac{K_c\left(\tau_d\tau_i s^2+\tau_i s+1\right)C\left\{\tau_m s+1\right\}Y_{set}(s)+X(s)\tau_i s\left(1+\tau_v s\right)\left\{\tau_m s+1\right\}kswC_p}{kswC_p\left(1+\tau_v s\right)(\tau s+1)\left\{\tau_m s+1\right\}\tau_i s+K_c\left(\tau_d\tau_i s^2+\tau_i s+1\right)C}$$

Hence, the closed-loop transfer functions are, for SP change alone

$$\frac{Y(s)}{Y_{set}(s)}=\frac{K_c\left(\tau_d\tau_i s^2+\tau_i s+1\right)C\left\{\tau_m s+1\right\}}{kswC_p\left(1+\tau_v s\right)\{\tau s+1\}\left\{\tau_m s+1\right\}\tau_i s+K_c\left(\tau_d\tau_i s^2+\tau_i s+1\right)C}$$

for load change alone

$$\frac{Y(s)}{X(s)}=\frac{X(s)\tau_i s\left(1+\tau_v s\right)\left\{\tau_m s+1\right\}kswC_p}{kswC_p\left(1+\tau_v s\right)\{\tau s+1\}\left\{\tau_m s+1\right\}\tau_i s+K_c\left(\tau_d\tau_i s^2+\tau_i s+1\right)C}$$

(b) for a unity step disturbance of SP alone, $Y(s)$ is given as

$$Y(s)=\frac{K_c\left(\tau_d\tau_i s^2+\tau_i s+1\right)C\left\{\tau_m s+1\right\}}{s\left[kswC_p\left(1+\tau_v s\right)\{\tau s+1\}\left\{\tau_m s+1\right\}\tau_i s+K_c\left(\tau_d\tau_i s^2+\tau_i s+1\right)C\right]}$$

Hence, by final value theorem,

Lim $Y(t)$ as $t\to\infty$ = Lim $s\,Y(s)$ as $s\to 0 = K_c\,C/K_c\,C = 1$

Hence, offset = $Y_{set}(\infty) - Y(\infty) = 1 - 1 = 0$ for load change by a unity step disturbance alone

$$Y(s)=\frac{X(s)\tau_i s\left(1+\tau_v s\right)\left\{\tau_m s+1\right\}kswC_p}{kswC_p\left(1+\tau_v s\right)\{\tau s+1\}\left\{\tau_m s+1\right\}\tau_i s+K_c\left(\tau_d\tau_i s^2+\tau_i s+1\right)C}$$

By final value theorem,

$$Y(\infty) = 0 \text{ hence, offset} = Y_{set}(\infty) - Y(\infty) = 0 - 0 = 0.$$

This indicates that the offset becomes zero in PID control both during SP and load change events. The offsets during a proportional control and PI control as well can be determined similarly. It can be proved that the offset exists in the proportional control; whereas, offset vanishes in the PI control.

However, it is to be noted that the applicability of the final value theorem fails when the denominator goes to infinity for any value of s. In that case, tedious partial fractionation followed by inversion of the transform is essential to evaluate the value of $Y(\infty)$.

EXERCISE 5.6

While a level control system as shown in Figure 5.19 is a practical control system, Laplacian analysis is difficult. Consider a level-control system with a control valve at the influent flow line and the effluent flow takes place by gravity where a disturbance to the level is caused by a second influent stream. This theoretical control system is shown in Figure 5.27.

The influent rate F_1 is manipulated by the controller while the level fluctuates because of the fluctuation of the flow F_2 and gravity flow of the effluent F_3. If R is the resistance in the exit pipe, $F_3 = h/R$. The differential equation for the material balance relating the level h of the tank is presented as

$$F_1 + F_2 - h/R = A\, dh/dt \tag{5.48}$$

and at steady initial condition,

$$F_{1s} + F_{2s} - hs/R = 0 \tag{5.49}$$

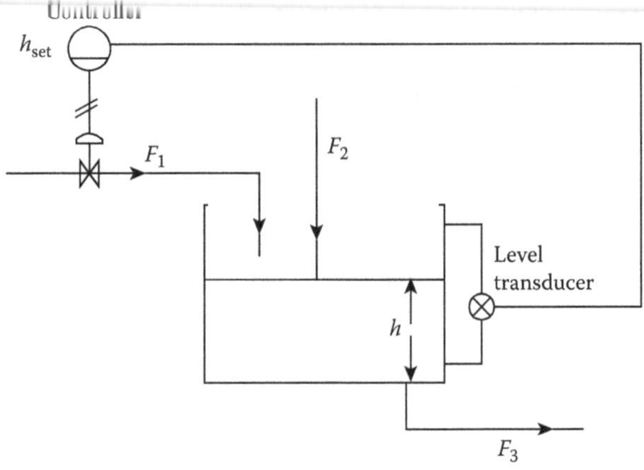

FIGURE 5.27 Level-control system with control valve at the inlet; effluent is by gravity flow.

subtracting Equation 5.49 from Equation 5.48,

$$F_1 - F_{1s} + F_2 - F_{2s} - (h - hs)/R = A\ d(h - hs)/dt \tag{5.50}$$

or

$$X_1(t) + X_2(t) - H(t)/R = A\ dH/dt \tag{5.51}$$

where $X_1(t)$, $X_2(t)$, and $H(t)$ are deviation forms of the instantaneous flow entities F_1, F_2, and h, respectively.

Thus taking the Laplace transformation,

$$H(s) = \{X_1(s) + X_2(s)\}\ R/(\tau s + 1). \tag{5.52}$$

The closed-loop Laplacian block diagram is then constructed as shown in Figure 5.28. The closed-loop relationship is then obtained as

$$\begin{aligned} H(s) &= \frac{\dfrac{G_c G_v R}{(\tau s+1)} H_{set}(s) + \dfrac{R}{(\tau s+1)} X_2(s)}{1 + \dfrac{G_c G_v G_m R}{(\tau s+1)}} \\ &= \frac{G_c G_v R H_{set}(s) + R X_2(s)}{(\tau s+1) + G_c G_v G_m R}. \end{aligned} \tag{5.53}$$

The overall closed-loop transfer function during a SP change alone is given as

$$\frac{H(s)}{H_{set}(s)} = \frac{G_c G_v R}{(\tau s+1) + G_c G_v G_m R} \tag{5.54}$$

and during load change alone

$$\frac{H(s)}{X_2(s)} = \frac{R}{(\tau s+1) + G_c G_v G_m R} \tag{5.55}$$

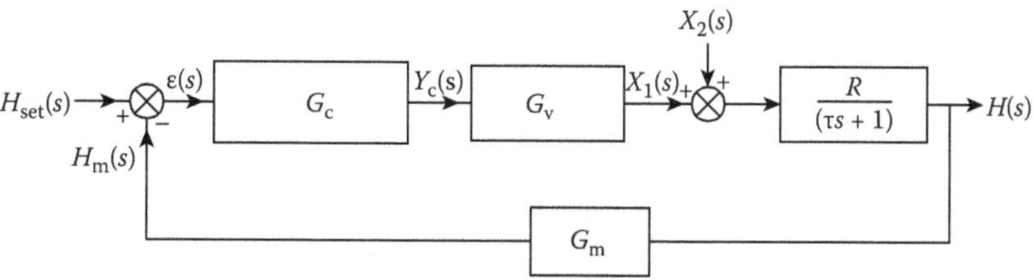

FIGURE 5.28 Closed-loop Laplacian block diagram of a level-control system corresponding to the process instrumentation diagram in Figure 5.27.

Now, the analysis of the loop for the response of the level, H, the control variable can be analyzed for any change either in the load or SP as done for the control systems described in the previous exercises.

EXERCISE 5.7

A pressure-control system, where, practically, a control valve is installed at the effluent line cannot be analyzed in the Laplacian feedback system. In order to make the analysis applicable, a theoretical pressure loop may be assumed with a control valve at the influent line. This is shown in Figure 5.29.

The material balance equation for the gas at unsteady and steady conditions are presented as

$$w_1 - \frac{(p - p_2)}{R} = VM/R_g T \frac{dp}{dt} \tag{5.56}$$

$$w_{1s} = \frac{(p_s - p_2)}{R} = 0 \tag{5.57}$$

where p and p_2 are the pressure within the vessel and at the exit line, respectively. The exit pressure at p_2 is the discharged pressure considered to be unchanged with time, and p_s is the steady pressure initially in the vessel. Subtracting Equation 5.57 from Equation 5.56 and expressing the variables in deviation form,

$$Q(t) = w_1(t) - w_{1s} \tag{5.58}$$

$$Y(t) = p(t) - p_s \tag{5.59}$$

$$Q(t) - \frac{Y(t)}{R} = VM/R_g T \frac{dY(t)}{dt} \tag{5.60}$$

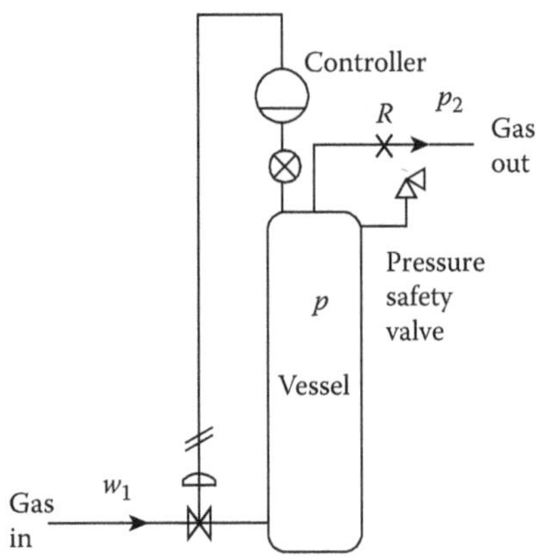

FIGURE 5.29 A pressure control system with a control valve at the inlet.

where R includes the resistance, area of the exit pipe, and density terms of the mass flow rate, and R_g is the universal gas constant, assuming the gas is behaving as an ideal gas, V is the volume of the vessel, and M is the molecular weight of the gas.

The Laplace transformation is obtained as

$$Y(s)/Q(s) = R/(\tau s + 1) \tag{5.61}$$

where $\tau = VMR/R_gT$ is the time constant.

The equivalent closed-loop Laplacian block diagram is presented as

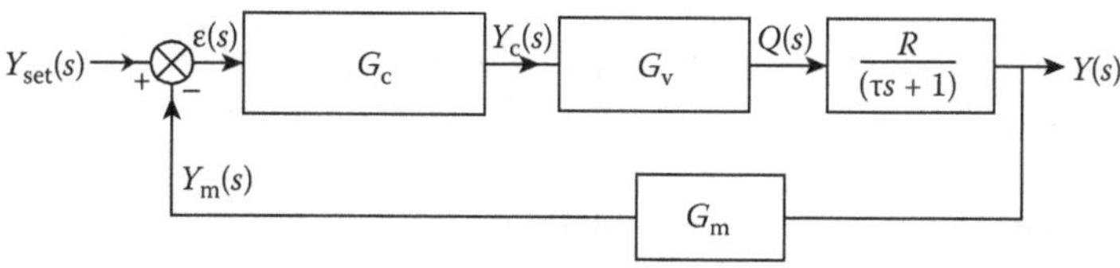

FIGURE 5.30 Laplacian block diagram for pressure-control system for instrumentation diagram in Figure 5.29.

Analysis of the response of the controlled variable for SP change can be carried out in a similar fashion as done for temperature- and level-control systems.

EXERCISE 5.8

Consider a temperature-control system as shown in Figure 5.31, where the level and temperature are simultaneously controlled. Present a Laplacian block diagram of the control system.

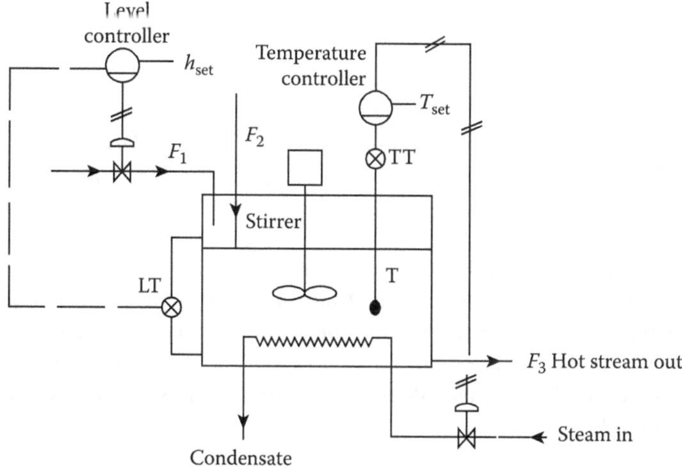

FIGURE 5.31 Simultaneous temperature and level control system.

The simultaneous material and heat balance equations for this system are given as

Material balance

$$F_1 + F_2 - \frac{h}{R} = A\frac{dh}{dt} \tag{5.62}$$

where $F_3 = h/R$, and R is the resistance at the exit pipe.
Heat balance,

$$F_1C_p\rho T_1 + F_2C_p\rho T_2 + q(t) - \frac{h}{R}C_pT = \rho C_pA\frac{d(hT)}{dt}. \tag{5.63}$$

Corresponding steady-state relationships are

$$F_{1s} + F_{2s} - \frac{h_s}{R} = 0 \tag{5.64}$$

$$F_{1s}C_p\rho T_{1s} + F_{2s}C_p\rho T_{2s} + q_s - \frac{h_s}{R}C_pT_s = 0 \tag{5.65}$$

where F_{1s}, F_{2s}, T_{1s}, T_{2s}, and T_s are the steady-state initial values of the F_1, F_2, T_1, T_2, and T. Subtracting the steady-state Equations 5.64 and 5.65 from Equations 5.62 and 5.63, respectively, we get

$$F_1 - F_{1s} + F_2 - F_{2s} - \frac{h - h_s}{R} = A\frac{dh}{dt} \tag{5.66}$$

$$F_1T_1 - F_{1s}T_{1s} + F_2T_2 - F_{2s}T_{2s} + \frac{q(t) - q_s}{C_p\rho} - \frac{\left(hT - h_sT_s\right)}{R} = A\frac{d\left(hT - h_sT_s\right)}{dt}. \tag{5.67}$$

The Equation 5.67 contains the nonlinear terms F_1T_1, F_2T_2, and hT, which cannot be transformed directly by the Laplace transformation method, which can handle only linear relationships. Hence, the nonlinear entities are converted to linearized entities using a Taylor series expansion and taking the differences of the variables from their corresponding steady values as negligibly small, so the higher-order terms become negligible.

$$F_1T_1 = F_{1s}T_{1s} + (F_1 - F_{1s})T_{1s} + (T_1 - T_{1s})F_{1s} \tag{5.68}$$

$$F_2T_2 = F_{2s}T_{2s} + (F_2 - F_{2s})T_{2s} + (T_2 - T_{2s})F_{2s} \tag{5.69}$$

$$hT = h_sT_s + (h - h_s)T_s + (T - T_s)h_s \tag{5.70}$$

so

$$\frac{dhT}{dt} = T_s\frac{dh}{dt} + h_s\frac{dT}{dt}. \tag{5.71}$$

Expressing in deviation variables with respect to the steady values,

$X_1(t) = F_1(t) - F_{1s}$, $X_2(t) = F_2(t) - F_{2s}$, $Y_1(t) = T_1(t) - T_{1s}$, $Y_2(t) = T_2 - T_{2s}$, $Y(t) = T - T_s$,

and $Q(t) = q - q_s$, the equations become

$$X_1(t) + X_2(t) - \frac{H(t)}{R} = A\frac{dH}{dt} \tag{5.72}$$

$$T_{1s}X_1(t) + T_{2s}X_2(t) + F_{1s}Y_1(t) + F_{2s}Y_2(t) + \frac{Q(t)}{C_p\rho} - \frac{T_sH(t) + h_sY(t)}{R}$$

$$= A\tau_s\frac{dH(t)}{dt} + Ah_s\frac{dT(t)}{dt}. \tag{5.73}$$

Taking the Laplace transformations, the above equations become

$$H(s) = \frac{R\left(X_1(s) + X_2(s)\right)}{(1 + \tau s)} \tag{5.74}$$

and

$$Y(s) = \frac{R}{(1 + \tau_s)h_s}\left[\left(T_{1s} - T_s\right)X_1(s) + \left(T_{2s} - T_s\right)X_2(s) + F_{1s}Y_1(s) + F_{2s}Y_2(s) + \frac{Q(s)}{C_p\rho}\right]. \tag{5.75}$$

The Laplacian block diagram for the control system is presented in Figure 5.32.

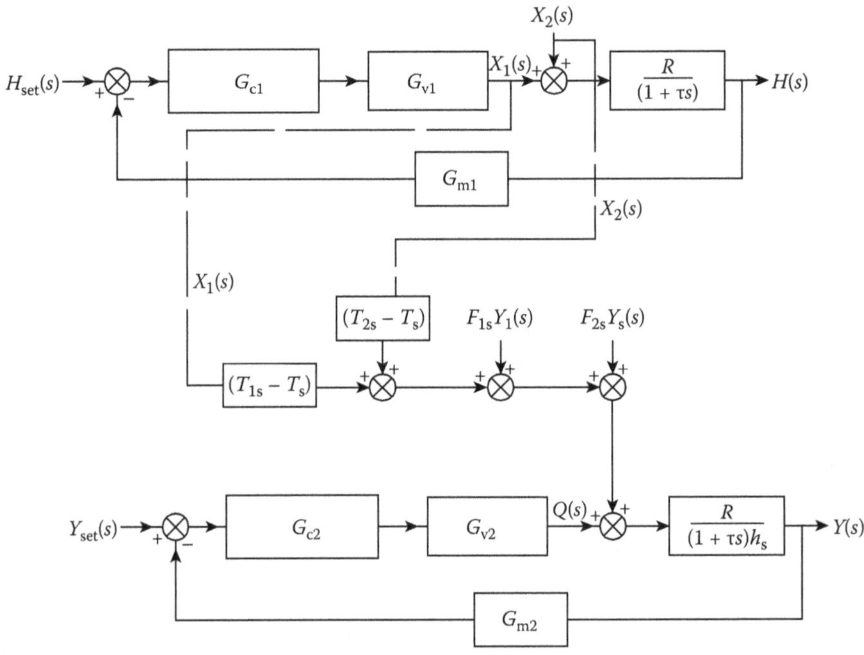

FIGURE 5.32 Laplacian block diagram for the simultaneous level and temperature control in a steam-heated tank.

where the transfer functions of the level controller, transducer, and control valve are G_{c1}, G_{m1}, and G_{v1}, respectively. The transfer functions of the temperature controller, temperature transducer, and steam control valve are G_{c2}, G_{m2}, and G_{v2}, respectively. It is to be noted that SP change in the level controller will affect both the level and temperature simultaneously. Similarly, SP change in the temperature control will not affect the level. If the level controller efficiently works such that the level is unchanged at a constant value, then the control loop will be same as that shown in Figure 5.14 where the mass flow rate is constant, and the liquid in the tank has a constant volume, i.e., level is unchanged.

EXERCISE 5.9

Select the controller parameters and process parameters from the following list: SP, proportional band, PV, bias, integration time, derivative time, and first-order time constant.

Answer:

Controller parameters are SP, proportional band, bias, integration time, and derivative time.

Process parameters are PV and first-order time constants.

EXERCISE 5.10

The SP of a temperature control system is 80°C, and the process temperature is 30°C. What will be the process temperature for a perfectly controlled system, and what will be the time required for this?

Answer:

For a perfect controlled system, the process temperature will be equal to the set point, 80°C, at no time.

6 Tuning of PID Controllers

6.1 CONTROLLER PARAMETERS AND PERFORMANCE

The equation of a proportional–integral–derivative (PID) controller is known as

$$O_c = A + K_c\varepsilon + \frac{K_c}{\tau_i}\int \varepsilon\, dt + K_c\tau_d \frac{d\varepsilon}{dt} \tag{6.1}$$

where the parameters K_c, τ_i, and τ_d already explained in the earlier chapters have a range of values typically from 0.5 to 500 in their respective units. Proper selection of a set of values from these three parameters for the best-controlling performance is known as tuning, i.e., tuning of controller parameters. Of course, trial values can be selected, and control action is observed. If the performance is satisfactory, then these parameters are selected; alternatively, another set is selected, and the performance test has to be repeated. But this type of hit-or-miss method is not suggested as, during the performance the test, there may be too much stress and strain over the equipment, which may cause accidents or damage to the plant and equipment. However, at this point, it is also necessary to know how the performance of a controller is judged.

In any closed-loop control system, the response of the control variable must be observed as a function of time as shown in Figure 6.1.

A desirable response should satisfy the following criteria with reference to Figure 6.1:

1. Overdamping ratio: The overdamp B should not be large. For example, in case the controlled variable Y was temperature, the overshoot should not be above the safe limit for temperature of use of the equipment.
2. Decay ratio: The decay C should be lower than B and should be as small as possible. Ideally a quarter decay ratio (i.e., $C/B <= 1/4$) is the maximum limit allowable for a good performance.
3. Rise time: The smaller the rise time, the quicker the response. For good performance, this time should be as small as possible.
4. Settling time: Settling time is the time to reach the set point. The control variable should be within ±5% of the set point.
5. Offset: The difference between the set point and the control variable while the settling time elapses. Theoretically, it is achieved in infinite time. The smaller the value of the offset, the better the achievement of the controlled system.

Ideally, a control system should have zero overshoot, zero decay, the smallest rise and settling times, and zero offset. But such ideal achievement is rarely found.

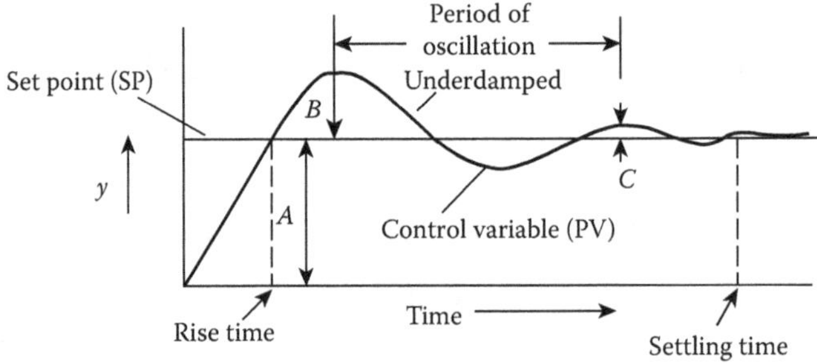

FIGURE 6.1 Example of response of controlled system.

However, a response, such as critically damped or overdamped systems, as shown in Figure 6.2 are usually not desirable when the settling time is long and the speed of response is slow compared to the underdamped-type response as shown in Figure 6.1. However, a critically damped response is ideal when the settling time is very short. An overdamped response is also acceptable if both the settling time and the offset are negligibly small.

However, a response with ever-increasing amplitude with or without oscillations, as shown in Figure 6.3, are uncontrollable responses and may lead to accidents.

These uncontrollable responses indicate unstable situations, i.e., the control variable will never be convergent.

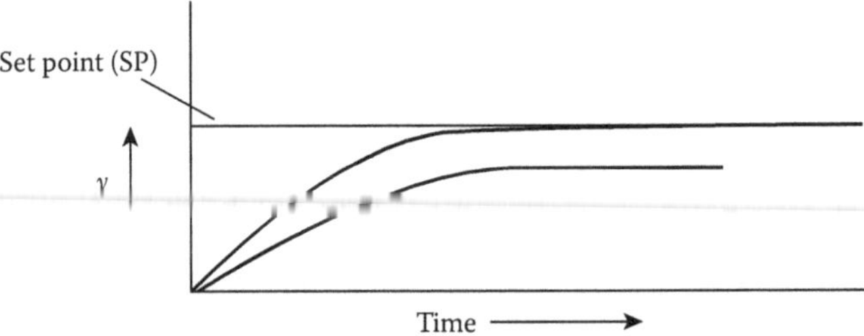

FIGURE 6.2 Critically damped or overdamped responses.

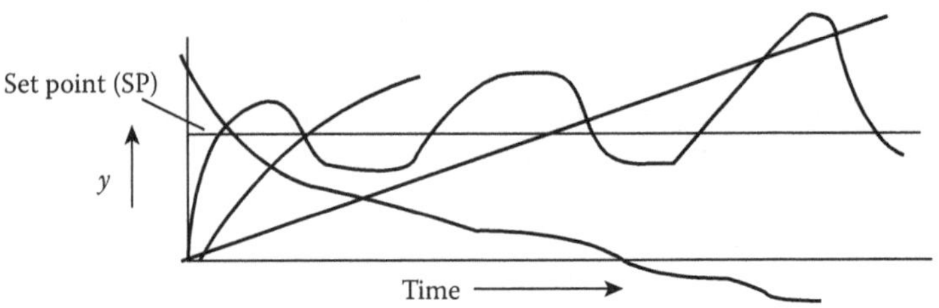

FIGURE 6.3 Uncontrollable responses.

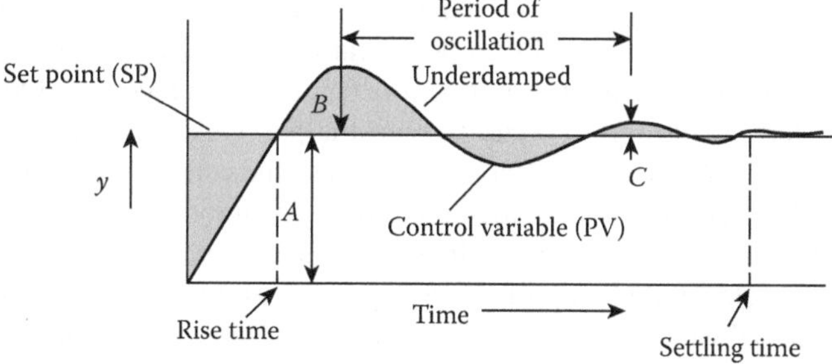

FIGURE 6.4 Area of the error integration in the shaded region.

Besides the above criteria, the following additional performance criteria are also taken into consideration:

(a) The area under the curve as shown in Figure 6.4 as shaded. The smaller this area, the better the performance. This is mathematically equal to

$$\int edt. \tag{6.2}$$

(b) Alternatively, the sum of the square of the error integrated over the time period is also taken as a measure of the performance.

$$\int e^2dt. \tag{6.3}$$

The smaller the value, the better is the performance.

6.2 EXPERIMENTAL TUNING

Tuning of a PID controller is carried out experimentally either by the Ziegler–Nichols or Cohen–Coon method.

6.2.1 Ziegler–Nichols Method

In this method, the following steps are used for determining the tuned parameters:

Step 1: The controller is set in proportional-control mode. Modern PID controllers can be used in proportional, integral, or derivative modes as desired. In these controllers, proportional mode is selected by turning the integration time to the highest possible value (such that the inverse of it nears zero) and the derivative time to zero. The value of the proportional gain must be noted.

Step 2: The system should be allowed to be in a steady-state condition, i.e., when the set point equals the control variable at the beginning of the experiment.

Step 3: The set point is suddenly changed to a small value, and the response is recorded. If the response of the control variable is found to be sinusoidal in nature with the time as shown in Figure 6.5, it is then recorded. Otherwise, the system is allowed to come back to the previous steady condition followed by a change in the proportional gain of the controller to another new value. The set point is then disturbed to record the response curve. A few trials may suffice to obtain the sinusoidal response record.

Step 4: After step 3, the value of the proportional gain is noted as the ultimate proportional gain of the controller K_u, and the period of the sine curve from the record is noted as the ultimate period P_u.

Step 5: The values of the parameters of the controller are then set with the values of K_u and P_u according to the Ziegler–Nichols chart.

Ziegler–Nichols Chart for Tuning

Control Mode	Proportional Gain, K_c	Integration Time, τ_i	Derivative Time, τ_d
Propotional	$0.50\ K_u$	–	–
Proportional–integral	$0.45\ K_u$	$P_u/1.2$	
Proportional–integral–derivative	$0.60\ K_u$	$P_u/2$	$P_u/8$

The Ziegler–Nichols method is also known as closed-loop tuning as the response has to be obtained in the closed-loop system with the controller.

However, the major drawback of this method is that, during tuning, the equipment may be damaged while a sinusoidal oscillating response is searched for. In this tuning method, the proportional–derivative control mode is absent; however, the settings for PID mode values may be used.

To avoid these disadvantages, the Cohen–Coon method is preferred as described next.

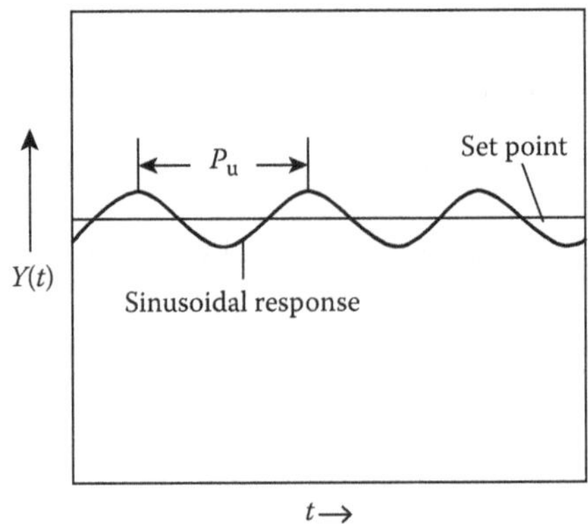

FIGURE 6.5 Sinusoidal response of the control variable $Y(t)$.

6.2.2 Cohen–Coon Method

In this method, the experiment is carried out in the open-loop condition, i.e., the feedback action of the controller is not allowed. Such an open-loop system is schematically shown in Figure 6.6. In fact, an open-loop system can be made by selecting the manual mode of the controller, and the control signal will be equal to the value of the set point selected. Thus, the output signal of the controller will not be a function of the error, i.e., bypassing the controller equation. The control valve is, therefore, opened or closed directly by increasing or decreasing the set point. A small step change or valve opening is done, and the record of the variable under study is obtained. Usually what's recorded is found to be a sigmoidal one as shown in Figure 6.7. This response is also known as the reaction curve.

From the sigmoidal response curve, three regions are observed. In the beginning, there is a slow rise of the curve followed by a sharp increase, almost a linear rise, followed by a slow rise to reach the final or ultimate value B. If a straight line is drawn along the linear rise portion of the curve as shown in Figure 6.8, the line intersecting the time axis is considered as the dead time or exponential lag time τ_D, and the slope of the line (slope) will indicate the fastest rate of rise. According to Cohen–Coon, the transfer function of the various processes resembles this type of sigmoidal response, and the transfer function of the processes can be approximated as

$$G_p(s) = \frac{K_p e^{-\tau_D s}}{\tau_p s + 1} \tag{6.4}$$

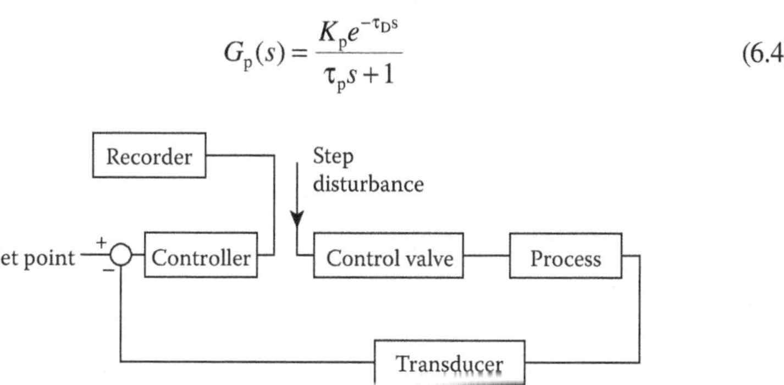

FIGURE 6.6 Open-loop system for tuning using Cohen–Coon method.

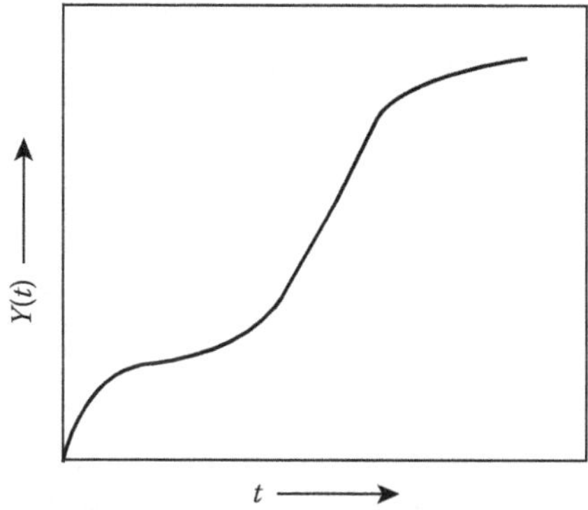

FIGURE 6.7 Sigmoidal response or reaction curve of the process variable in open-loop tuning.

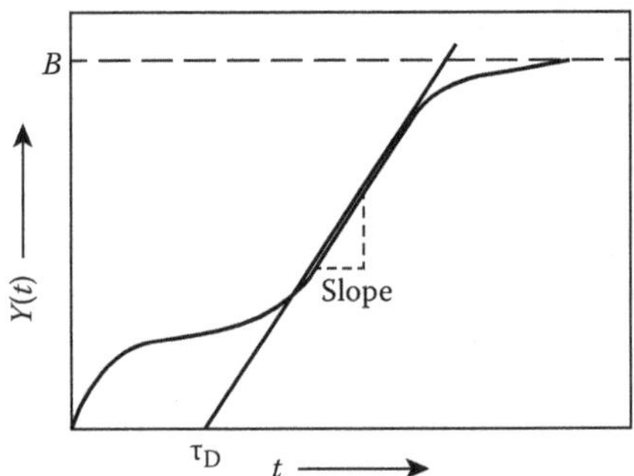

FIGURE 6.8 Parameters from the sigmoidal response of a process.

where

Process gain: $K_p = B/M$ where B is the ultimate value, and M is the magnitude of the step disturbance or controller opening.

Process dead time: τ_D

Process first-order time constant: $\tau_p = B/\text{slope}$

Using these parameters, Cohen–Coon's tuning parameters for various control modes are presented in the following table.

Cohen–Coon's Tuning Parameters

Modes of Control	Proportional Gain of Controller, K_c	Integration Time, τ_i	Derivative Time, τ_d
P	$K_c = \frac{\tau_p}{k_p\tau_D}\left(1 + \frac{\tau_D}{3\tau_p}\right)$		
PI	$K_c = \frac{\tau_p}{k_p\tau_D}\left(0.9 + \frac{\tau_D}{12\tau_p}\right)$	$\tau_i = \tau_D\left(\frac{30 + 3\frac{\tau_D}{\tau_p}}{9 + 20\frac{\tau_D}{\tau_p}}\right)$	
PD	$K_c = \frac{\tau_p}{k_p\tau_D}\left(1.25 + \frac{\tau_D}{6\tau_p}\right)$		$\tau_d = \tau_D\left(\frac{6 - 2\frac{\tau_D}{\tau_p}}{22 + 3\frac{\tau_D}{\tau_p}}\right)$
PID	$K_c = \frac{\tau_p}{k_p\tau_D}\left(1.33 + \frac{\tau_D}{4\tau_p}\right)$	$\tau_i = \tau_D\left(\frac{32 + 6\frac{\tau_D}{\tau_p}}{13 + 8\frac{\tau_D}{\tau_p}}\right)$	$\tau_d = \tau_D\left(\frac{4}{11 + 2\frac{\tau_D}{\tau_p}}\right)$

6.3 THEORETICAL TUNING

Theoretical tuning is possible if the proper mathematical model of the controlled system, consisting of the process, transducer, controller, and control valve, is available. With the help of this model, a set of controller parameters are selected that yield a stable control system.

6.3.1 STABILITY ANALYSIS OF A CONTROL SYSTEM

If the transfer functions of the process G_p, transducer G_m, and control valve G_v are known, theoretical tuning is possible. There are various methods applicable, such as the Routh–Hurwitz method, Bode's stability criteria, Nyquist stability criteria, etc. A stable control system is defined when the variable to be controlled comes as close as possible to the set point. If the deviation from the set point reaches a zero value, i.e., when the offset becomes zero, the control variable equals the set point and is maintained there as long as there is no further disturbance. However, the deviation may be constant and is unchanged until further disturbance. These systems when the deviation ultimately (after a long time) becomes either zero (without offset) or a constant value (offset exists) are stable control systems. But if the deviation goes on increasing or decreasing above or below the set point without any stopover as shown in Figure 6.3, the systems are unstable. Mathematically, if any variable as a function of time becomes infinity (positive or negative) as time goes to infinity, it becomes a divergent or unstable entity. Alternatively, when the variable reaches a constant value or zero, it is a convergent or stable entity. Hence, if the ultimate error or offset is convergent, then it is a stable system. Theoretically, the control variable can be determined as a function of time from the closed-loop transfer function by the partial fraction method as described next.

The open-loop transfer function is the transfer function of the loop without the feedback as shown in Figure 6.9.

The open-loop transfer function G_{ol} is the product of all the transfer functions in the control loop, i.e., $G_{ol} = G_p G_m G_v G_c$. If the feedback loop was closed, the closed-loop transfer function (for set point change only) will be

$$G(s) = G_c G_v G_p/(1 + G_c G_v G_p G_m)$$

$$= G_c G_v G_p/(1 + G_{ol}). \tag{6.5}$$

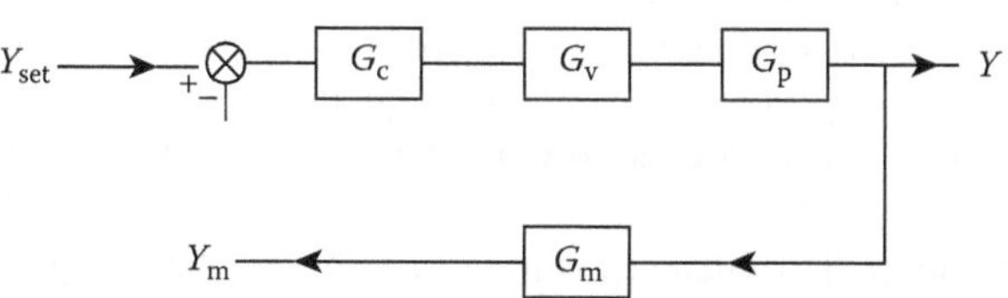

FIGURE 6.9 Open-loop system.

The denominator of the closed-loop transfer function influences (characterizes) the nature of the output variable's response to a change in set point. Factorization of the denominator is essential for getting the output variable (response) as a function of time. Factorization of any polynomial is possible only by knowing the roots of the equation by equating the denominator as zero and solving for the roots of the equation. Thus, the equation is

$$1 + G_{ol} = 0. \tag{6.6}$$

This is the characteristic equation. Roots of such a characteristic equation are utilized for getting the response as described next.

6.3.2 Roots of a Characteristic Equation

$$\begin{aligned} G(s) = Y(s)/X(s) &= \frac{1}{a_0 s^n + a_1 s^{n-1} + a_2 s^{n-2} \ldots\ldots\ldots\ldots\ldots\ldots + a_n} \\ &= \frac{1}{(s-\alpha)(s-\beta)(s-\gamma)(s-\delta)\ldots\ldots\ldots\ldots\ldots\ldots} \end{aligned} \tag{6.7}$$

where α, β, γ, δ, …, are the roots of the characteristic equation

If $X(s) = 1/s$,

$$Y(s) = \frac{1}{s(s-\alpha)(s-\beta)(s-\gamma)(s-\delta)\ldots} \tag{6.8}$$

and by partial fraction,

$$Y(s) = \frac{A}{s} + \frac{B}{(s-\alpha)} + \frac{C}{(s-\beta)} + \frac{D}{(s-\gamma)} + \frac{E}{(s-\delta)} + \ldots$$

Inverting to time functions,

$$Y(t) = A + Be^{\alpha t} + Ce^{\beta t} + De^{\gamma t} + Ee^{\delta t} + \ldots. \tag{6.9}$$

where *A*, *B*, *C*, *D*, *E*, …. are the constants.

The roots may be real, imaginary, or complex, depending on the values of the coefficients a_0, a_1, a_2, a_n, of the characteristic equation.

If any of the roots is complex, say, $\alpha = a + jb$ where *a* is the real and *b* is the imaginary part where $j = \sqrt{-1}$. Further, if *a* is positive, the value of $e^{(a+bj)t}$ will be infinite as *t* tends to infinity. Hence, *Y(t)* will tend to infinity, i.e., *Y(t)* will be divergent. Alternatively, if *a* is negative, the value of $e^{(a+b\,j)t}$ will be zero as *t* tends to infinity. Thus, if all the roots have a negative real part, *Y(t)* will be convergent, and the system will be stable. Hence, the stability or instability of a control system can be predicted from the roots of the characteristic equation. Of course, a stable system may become unstable because of an unstable cause or input. If the input has limits or bounds, the response of a stable system will generate responses that will also be bounded. Thus, a stable system is, therefore, defined as the system where a bound input produces a bound output or response.

6.3.3 Routh–Hurwitz Test for Stability

If the characteristic equation can be written as a polynomial of *n*th degree as

$$a_0 s^n + a_1 s^{n-1} + a_2 s^{n-2} + \dots\dots\dots\dots + a_n = 0 \tag{6.10}$$

then the coefficients a_0, a_1, a_2, a_n will be used to determine the stability of the closed-loop control system. For this, an array is to be constructed that will have $n + 1$ number of rows with the column elements as given below in the Routh–Hurwitz matrix.

$$\begin{vmatrix} a_0 & a_2 & a_4 & \cdots \\ a_1 & a_3 & a_5 & \cdots \\ b_1 & b_2 & b_3 & \\ c_1 & c_2 & c_3 & \\ \vdots & \vdots & \vdots & \end{vmatrix}$$

where the values of b_1, b_2, b_3, c_1, c_2, are determined from the elements of the previous two rows and are obtained as

$$b_1 = \frac{a_1 a_2 - a_0 a_3}{a_1}, \quad b_2 = \frac{a_1 a_4 - a_0 a_5}{a_1}, \quad \dots\dots$$

$$c_1 = \frac{b_1 a_3 - a_1 b_2}{b_1}, \quad c_2 = \frac{b_1 a_5 - a_1 b_3}{b_1}, \quad \dots\dots$$

and so on. Thus, the Routh–Hurwitz array is constructed. If all the elements of the first column are found to be positive and nonzero, the closed-loop control system will be stable. Alternatively, if any of the elements in the first column of this array is found to be negative or zero, the control system will be closed-loop unstable. This conclusion of the control system is based on the following facts:

1. If all the elements of the first column are positive, none of the roots of the characteristic equation will have a positive real part; rather, all the roots will have a negative real part. This means the roots of the characteristic equation will lie on the left half of the complex plane as shown in Figure 6.10. As a result, the response of the control variable will be convergent to a value or zero. In other words, the control system will be stable.

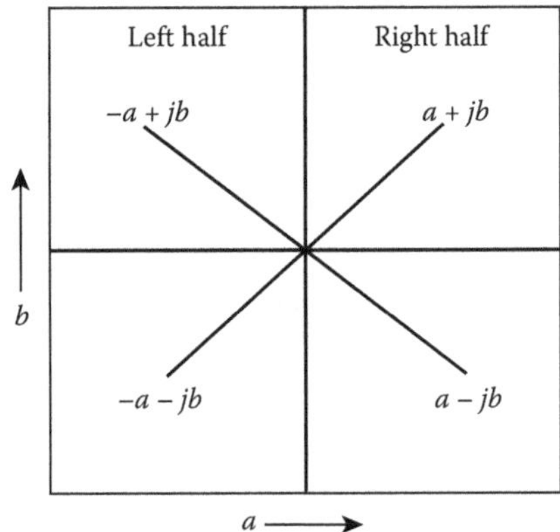

FIGURE 6.10 Complex plane and complex roots.

2. If any of the columns are found to be negative, the number of roots having a positive real part will be equal to the number of sign changes, i.e., those lying on the right-hand side of the complex plane. These roots will then tend to make the control system unstable as the control variable will be divergent.
3. If the nth row elements vanish but the $(n - 1)$th row exists, the system will be unstable with two roots lying on the imaginary axis. The control system will be oscillatory like a sine curve and will not converge.

These two roots are determined as

$$0.0 \pm j\sqrt{(D/C)} \tag{6.11}$$

where C and D are the elements of the $(n - 1)$th row read from left to right.

With these Routh–Hurwitz rules, it is possible to tune a controller. This will be demonstrated in the question and answer section later. It is also to be noted that this analysis has to be carried out in a closed-loop system. However, if the transfer function of any of the elements in the control loop includes a transportation lag (i.e., exponential lag), this method will not be applicable. In this case, the frequency response methods of Bode stability criteria must be used for tuning.

6.3.4 Frequency Response Analysis

In this method, a predetermined sine disturbance is introduced in the open-loop system, and the response is determined. The response is recorded after some time, theoretically at infinite time. This is known as the ultimate response, which also resembles a sine curve. Thus, if the input disturbance is $A \sin \omega t$, then the ultimate response is $B \sin (\omega t + \Phi)$, where A and B are the amplitudes of the input and output sine response and Φ is the phase difference between them. The amplitude and phase difference of the ultimate response are then analyzed for the determination of a stable control system. Bode stability and Nyquist stability methods are used for tuning controllers by analyzing the ultimate responses.

6.3.5 Bode Stability Analysis

In this method, the open-loop transfer function $G(s)$ is determined. In order to determine the ultimate response, the following steps are followed:

1. The Laplacian operator s is then replaced by $j\omega$ where $j = \sqrt{-1}$, and the transfer function $G(j\omega)$ becomes a function of angular frequency ω. Thus, the transfer function becomes a complex number.
2. The magnitude $|G(j\omega)|$ and the angle $< G(j\omega)$ of the transfer function are determined.
3. The amplitude ratio of the output sine response to its sine disturbance is the magnitude $| G(j\omega)|$ i.e. $| G(j\omega)| =$ amplitude ratio $= B/A$. Hence, $B = A \mid G(j\omega)|$.
4. If the phase difference $\Phi = -180°$ or $-\Pi^{\text{radian}}$ for a radian frequency ω_{co} (known as the crossover frequency) and the amplitude ratio determined at this value is found to be less than unity, the closed-loop controlled system will be stable. This phenomenon can be obtained by analyzing the control system as shown in Figure 6.11, where a sine disturbance of amplitude A and radian frequency ω is introduced in the open-loop system.

The output is also a sine function with its amplitude equal to $A \times AR$.

AR is the amplitude ratio of the amplitudes of output function to those of input.

If the phase difference Φ is made to $-180°$ by varying the radian frequency ω, the output will be changed to $A \times AR \sin(\omega t - 180) = -A \times AR \sin \omega t$. Now, if this output is fed back to the system and, simultaneously, the input sine disturbance is withdrawn, the output will continue to be oscillating. This is shown in Figure 6.12.

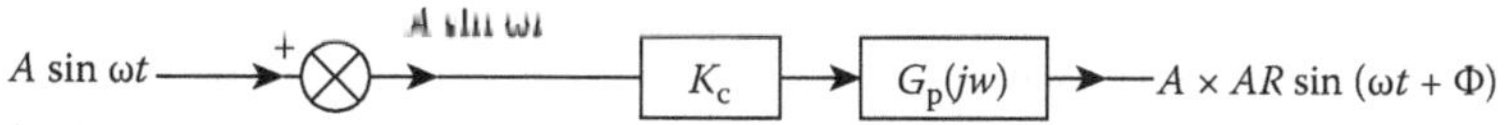

FIGURE 6.11 Open loop disturbance using sine input.

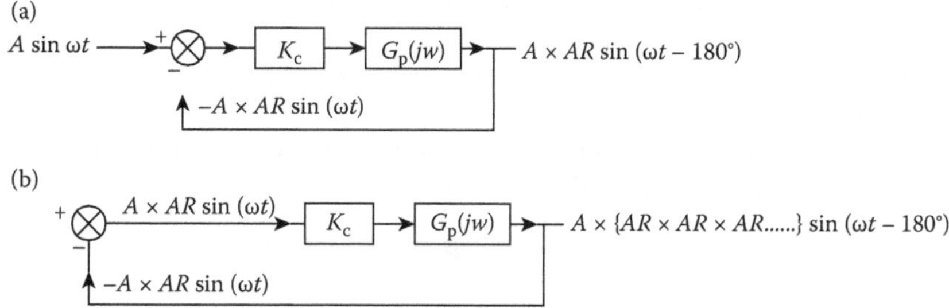

FIGURE 6.12 Output of system at $\Phi = -180°$ (a) before closing the loop with the sine input and (b) after closing the loop without the sine input.

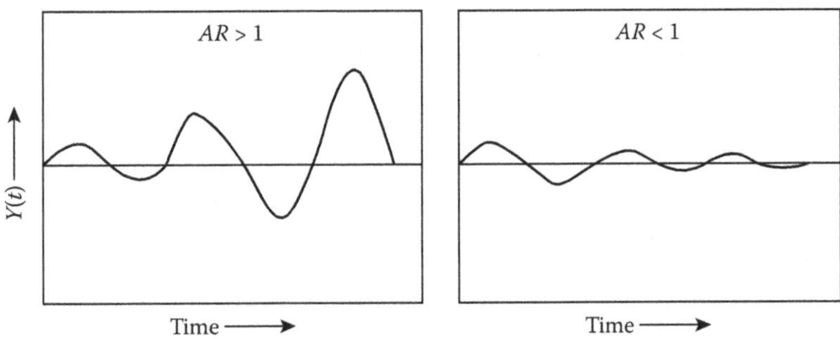

FIGURE 6.13 Sine response curve with (a) increasing amplitude and (b) decreasing amplitude.

This will indicate that at the phase difference of −180°, the output sine function will continue to exist in the closed loop even though the input sine disturbance is absent. The amplitude of the output function will be AR^n where n will be equal to the number of times the controller sends signals. For example, at the first, second, third, nth instances of signals, n will be 2,3,4, n, respectively. If the value of amplitude ratio AR is >1, the amplitude of the output sine function will continue to grow and may lead to an unstable system. Alternatively, if AR is <1, the amplitude of the output will come down to the low value, which may even be less than the amplitude of the input sine function. This is explained in Figure 6.13. Thus, a stable system will be obtained. Hence, it can be concluded that a system will be stable if the amplitude ratio is less than unity at a phase lag of 180°

$$(\Phi = -180°).$$

6.3.6 Bode Plots

Bode plots are the plots of amplitude ratio AR and phase difference Φ against radian frequency ω.

Bode plot of a first-order system:
The Laplacian transfer function for a first-order system is

$$G(s) = 1/(\tau s + 1). \tag{6.12}$$

Substituting s with $j\omega$, $G(j\omega) = 1/(j\tau\omega + 1) = (1 - j\tau\omega)/(1 + \tau^2\omega^2) = a + bj$, where $a = 1/(1 + \tau^2\omega^2)$ and $b = -\tau\omega/(1 + \tau^2\omega^2)$.

Hence, amplitude of $G(j\omega) = \sqrt{(a^2 + b^2)} = 1/\sqrt{(1 + \tau^2\omega^2)}$, and the phase difference is given as $\angle G(j\omega) = \tan^{-1}(-\omega\tau)$.

If a proportional controller of gain K_c is used, the open-loop transfer function becomes $K_cG(s)$, and the amplitude ratio and the phase difference will be given as

$$AR = \frac{K_c}{\sqrt{(\tau^2\omega^2 + 1)}} \tag{6.13}$$

$$\Phi = \tan^{-1}(-\omega\tau). \tag{6.14}$$

Amplitude ratio and phase difference of some of the transfer functions are similarly evaluated and listed below:

Transfer Function	Amplitude Ratio, $G(jw)$	Phase Difference, Φ
$\frac{K_c}{(\tau_1 s+1)}$	$\frac{K_c}{\sqrt{(\tau_1^2\omega^2+1)}}$	$\Phi = \tan^{-1}(-\omega\tau_1)$
$\frac{K_c}{(\tau_1 s+1)(\tau_2 s+1)}$	$\frac{K_c}{\sqrt{(\tau_1^2\omega^2+1)}\sqrt{(\tau_2^2\omega^2+1)}}$	$\tan^{-1}(-\omega\tau_1) + \tan^{-1}(-\omega\tau_2)$
$\frac{K_c}{(\tau_1 s+1)(\tau_2 s+1)(\tau_3 s+1)}$	$\frac{K_c}{\sqrt{(\tau_1^2\omega^2+1)}\sqrt{(\tau_2^2\omega^2+1)}\sqrt{(\tau_3^2\omega^2+1)}}$	$\tan^{-1}(-\omega\tau_1) + \tan^{-1}(-\omega\tau_2) + \tan^{-1}(-\omega\tau_3)$
$\frac{K_c K_p e^{-\tau_d s}}{\tau_p s+1}$	$\frac{K_c K_p}{\sqrt{(\tau^2\omega^2+1)}}$	$\Phi = \tan^{-1}(-\omega\tau) - \omega\tau_d$

Bode plots are obtained by plotting AR vs. ω as the upper plot and Φ vs. ω as the lower plot. These two plots are jointly called as the Bode plots. This is shown in Figure 6.14.

Bode plots for a first-order system coupled with a transportation or exponential lag are presented in Figure 6.14.

6.3.7 Gain and Phase Margin Method

If the amplitude ratio at a phase lag of 180° is a, then the ratio of $1/a$ is the gain margin (GM). The smaller the value of a, the larger the GM and the safer the controlled process is as far as stability is concerned. But too low a value of a will decrease the

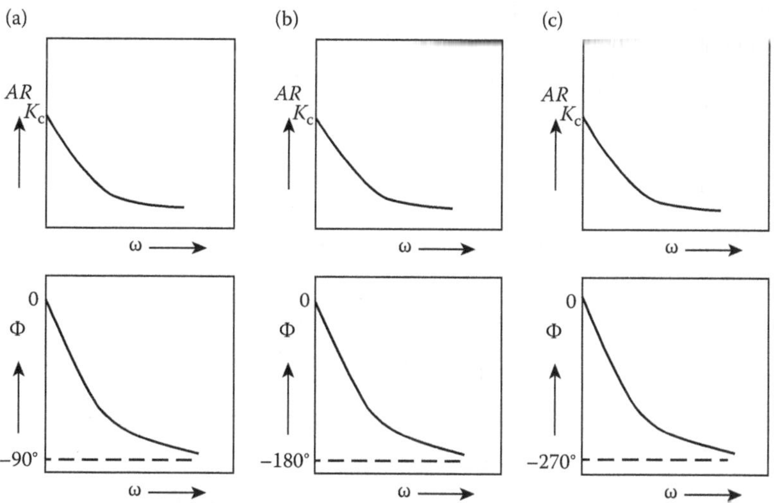

FIGURE 6.14 Bode plots of (a) first-order, (b) second-order, and (c) third-order systems.

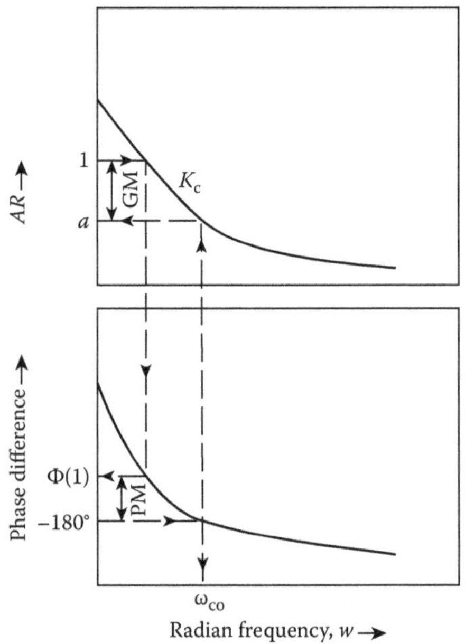

FIGURE 6.15 Gain and phase margins.

speed of response for lower K_c. Practically, a value of the GM is compromised to 1.7 and more. It is also understood that operating the control system when the amplitude ratio equals unity (one) is the maximum limit for stable operation. Alternatively, the phase difference at this unity amplitude ratio is determined as Φ(1), which should be far from the crossover phase lag of −180°. This difference, Φ(1) – (−180°) or (180° + Φ(1)) is the phase margin (PM). The greater the margin, the safer it is from instability. In practice, the PM should be greater than 30°. GM and PM on Bode plots are explained in Figure 6.15. Theoretical tuning of controllers using the Ziegler–Nichols method, Bode plots, and gain and phase margin methods are explained in the following example.

Tuning Example

In a control loop, the transfer function of the process is given in the block diagram shown in Figure 6.16.

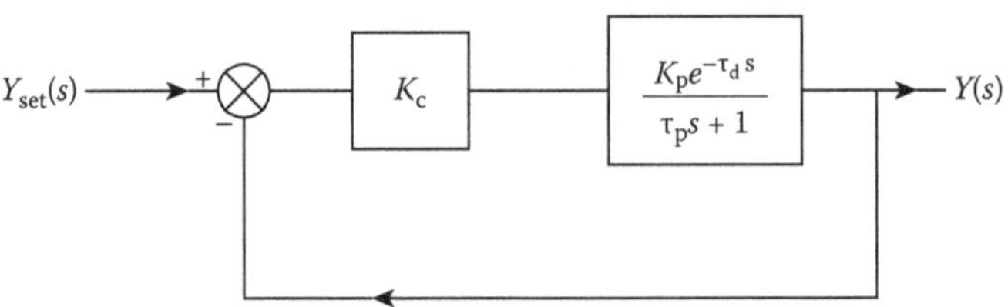

FIGURE 6.16 Control loop to be studied.

(a) Determine the open-loop transfer function
(b) Determine the amplitude ratio and phase difference using frequency response
(c) Determine the maximum value of K_c according to Bode stability criteria
(d) Determine the value of P_u
(e) Tune the controller for P, PI, and PID modes using Ziegler–Nichols settings
(f) Determine the gain and phase margins for the above settings
(g) Retune the controller using the gain and phase margins only

given that $K_p = 3$, $\tau_p = 1$, $\tau_d = 0.5$.
Answer:

(a) The open-loop transfer function is

$$G_{ol} = \frac{K_c K_p e^{-\tau_d s}}{\tau_p s + 1}. \tag{6.15}$$

(b) Amplitude ratio is

$$AR = \frac{K_c K_p}{\sqrt{(\tau_p^2 \omega^2 + 1)}}. \tag{6.16}$$

And phase difference is

$$\Phi = \tan^{-1}(-\omega\tau_p) - \omega\tau_d. \tag{6.17}$$

(c) Bode plots are presented in Figure 6.17.
Numerically, radian frequency, AR/K_c and phase difference are determined as listed in the following table.

ω	AR/K_c	Φ	ω	AR/K_c	Φ	ω	AR/K_c	Φ
0.1	2.99	−8.58	0.2	2.94	−17.0	0.3	2.87	−25.3
0.4	2.79	−33.2	0.5	2.68	−40.9	0.6	2.57	−48.1
0.7	2.46	−55.0	0.8	2.34	−61.6	0.9	2.23	−67.8
1.0	2.12	−73.6	1.1	2.02	−79.2	1.2	1.92	−84.6
1.3	1.83	−89.7	1.4	1.74	−94.6	1.5	1.66	−99.3
1.6	1.59	−103.8	1.7	1.52	−108.2	1.8	1.46	−112.5
1.9	1.4	−116.7	2.0	1.34	−120.7	2.10	1.29	−124.7
2.2	1.24	−128.65	2.3	1.2	−132.4	2.4	1.15	−136.2
2.5	1.11	−139.8	2.6	1.08	−143.5	2.7	1.04	−147.1
2.8	1.01	−150.6	2.9	0.98	−154.1	3.0	0.95	−157.5
3.1	0.92	−161.0	3.2	0.89	−164.4	3.3	0.87	−167.7
3.4	0.85	−171.1	3.5	0.82	−174.4	3.6	0.80	−177.7
3.7	0.78	−180.96	3.8	0.76	−184.2	3.9	0.75	−187.4

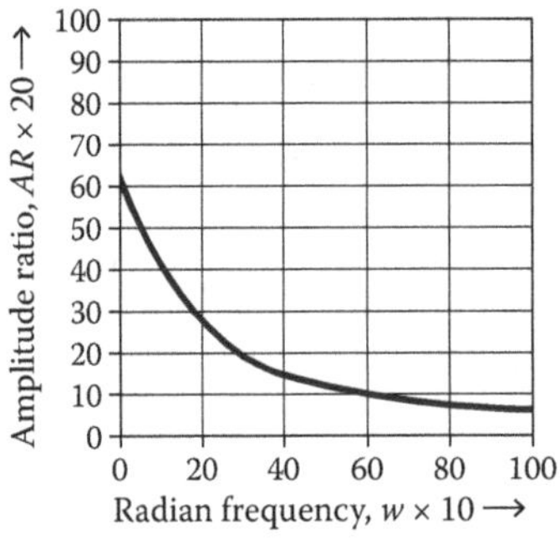

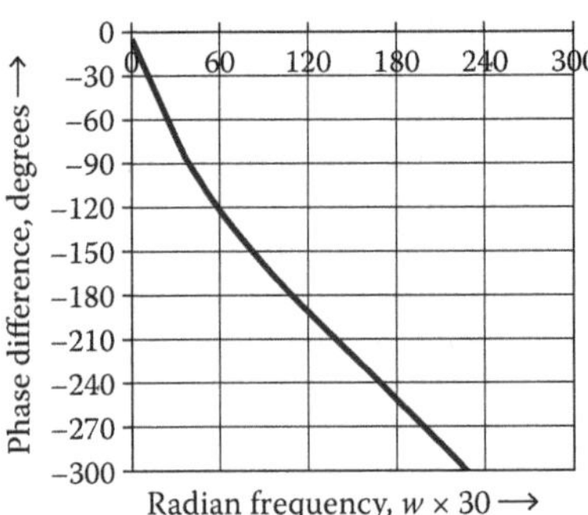

FIGURE 6.17 Bode plots of Exercise 6.1.

From the phase plot, the value of the crossover frequency is $\omega_{co} = 3.7$ at $\Phi = -180$. The value of AR/K_c at this frequency is obtained as 0.78 from the amplitude ratio plot. Hence, the maximum value of K_c (ultimate value K_u) when $AR = 1$ is 1/0.78 = 1.28, i.e., $K_u = 1.28$.

(d) Ultimate period $P_u = 2 \times \Pi/\omega_{co} = 6.28/3.7 = 1.69$.

(e) Tuned parameters as per Ziegler–Nichols settings based on K_u and P_u are presented in the following table.

Ziegler–Nichols Settings Determined

Control Mode	Proportional Gain, K_c	Integration Time, τ_i	Derivative Time, τ_d
Proportional	$0.50\ K_u = 0.64$	–	–
Proportional–integral	$0.45\ K_u = 0.576$	$P_u/1.2 = 1.4$	–
Proportional–integral–derivative	$0.60\ K_u = 0.768$	$P_u/2 = 0.845$	$P_u/8 = 0.2115$

(f) Gain margin and phase margin are determined for the above settings. Considering the proportional mode, $K_c = 0.64$: hence, the AR and phase difference are given as at $\omega = \omega_{co} = 3.7$, AR is obtained as

$$AR = \frac{K_c\, K_p}{\sqrt{(\tau_p^2 \omega^2 + 1)}} = \frac{0.64 \times 3}{\sqrt{(3.7^2 + 1)}} = 0.50. \tag{6.18}$$

Hence, GM = 1/0.5 = 2 > 1.7.

When $AR = 1$, ω is obtained from the equation

$$1 = \frac{0.64 \times 3}{\sqrt{(\omega^2 + 1)}}. \tag{6.19}$$

Solving, ω = 1.50, and the value of the phase difference at this ω is obtained from the relationship

$$\Phi = \tan^{-1}(-\omega\tau_p) - \omega\tau_d$$

$$\Phi(1) = \tan^{-1}(-1.5) - 1.5 \times 0.5 \times 180/3.14$$

$$= -99.33°.$$

Hence, the phase margin is given as

$$\text{PM} = 180 - 99.33 = 80.67$$

which is greater than 30°.

The margins for the PI control for the Z–N settings are evaluated in the following ways: The open-loop transfer function with the controller is

$$G_{ol} = \frac{K_c\left(1+\dfrac{1}{\tau_i s}\right)K_p\, e^{-\tau_d s}}{\tau_p s+1} \tag{6.20}$$

and the amplitude ratio and phase difference transfer function in frequency ω is given as

$$AR = \frac{K_p\, K_c\sqrt{1+\dfrac{1}{\tau_i^2\omega^2}}}{\sqrt{\tau_p^2\omega^2+1}} \tag{6.21}$$

$$\Phi = \tan^{-1}(-1/\tau_i\omega) + \tan^{-1}(-\tau_p\omega) - \tau_d\omega. \tag{6.22}$$

Substituting the values of K_p, τ_p, and τ_d, and Z–N values for $K_c = 0.576$ and $\tau_i = 1.4$, the Bode plots are drawn to determine the crossover frequency and the corresponding *AR* (Figure 6.18).

At ω = 3.30, Φ = –180, and *AR* = 0.51; hence, GM = 1/0.51 = 1.96 > 1.7, and at *w* = 1.50, *AR* = 1 and Φ = –124.65°; hence, PM = 180 – 124.65 = 55.35 >30°.

Similarly, the open-loop transfer function with a PID controller is

$$G_{ol} = \frac{K_c\left(1+\dfrac{1}{\tau_i s}+\tau_D s\right)K_p\, e^{-\tau_d s}}{\tau_p s+1}. \tag{6.23}$$

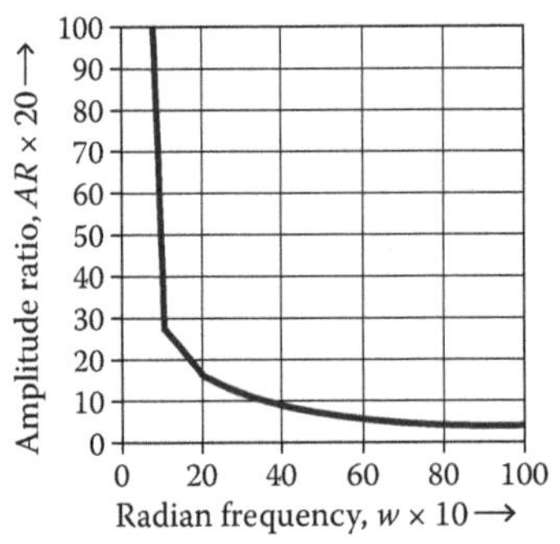

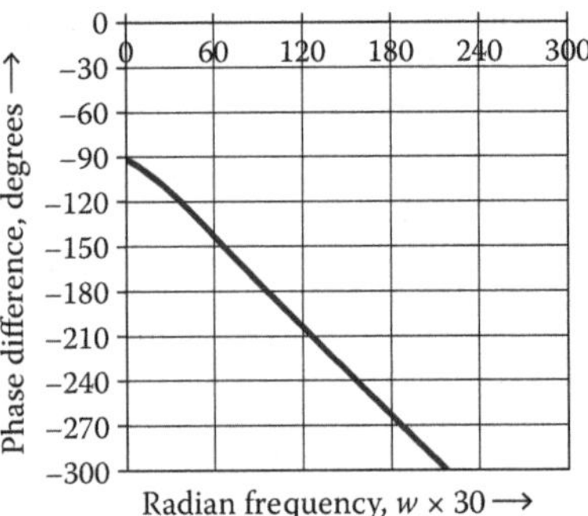

FIGURE 6.18 Bode plots for PI controller with process.

Amplitude ratio and phase difference are given as

$$AR = \frac{K_p\ K_c\sqrt{1+\left(\tau_D\omega - \dfrac{1}{\tau_i\omega}\right)^2}}{\sqrt{\tau_p^2\omega^2+1}} \tag{6.24}$$

$$\Phi = \tan^{-1}(\tau_D\omega - 1/\tau_i\omega) + \tan^{-1}(-\tau_p\omega) - \tau_d\omega. \tag{6.25}$$

Substituting the values of K_p, τ_p, and τ_d, and Z–N values for $K_c = 0.768$, $\tau_i = 0.848$, and $\tau_D = 0.212$. The Bode plots are drawn to determine the crossover frequency, and the corresponding AR are presented in Figure 6.19.

At $\omega = 4.9$, $\Phi = -180$, and $AR = 0.59$; hence, GM = 1/0.59 = 1.695 ≈ 1.7, and at $\omega = 2.10$, $AR = 1$, and $\Phi = -131.40°$; hence, PM = 180 – 131.40 = 48.6 > 30°.

Hence, tuned parameters for all the control modes according to Ziegler–Nichols settings satisfy gain and phase margins.

(g) Tuning of the controller in all the P, PI, and PID modes can also be done using Bode stability criteria and margins in the following way:

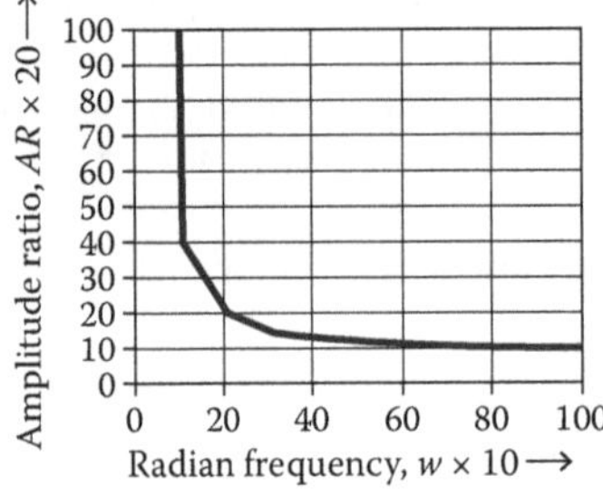

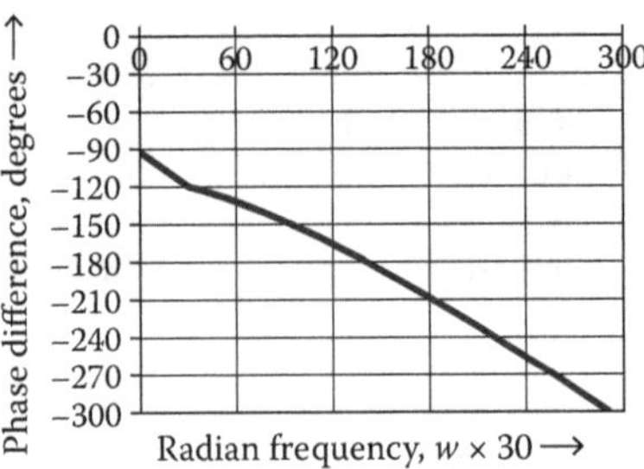

FIGURE 6.19 Bode plots for PID controller with process.

Proportional control:
Taking GM = 1.7, $AR = 1/1.7 = 0.588$, at $\Phi = -180°$. Also, AR is related to K_c as

$$AR = \frac{K_c\, K_p}{\sqrt{(\tau_p^2\omega^2+1)}} = \frac{K_c \times 3}{\sqrt{(3.7^2+1)}} = 0.588\,.$$

Hence, $K_c = 0.75$ at $\omega >= 3.7$ obtained from Figure 6.19.

PM = 30, so $180 + \Phi(1) = 30$, i.e., $\Phi(1) = -150°$. Now, from the phase plot of Figure 6.14, the corresponding $AR/K_c = 1.01$. AR should be unity as per the phase margin concept. Hence, $K_c = 1/1.01 = 0.99$.

Hence, $K_c = 0.99$ should be taken. This value will satisfy both the margins.

Similarly, the tuning is carried out using the margins for PI and PID modes.

Proportional–integral:
In this case, the transfer function will change from the previous proportional control system as

$$G_{ol} = \frac{K_c\left(1+\dfrac{1}{\tau_i s}\right)K_p\, e^{-\tau_d s}}{\tau_p s+1}\,.$$

and Bode plots are to be obtained.

GM = 1.7, PM = 30, and τ_i (=1.4) is assumed. AR/K_c and Φ are then plotted against ω as shown in Figure 6.20 where the amplitude ratio axis will indicate AR/K_c.

For $\Phi = -180$, $\omega = 3$, and $AR/K_c = 0.98$, so $K_c = AR/0.98 = (1/1.7)/0.98 = 0.60$ (AR should be $1/\text{GM} = 1/1.7$).

For PM = 30, $\Phi(1) = -150°$, $\omega = 2$, and $AR/K_c = 1.5$; hence, $K_c = 1/1.5 = 0.67$.

Hence, $K_c = 0.67$ should be taken for satisfying both the GM and PM.

However, if integration time is changed, K_c value has to be redetermined in a similar way.

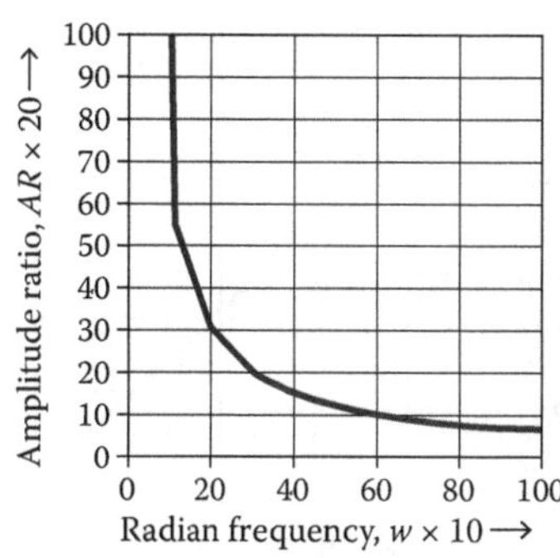

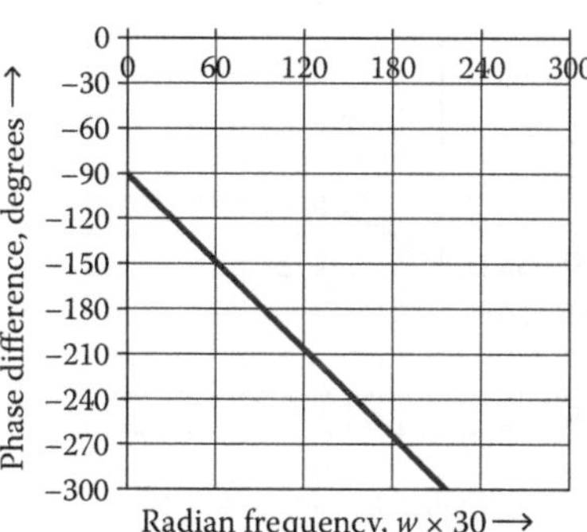

FIGURE 6.20 Bode plots for PI control with AR/K_c and Φ vs. ω.

Proportional–integral–derivative control:
Taking integration and derivative time as 0.8 and 0.2, respectively, PM = 30, and GM = 1.7, the transfer function will include the PID control mode and be given as

$$G_{ol} = \frac{K_c\left(1+\dfrac{1}{\tau_i s}+\tau_D s\right)K_p\, e^{-\tau_d s}}{\tau_p s+1}. \tag{6.26}$$

Bode plots are redrawn as shown in Figure 6.21.

Now, GM = 1.7, $AR/K_c = 0.75$ while $\omega = 5$; hence, $K_c = (1/1.7)/0.75 = 0.784$.

Also at $\Phi(1) = -150°$, $\omega = 3$, and $AR/K_c = 0.98$; hence, $K_c = 1/0.98 = 1.02$.

Hence, $K_c = 1.02$ should be taken to satisfy both the margins.

If integration and derivative times are each taken as unity, the Bode plots will be as shown in Figure 6.22.

This type of Bode plot will generate ambiguous values as more than two values may be obtained for AR/K_c. This type cannot be handled by the Bode stability method. Hence, the integration and derivative time should be set at lower values such that Bode plots similar to Figure 6.21 will be generated to make the Bode stability method applicable.

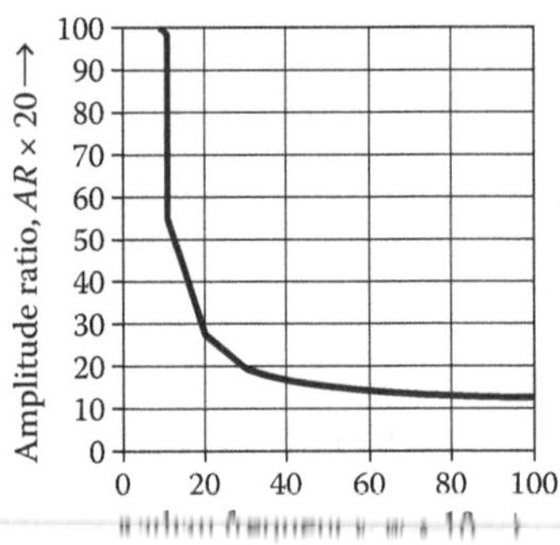

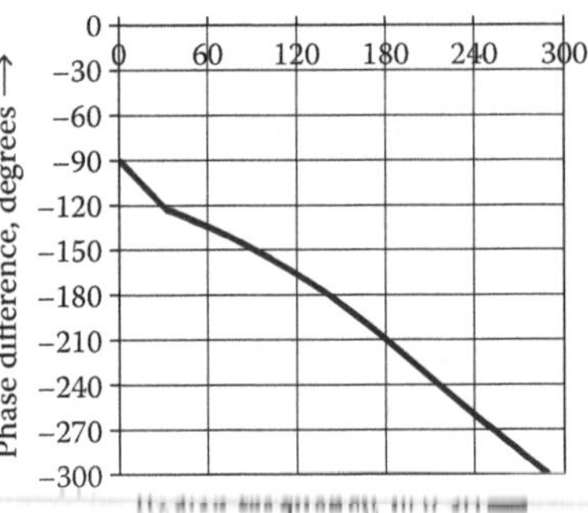

FIGURE 6.21 Bode plots for PID control system.

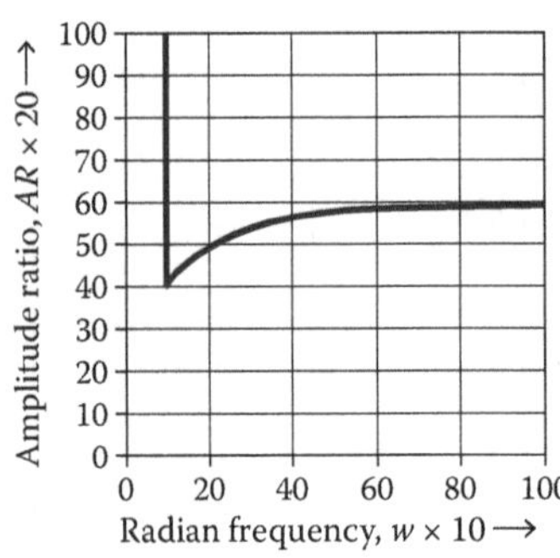

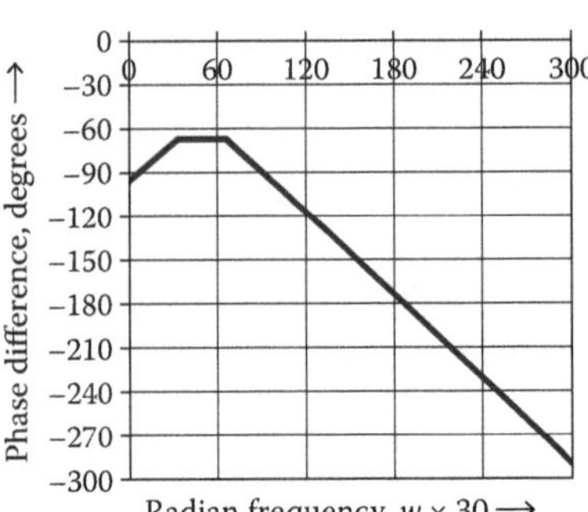

FIGURE 6.22 Bode plots for PID control with unity integration and derivative times.

6.3.8 Limitations of Bode Stability Analysis

Bode stability criteria is applicable when the amplitude ratio and the phase difference decrease or increase continuously (mono-modal curves) where the *AR* and Φ each assumes only a single value at a particular value of radian frequency (ω). These types are already shown in Figures 6.17 through 6.21. However, in Figure 6.22, the Bode plots are not unimodal and yield two values of ω at the same value of *AR* or Φ in certain regions of the curve where Bode stability analysis cannot be made. Other examples of plots of amplitude ratio and phase difference where Bode stability analysis cannot be made are shown in Figure 6.23.

In such cases, Nyquist stability analysis must be used.

6.3.9 Nyquist Plots

A Nyquist plot (or polar plot) is a single plot of the transfer function on a complex plane where the points are located with the help of the *AR* and Φ with respect to the origin as shown in Figure 6.24. A point $(a + bj)$ is located such that $r = \sqrt{(a^2 + b^2)}$, and

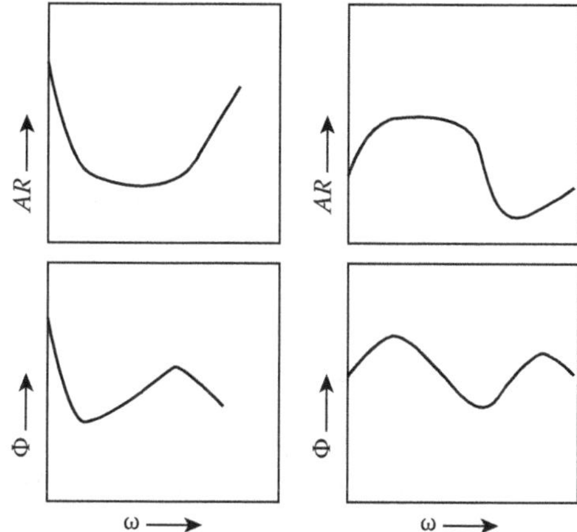

FIGURE 6.23 Multimodal plots where Bode stability analysis is not applicable.

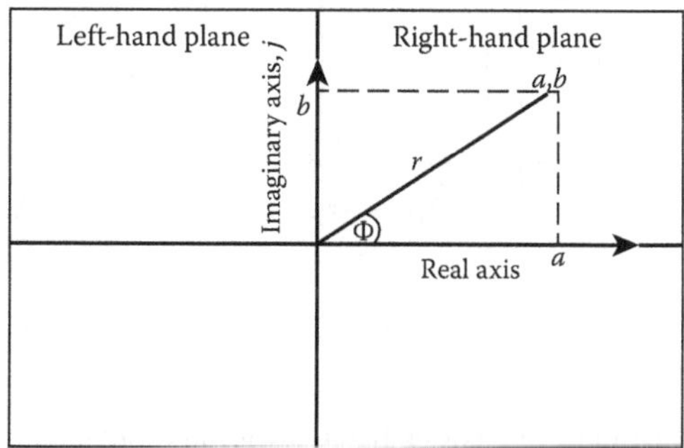

FIGURE 6.24 Location of a point on a complex plane.

the angle is Φ where r is the AR, and Φ is the phase difference between the input and response sine functions as defined earlier in the Bode plots.

Nyquist plot of a first-order transfer function:

$$AR = \frac{K_c}{\sqrt{\tau^2\omega^2 + 1}}$$

$$\Phi = \tan^{-1}(-\omega\tau).$$

Taking $K_c = 1$ and $\tau = 1$, then from the AR relationship, $\omega = 0$, $AR = 1$, and $\Phi = 0$; thus, the point is located as $1 + j0$ along the real axis. Similarly, as $\omega = 1$, $AR = 1/\sqrt{2}$, and $\Phi = -45°$. Thus, various points are determined until $\omega \to \infty$ when $AR = 0$ and $\Phi = -90°$. The values are listed below for a few points. The Nyquist plot is shown in Figure 6.25a.

ω	*AR*	Φ
0	1	0
1	$1/\sqrt{2}$	−45
2	$1/\sqrt{5}$	−63.4
10	$1/\sqrt{101}$	−84.28
20	$1/\sqrt{401}$	−87.13
∞	0	−90

Nyquist plots of other transfer functions are similarly plotted as shown in Figure 6.25b through 6.25f.

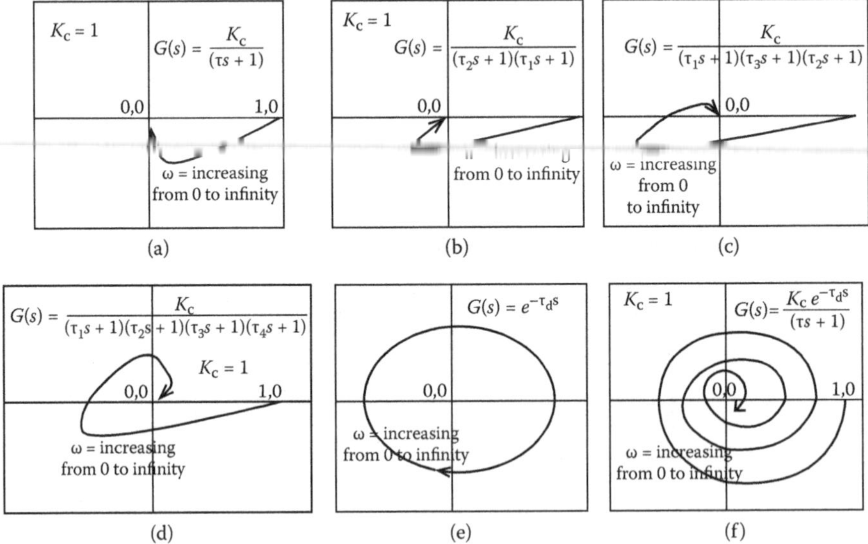

FIGURE 6.25 Nyquist plots of transfer functions (a) first order, (b) second order of two first orders in series, (c) third order, (d) fourth order, (e) transportation lag, and (f) first order coupled with transportation lag.

6.3.10 Nyquist Stability Analysis

In order to apply Nyquist stability law, the following points must be remembered:

(a) Phase angle: As shown in Figure 6.25, the plots for first- and second-order systems lie on the lower right and lower left of the complex plane. However, the third-order and higher-order systems, and when the transportation lag is present, the plots extend to the top left and top right of the plane. Imagine a point is moving in a clockwise direction around the origin. The angular position of the point lying on the right half of the plane is zero degrees and on the left half, it is –180° with respect to the origin. Thus, a point lying on the imaginary axis in the lower half is –90° and in the upper half is –270°. If the point continues to rotate further from the real axis on the upper right half, the angle will be –360°, and on the left half, it will be –540°. Thus, the locus of a point representing the transportation lag as shown in Figure 6.25e is encircling the origin at 0°, –90°, –180°, –270°, –360°, and –450°, approaching infinity by endless encircling. This is also true for a transfer function coupled with a transportation lag as shown in Figure 6.25f. The various angles of encirclements in a single rotation of the point are shown in Figure 6.26. However, these angles will be taken as positive in a counter-clockwise direction with respect to the origin.

(b) Poles and zeroes:
From the closed-loop characteristic equation, the roots are determined at two values of K_c at zero and infinity, and these roots are known as the poles and zeroes, respectively. This can be understood from the following characteristic equation:

$$1 + K_c N/D = 0 \tag{6.27}$$

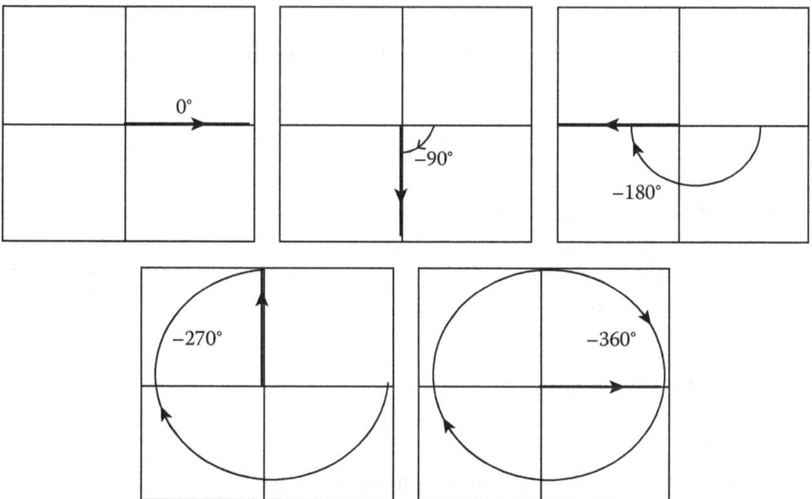

FIGURE 6.26 Various angular positions of any point on the axes.

where N is the numerator and D is the denominator of the open-loop transfer function. If $K_c = 0$, then $D = 0$, and the roots are known as poles. If $K_c = \infty$, then $N = 0$ and the roots are zeroes. For instance, if the open-loop transfer function is

$G(s) = K_c(s-1)(s-2)/(s-3)(s-4)(s-5)$, i.e., $N = (s-1)(s-2)$ and $D = (s-3)(s-4)(s-5)$

Hence, poles are 3, 4, and 5, and zeroes are 1 and 2. Note that all of these roots are lying on the right half of the plane (RHP). It is also to be noted that if poles are lying on the RHP, it means that the process is open-loop unstable. But the presence of zeroes on the RHP is responsible for the instability of the highest value of K_c in the closed-loop control system. Hence, determination of zeroes (Z) is vital for ascertaining the stability of a controlled system. Of course, the number of poles (P) on the RHP will be helpful for determining the number of zeroes on the RHP by the Nyquist encirclements as described next.

(c) Encirclement of the point (1,0):

If a Nyquist plot of an open-loop transfer function is drawn as explained earlier, and if it is found that the locus encircles the point on the real axis (left half) at the location of coordinates (−1,0), then it is taken as an encirclement. If the said encirclement takes place once in the clockwise direction as ω increases from 0 to ∞, then the number of encirclements is taken as 1. If this occurs in a counterclockwise direction, the number of encirclements will be taken as −1. If the number of encirclement occurs N times around the point (−1,0), then these will be taken as $+N$ (clockwise) and $-N$ (counterclockwise).

(d) Stability rules:

After ascertaining the number of encirclements ($+N$) and the number of poles lying on the right half of the plane (P), then the number of zeroes (Z) lying on the right half of the plane is given as

$$\mathrm{Z} = \mathrm{P} + N \tag{6.28}$$

Thus, Z in Equation 6.28 equals the number of roots having positive real parts for very high K_c. If Z is equal to one or more, then the control system will be unstable. Alternatively, if Z = 0 or is negative, then the system will be stable at the highest value selected for K_c.

6.4 QUESTIONS AND ANSWERS

EXERCISE 6.1

Consider a third-order system of three first-order systems in series having time constants of 1, 2, and 3 minutes, respectively. If a proportional controller is used, determine the maximum value of K_c that should be used applying Nyquist stability laws.

Solution:
The open-loop transfer function is

$$G(s) = \frac{K_c}{(\tau_1 s + 1)(\tau_3 s + 1)(\tau_2 s + 1)}$$

where $\tau_1 = 1$, $\tau_2 = 2$, and $\tau_3 = 3$.

The Nyquist plots are drawn for $K_c = 5$ and $K_c = 10$, shown in Figure 6.27.

From Figure 6.27, it is seen that the point (–1,0) is encircled clockwise when K_c is greater than or equal to 10 because P = 0 and N = +1. Therefore, Z = P + N = 0 + 1 > 0, which indicates an unstable system. Hence, the system will be stable as long as $K_c < 10$.

The same can also be supported by the Bode stability criterion. As at $\Phi = -180$, $AR = 0.978$ at $K_c = 10$. Hence, AR is nearly unity, which is vulnerable to instability.

According to the closed-loop characteristic equation,

$$1 + \frac{K_c}{(s+1)(2s+1)(3s+1)} = 0$$

or $6s^3 + 11s^2 + 6s + (K_c + 1) = 0$.

Using the Routh–Hurwitz array,

$$\begin{vmatrix} 6 & 6 \\ 11 & (K_c + 1) \\ \dfrac{66 - 6(K_c + 1)}{11} & \\ (K_c + 1) & \end{vmatrix}$$

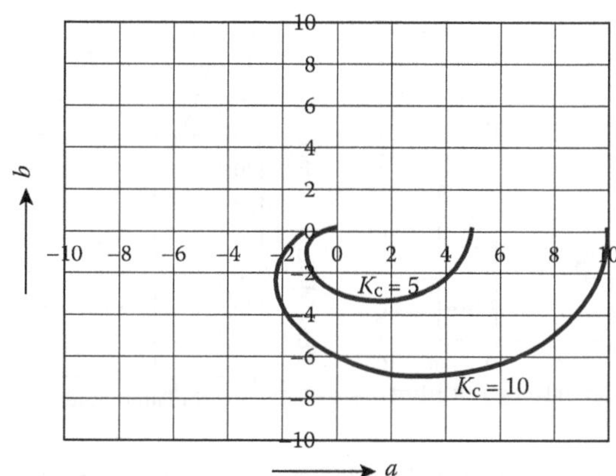

FIGURE 6.27 Nyquist plots for $K_c = 5$ and 10.

To make the third element in the first column positive, K_c must be less than 10. Hence, for a stable closed-loop control system, K_c must be selected below 10.

EXERCISE 6.2

Tune a proportional controller having the following transfer functions as shown in Figure 6.28.

The transfer functions of the control valve and the transducer are taken as unity. Take $\tau = 1$ and $\tau_d = 0.5$.

The Routh–Hurwitz method is not applicable for the presence of exponential lag, which is a nonlinear term. Hence, frequency response methods will be attempted.

(a) Bode stability method: Bode plots are shown in Figure 6.29.
(b) Nyquist stability method: Nyquist plots for $K_c = 1$ and 4 are shown in Figure 6.30.

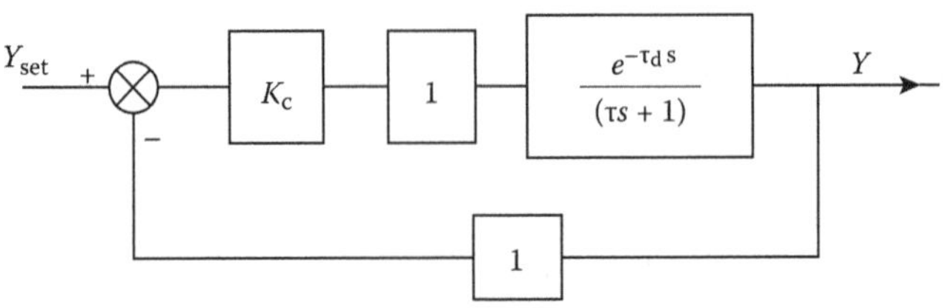

FIGURE 6.28 Transfer functions and the closed loop.

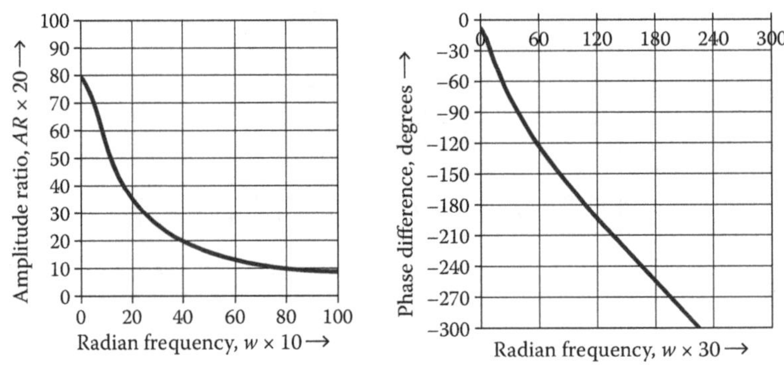

FIGURE 6.29 Bode plots of the problem in Exercise 6.1 where $K_c = 4$ is the maximum value as $AR = 1$ at $\Phi = -180°$.

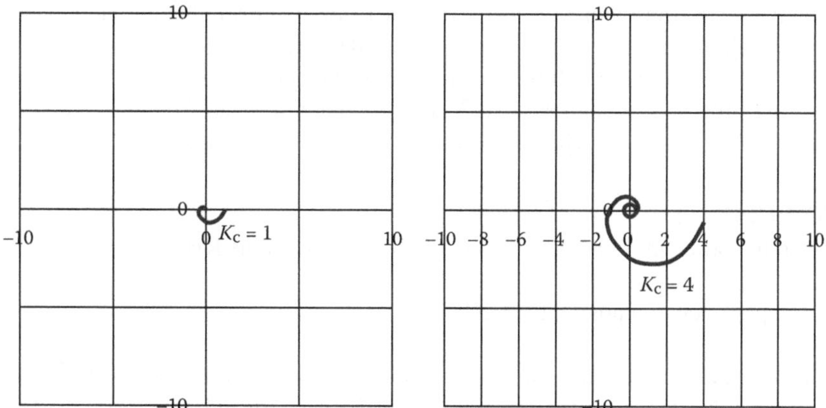

FIGURE 6.30 Nyquist plots of the problem in Exercise 6.1 for $K_c = 1$ and $K_c = 4$. At $K_c = 4$, the plot encircles the point (–1,0). Hence, $K_c < 4$ should be selected.

EXERCISE 6.3

Determine the values of proportional gain of the controller for which the systems will be stable for the Nyquist plots for $K_c = 1$ given in Figure 6.31a, b, c, and d if the number of poles on the RHP is absent.

Solution:

(a) $K_c > 10$ will cause an encirclement clockwise; hence, $Z > 0$ and is unstable. Hence, $K_c < 10$ is stable.

(b) For $K_c > 2$ and $K_c < 50$, there will be one encirclement clockwise, which, hence, is unstable. Hence, $K_c < 2$ and >50 are stable.

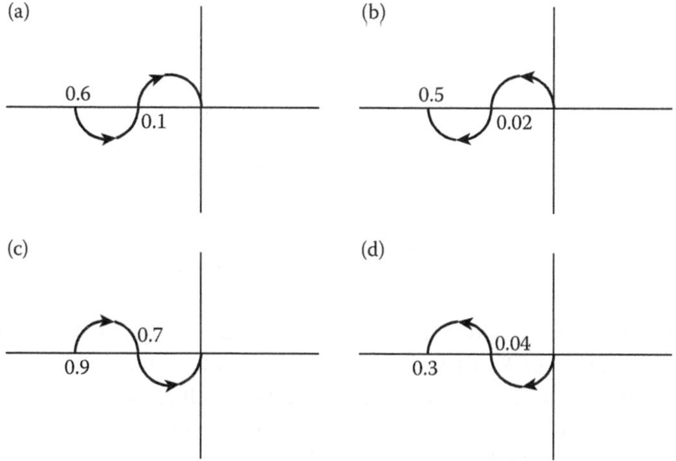

FIGURE 6.31 Various Nyquist plots of open-loop transfer functions.

(c) For $K_c > 1/0.9$ and $K_c < 1/0.7$, there will be one encirclement clockwise, which, hence, is unstable. Hence, $K_c < 1/0.9$ and $K_c > 1/0.7$ are stable.

(d) For $K_c > 25$, there will be one encirclement clockwise, so $Z = 1 > 0$ and is unstable. $K_c < 25$ and $K_c > 10/3$, $N = -1$, and $Z = -1$ are stable as are $K_c < 1/0.3$, $N = 0$, $Z = 0$. Hence, at $K_c < 25$, the system will be stable.

EXERCISE 6.4

Repeat the problem if one pole is found to be present on the RHP.

Solution:

For the presence of a pole in the RHP,

(a) $K_c > 10$ will cause an encirclement clockwise; hence, $Z = P + N = 0 + 1 > 0$ and is unstable. Hence, for $10/6 < K_c < 10$, $N = -1$ and $Z = 1 - 1 = 0$ are stable. For $K_c < 10/6$, $N = 0$ and $Z = 1 + 0 = 1 > 0$, which is unstable. Hence, K_c should be greater than 10/6 but below 10 for a stable system.

(b) For $K_c > 2$ and $K_c < 50$, there will be one encirclement clockwise; hence, $N = 1$, $Z = 1 + 1 = 2$, which is unstable. Also, for $K_c < 2$, $N = 0$, but $Z = 1 + 0 > 0$ is unstable, and $K_c > 50$, $N = -1$, $Z = 1 - 1 = 0$ is stable. Hence, only when $K_c > 50$ will the system be stable.

(c) For $K_c > 1/0.9$ and $K_c < 1/0.7$, there will be one encirclement clockwise; hence, $N = 1$, $Z = 1 + 1 = 2$ and is unstable. For $K_c < 1/0.9$, $N = 0$, $Z = 1 + 0 = 1 > 0$ and is unstable, and $K_c > 1/0.7$, $N = -1$, $Z = 1 - 1 = 0$ is stable. So, K_c should be greater than 1/0.7 (= 1.42) for stable operation.

(d) For $K_c > 25$, there will be one encirclement clockwise, so $Z = 1 + 1 > 0$; hence unstable, and $K_c < 25$ and $K_c > 10/3$, $N = -1$ and $Z = 1 - 1 = 0$ is stable. But, $K_c < 10/3$, $N = 0$, $Z = 1 + 0 > 0$ is again unstable. Hence, $3.3 < K_c < 25$ is stable.

EXERCISE 6.5

A proportional control system has an open-loop function as

$$G_{ol} = K_c/(s - 1)(s + 2)(s + 3)$$

Determine the value of K_c for a stable, closed-loop system.

Solution:

$$G_{ol} = K_c/6\ (s - 1)(s/2 + 1)(s/3 + 1)$$

From this, it is understood that there is a pole on the right-hand half of the plane. The Nyquist plots for $K_c = 1$ and $K_c = 10$ are shown in Figure 6.32.

From the plots, it is seen that as long as $K_c > 6$ but <10, $N = -1$, so $Z = 1 - 1 = 0$, hence stable. But if $K_c < 6$, $N = 0$ but $P = 1$ as understood from the transfer function, $Z = 1 + 0 = 1$, hence unstable. Also, $K_c > 10$, $N = 1$, $Z = 1 + 1 = 2 > 0$ and hence unstable. Thus, the system will be stable if $6 < K_c < 10$ only. Because the transfer function does not contain exponential lag, the Routh–Hurwitz law can also be applied.

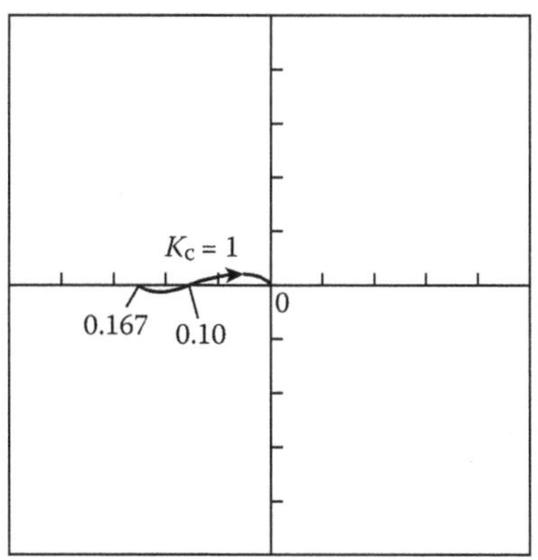

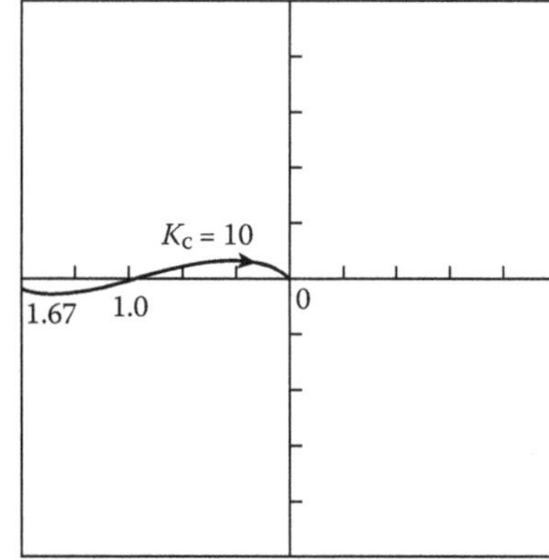

FIGURE 6.32 Nyquist plots for $K_c = 1$ and 10.

The closed-loop characteristic equation is

$$(s-1)(s+2)(s+3)+K_c=0$$

or

$$s^3+4s^2+s+(K_c-6)=0.$$

The Routh–Hurwitz array is

$$\left| \begin{array}{cc} 1 & 1 \\ 4 & K_c-6 \\ \dfrac{4-K_c+6)}{4} & \\ K_c-6 & \end{array} \right|.$$

From this, it is understood that $K_c < 10$ and $K_c > 6$ to have positive elements in the first column. Hence, for a stable control system, $6 < K_c < 10$.

This confirms the Nyquist stability findings above.

7 Advanced Control Strategies

7.1 CASCADE CONTROL

Cascade control is a control strategy that involves a master controller, which acts over a cascade of slave controllers. Two controllers are widely used: one is the master controller, which instructs the second controller, which is the slave controller. However, each controller controls different process variables. The benefit of such a control strategy over a single-controller system will be understood from the following example. Consider a steam-heated tank where steam flow is manipulated by a temperature controller. However, the temperature of the tank increases when the temperature of the steam coil is increased and vice versa. Hence, the hierarchy of change of temperature is from the steam-coil temperature to the tank temperature, i.e., the change in coil temperature is followed by a change in tank temperature. If the coil temperature slowly rises, the rise of the tank temperature will also be delayed. In order to reduce this delay, the tank temperature controller is connected with another controller that controls the coil temperature. This arrangement is shown in Figure 7.1 where the tank-temperature controller is the master controller, and it senses the tank temperature and sends the output signal as the set point for the coil temperature to the slave controller. The slave controller senses the coil temperature and generates an output signal based on the difference between the output signal of the master controller and the coil temperature and finally manipulates the steam flow accordingly. The cascade control system has two controllers and two sensors but one control valve.

The Laplacian block diagram of such a cascade control system can be developed in the following way:

A heat balance equation for the tank, assuming constant mass flow rate w for the liquid in and out, unchanged specific heat C_p, and density ρ, is given as

$$X(t) + Q(t)/(w\ C_p) = \tau\ dY(t)/dt + Y(t) \tag{7.1}$$

Heat balance over the coil

$$Q(s) = A_c U_c (Y_c(s) - Y(s))$$

or

$$Y_c(s) = Q(s)/(A_c U_c) + Y(s) \tag{7.2}$$

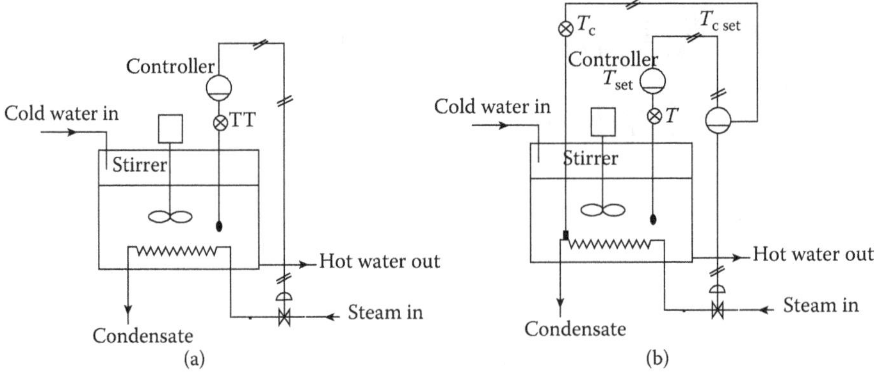

FIGURE 7.1 Temperature control in a steam-heated tank (a) single controller loop (b) cascade control system of two controllers where T and T_c are tank and coil temperatures, respectively.

where Y_c is the coil temperature in deviation form, and A_c and U_c are the surface area and heat transfer coefficient of the coil. Thus, the equivalent Laplacian block diagram of the control system is shown in Figure 7.2.

The master controller has the transfer function of G_{c1}, and that of the slave controller has G_{c2}. The master controller measures the tank temperature and generates the set point, $Y_{c\ set}$, of the slave controller. $Y(s)$ can be expressed in terms of $Y_{set}(s)$ and $X(s)$ after manipulations as

$$Y(s) = \frac{G_{c2}G_{c1}G_vU_cA_cY_{set}(s) + wC_p(U_cA_c + G_{c2}G_v)X(s)}{wC_p(1+\tau s)U_cA_c + G_{c2}G_uwC_p(1+\tau s) + G_{c2}G_v(1+\tau s)U_cA_c}. \tag{7.3}$$

The control response can be verified for set point or load disturbances for the given values of process and controller parameters.

For example, if proportional controllers are used with gains K_{c1} and K_{c2}, and the control valve has no lag, such that $G_v = 1$ and $wC_p = \tau = 1$, then the response for $Y(s)$ is obtained from Equation 7.3.

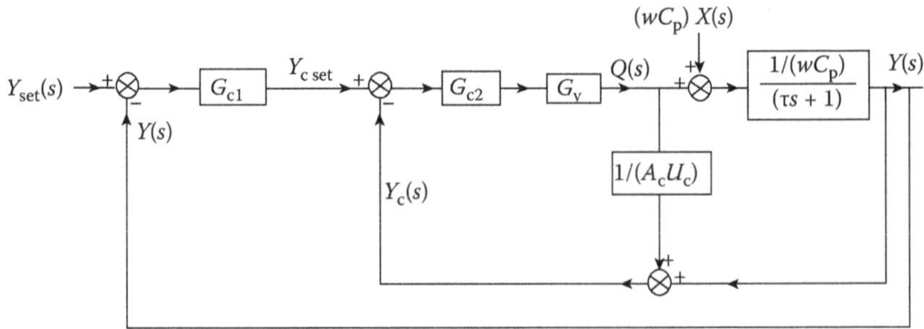

FIGURE 7.2 Laplacian block diagram of cascade control of tank temperature.

$$Y(s) = \frac{K_{c1}K_{c2}Y_{set}(s) + (1 + K_{c2})X(s)}{(1 + K_{c2})(s + 1) + (1 + K_{c1})K_{c2}} \tag{7.4}$$

This can be used to determine the responses during set point or load disturbance.

Example 7.1

Consider single-loop and cascade control systems as shown in Figures 7.3 and 7.4. Compare the performances of the control systems.
Single-control loop analysis:
From Figure 7.3, the transfer function is obtained as

$$\begin{aligned}\frac{Y_1(s)}{Y_{1set}(s)} &= \frac{K_1K_2\dfrac{1}{(\tau_1 s+1)}\dfrac{1}{(\tau_2 s+1)}}{1+K_1K_2\dfrac{1}{(\tau_1 s+1)}\dfrac{1}{(\tau_2 s+1)}} \\ &= \frac{K_1K_2}{(\tau_1 s+1)(\tau_2 s+1)+K_1K_2} \\ &= \frac{K_1K_2}{\tau_1\tau_2 s^2+(\tau_1+\tau_2)s+1+K_1K_2}.\end{aligned} \tag{7.5}$$

Applying the final value theorem, the ultimate value is determined as

$$Y(\infty) = \frac{K_1K_2}{1+K_1K_2} \tag{7.6}$$

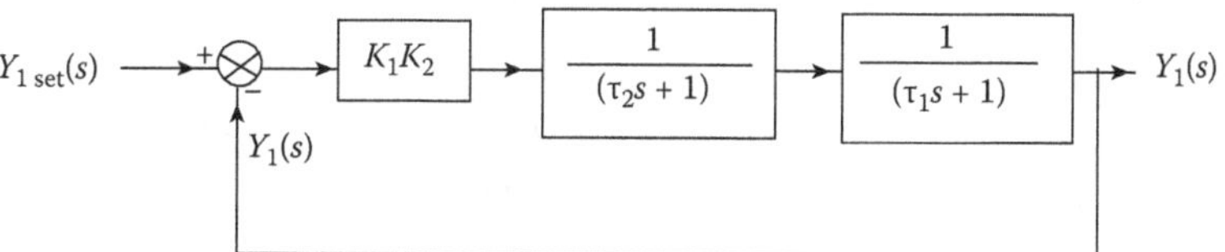

FIGURE 7.3 Single-loop feedback control system.

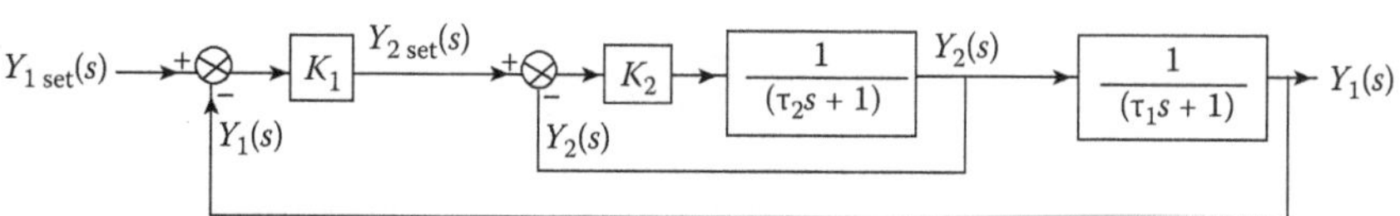

FIGURE 7.4 Cascade control for the same process as in Figure 7.3.

or

$$\text{offset} = \frac{1}{1+K_1K_2} \tag{7.7}$$

considering unit step disturbance in set point.

Also, comparing the generalized second-order system with a second-order time constant τ and a damping coefficient ξ with Equation 7.5,

$$\tau = \sqrt{\frac{\tau_1\tau_2}{1+K_1K_2}} \tag{7.8}$$

$$\zeta = \frac{\tau_1+\tau_2}{2\sqrt{\tau_1\tau_2(1+K_1K_2)}}. \tag{7.9}$$

From Figure 7.4, the transfer functions are presented as

$$\begin{aligned}\frac{Y_1(s)}{Y_{1\text{set}}(s)} &= \frac{K_1K_2}{\tau_1\tau_2s^2+(\tau_1+\tau_2+K_2\tau_1)s+(1+K_1K_2+K_2)} \\ &= \frac{K_1K_2/(1+K_1K_2+K_2)}{\dfrac{\tau_1\tau_2s^2}{(1+K_1K_2+K_2)}+\dfrac{(\tau_1+\tau_2+K_2\tau_1)s}{(1+K_1K_2+K_2)}+1}.\end{aligned} \tag{7.10}$$

Similarly the offset, second-order time constant, and damping coefficient are determined as

$$[\text{illegible}] \frac{K_1K_2}{(1+K_1K_2+K_2)} \tag{7.11}$$

$$\text{offset} = \frac{(1+K_2)}{(1+K_1K_2+K_2)} \tag{7.12}$$

$$\tau = \sqrt{\frac{\tau_1\tau_2}{(1+K_1K_2+K_2)}} \tag{7.13}$$

$$\zeta = \frac{\tau_1+\tau_2}{2\sqrt{\tau_1\tau_2(1+K_1K_2+K_2)}}. \tag{7.14}$$

A comparison of these values for both the single-loop control and cascade control are determined using numerical values of τ_1, τ_2, K_1, and K_2.

$\tau_1 = 1, \tau_2 = 2, K_1 = 1$, and $K_2 = 1$		
Parameters	**Single Loop**	**Cascade**
τ	1	0.816
ξ	0.75	0.816
Offset	0.50	0.67

$\tau_1 = 1, \tau_2 = 2, K_1 = 1$, and $K_2 = 2$		
Parameters	**Single Loop**	**Cascade**
τ	0.816	0.632
ξ	0.612	0.791
Offset	0.330	0.60

$\tau_1 = 1, \tau_2 = 2, K_1 = 2$, and $K_2 = 1$		
Parameters	**Single Loop**	**Cascade**
τ	0.816	0.707
ξ	0.612	0.707
Offset	0.33	0.500

$\tau_1 = 1, \tau_2 = 2, K_1 = 4$, and $K_2 = 1$		
Parameters	**Single Loop**	**Cascade**
τ	0.632	0.577
ξ	0.474	0.577
Offset	0.200	0.333

$\tau_1 = 1, \tau_2 = 2, K_1 = 10$, and $K_2 = 1$		
Parameters	**Single Loop**	**Cascade**
τ	0.426	0.408
ξ	0.320	0.408
Offset	0.091	0.167

From the above calculations, it is understood that the time constant for the cascade control is lower than that of a single-control system, i.e., cascade control is faster than a single-control system. Also, the damping coefficient is greater than that of a single-control system, which implies the cascade control is less oscillatory than a single-control system. The offset is greater in the cascade control but reduces when the gain of the master controller (K_1) is increased and is greater than that of the slave controller (K_2).

Example 7.2

Consider a control system of the block diagram shown. Determine the response of the control variable $c(t)$ as function of time (a) for a conventional PI control with $K_c = 1$ and $\tau_i = 0.63$ (b) cascade control with $K_c = 1.0$, $K_{c2} = 5$, and $\tau_i = 0.63$.
Solution:

(a) Conventional PI control

(a)

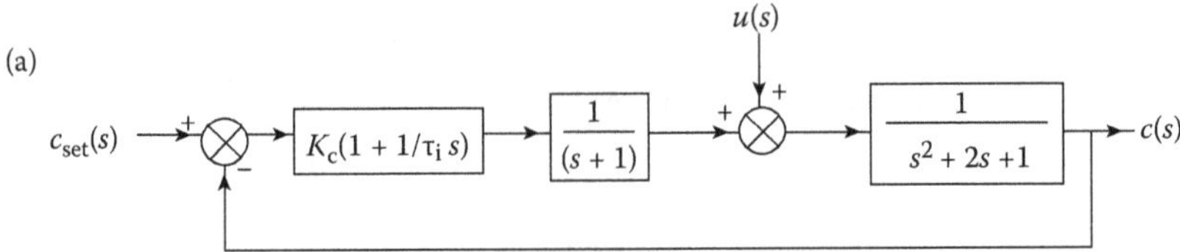

(b)

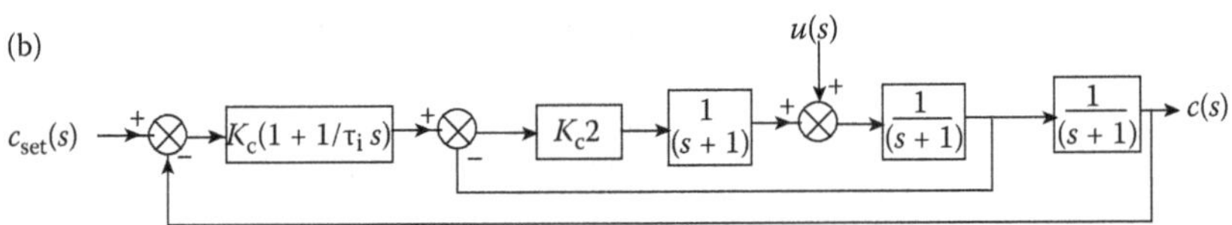

FIGURE 7.5 (a) Conventional PI control of the problem under study. (b) Cascade control of the problem under study.

The transfer function is

$$\frac{c(s)}{c_{set}(s)} = \frac{K_c(1+\tau_i s)}{\tau_i s(s^2+2s+1)(s+1)+K_c(1+\tau_i s)}.$$

The response of the system for a unit step change in set point only is obtained as shown in Figure 7.6.

(b) Cascade control

The transfer function $c(s)/c_{set}(s)$ is given as

$$\frac{c(s)}{c_{set}(s)} = \frac{K_{c2}\, K_c(1+\tau_i s)}{\tau_i s(s^2+2s+1+K_{c2})(s+1)+K_{c2}\, K_c(1+\tau_i s)}.$$

The response of this control system is presented in Figure 7.7, which clearly shows a quicker response as compared to the single-control system.

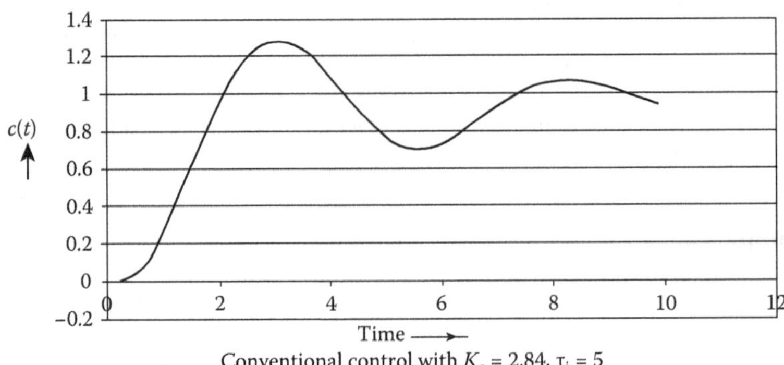

FIGURE 7.6 Response of conventional control system (plotted from a computer-aided solution).

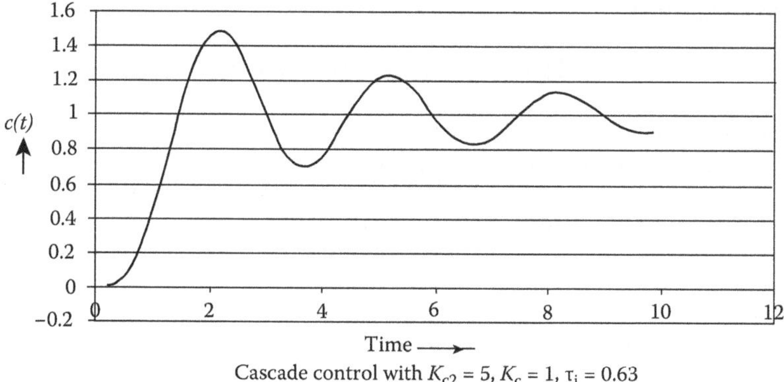

FIGURE 7.7 Response of cascade control system (plotted from computer-aided solution).

7.2 FEED-FORWARD CONTROL

Feed-forward control is applicable when the control variable is sluggish in nature, i.e., too slow to change, for example, the product composition in a reactor, the level in a boiler drum, etc. The controllers are different from the PID controller we have dealt with so far. A PID controller has three controller parameters: K_c, τ_I, and τ_d. But in the feed-forward controller, the parameters are dependent on the process to be controlled. As the process changes, the controller parameters will also have to be changed. Hence, the controller is, in fact, a digital computer where the control equation based on the process design will have to be entered. The output of this computer will then drive one or more appropriate control valves. A single computer may be sufficient to make feed-forward control programs for a number of processes and deliver the output signals to a number of control valves. A simplified example is discussed here for understanding the feed-forward strategy.

Consider a steam-heated tank problem as shown in Figure 7.8.

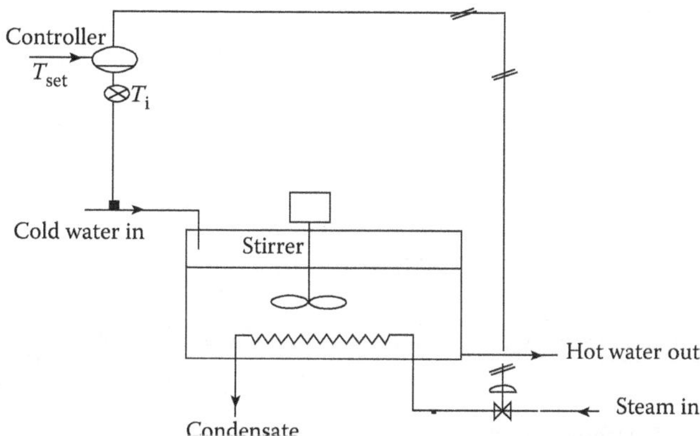

FIGURE 7.8 Feed-forward control of a steam-heated tank.

Process equation:

$$Y(s) = [Q(s)/wC_p + X(s)]/(\tau s + 1). \tag{7.15}$$

Controller equation:
To make the set point, $Y_{set}(s)$ equal to $Y(s)$,

$$Q(s) = wC_p\,[(\tau s + 1)\,Y_{set}(s) - X(s)]. \tag{7.16}$$

Hence,

$$Y(s) = [(\tau s + 1)\,Y_{set}(s) - X(s) + X(s)]/(\tau s + 1) = Y_{set}(s).$$

Hence, offset = 0 both for set point change and load change separately. However, the output of the controller $Q(s)$ with respect to its initial steady value will be changed in both cases.

The equivalent Laplacian block diagram is presented in Figure 7.9.

Note that the controller parameters involve process model, time constant, flow data w, and property C_p. Hence, feed-forward control cannot be achieved without precise information about the process.

Example 7.3

Consider a steam-heated tank process actually having a time constant of 2 seconds. Present how a feed-forward controller should perform.

(a) Conventional PI control is studied for $w = 10$, $C_p = 1$, $K_c = 1$, $\tau_i = 0.5$, and $\tau = 2$. This is shown in Figure 7.10.
The response of the controlled system for a unit step disturbance of the set point is presented in Figure 7.11.

(b) If the same process is controlled by a feed-forward controller as shown in Figure 7.12, the corresponding response to a unit step change in set point is given in Figure 7.13.

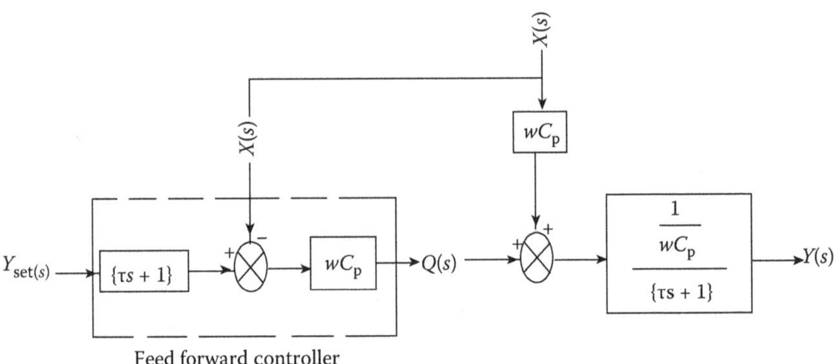

FIGURE 7.9 Feed-forward control of steam-heated tank in Figure 7.5.

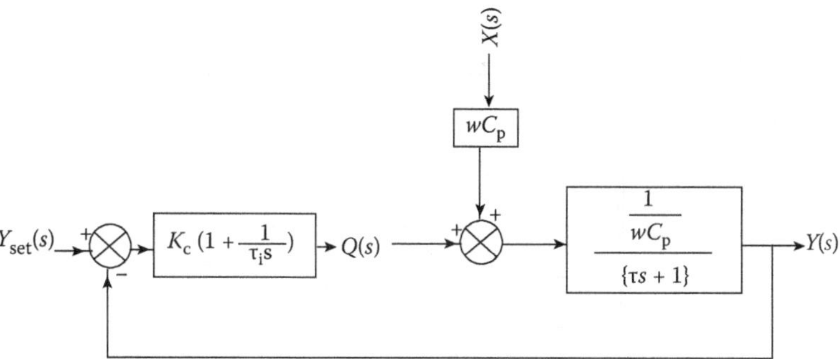

FIGURE 7.10 Feedback PI control of a steam-heated tank problem.

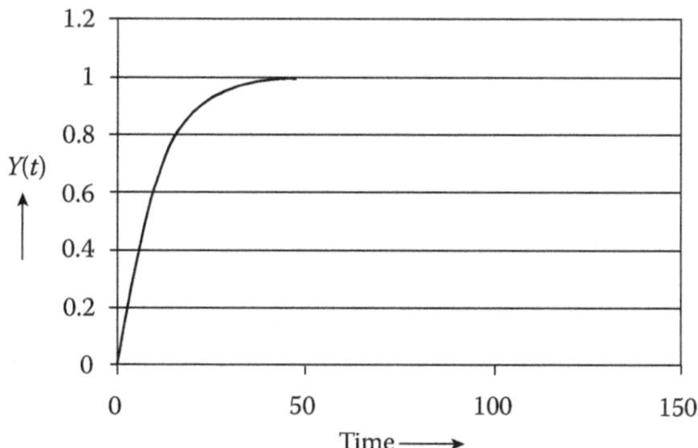

FIGURE 7.11 Response of PI control of the problem.

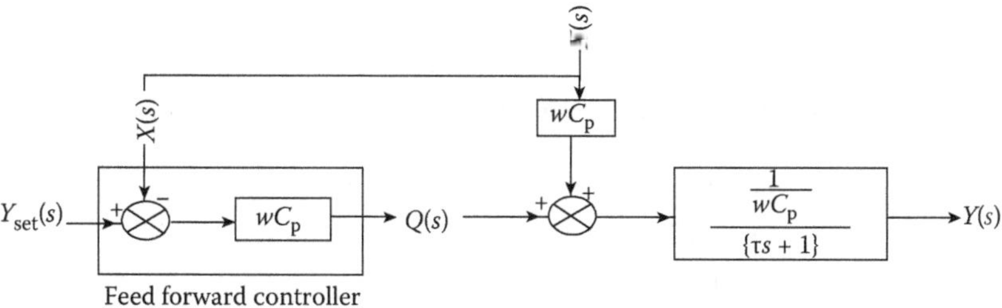

FIGURE 7.12 A feed-forward control of the steam-heated tank.

(c) If the feed-forward control system is fed back with the measured value of the control variable, the control system will be a feed forward–feedback control. This is shown in Figure 7.14, and the response is shown in Figure 7.15.

Thus the feed forward–feedback control system is faster than the conventional PI control system.

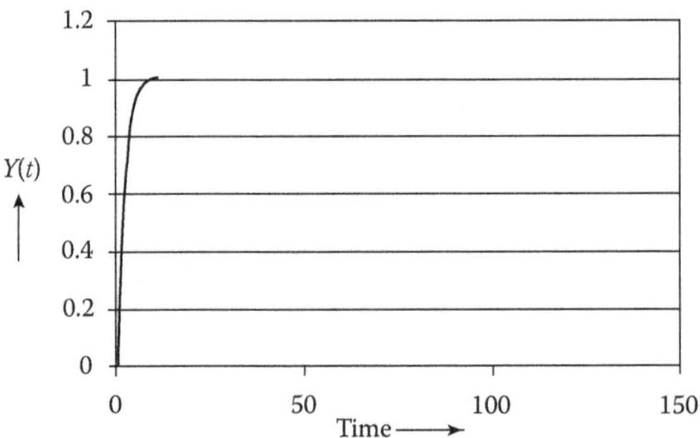

FIGURE 7.13 Response of feed-forward control of the problem.

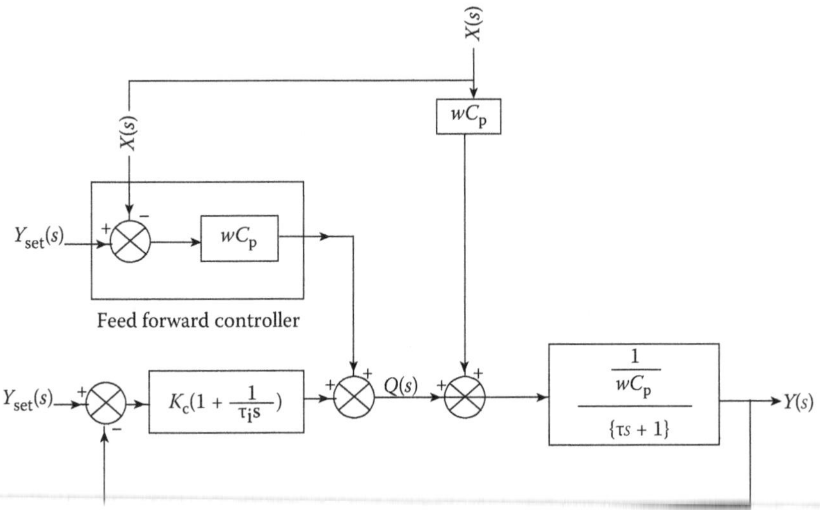

FIGURE 7.14 A feed forward–feedback control of the steam-heated tank problem.

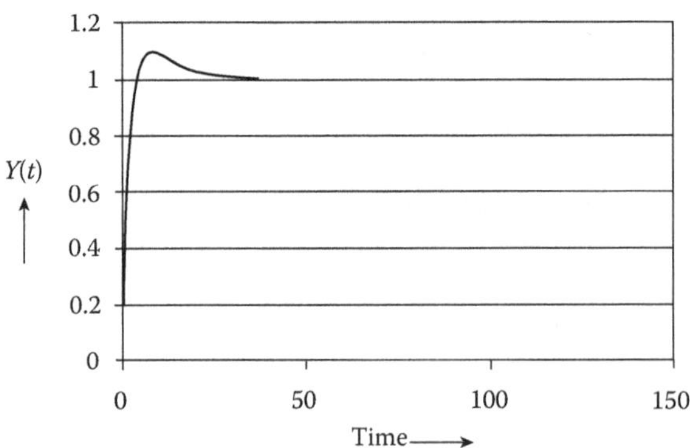

FIGURE 7.15 Response of the feed forward–feedback control of the steam-heated tank problem.

7.3 INTERNAL MODEL CONTROL

If the models of a process, transducer, and control valve are precisely known, then a control strategy can be devised to generate the necessary output signal as presented in Figure 7.16.

As shown in the control scheme, the model process has a time constant τ_2; whereas, the actual process has the time constant τ_1. The control variable of the actual process is $Y(s)$; whereas, the process model used in the IMC controller is $Z(s)$. Assuming the control valve and transducer have unity transfer function for the simplicity of explanation (of course, these elements must be taken into consideration for a more appropriate control scheme), in this example of an IMC scheme, the model differs from the actual time constant of the process only. Online correction is carried out by evaluating the difference between $Z(s)$ and $Y(s)$ as

$$\varepsilon = Z(s) - Y(s) \text{ and } \tau_2 = \tau_2 (1 + \varepsilon^2). \tag{7.17}$$

The response of the control variable to a unit step change in set point is presented in Figure 7.17 for a large difference in process time constants, $\tau_1 = 2$ and $\tau_2 = 10$, where the controller self adjusts for the correction of time constant, and set point is achieved in a shorter time.

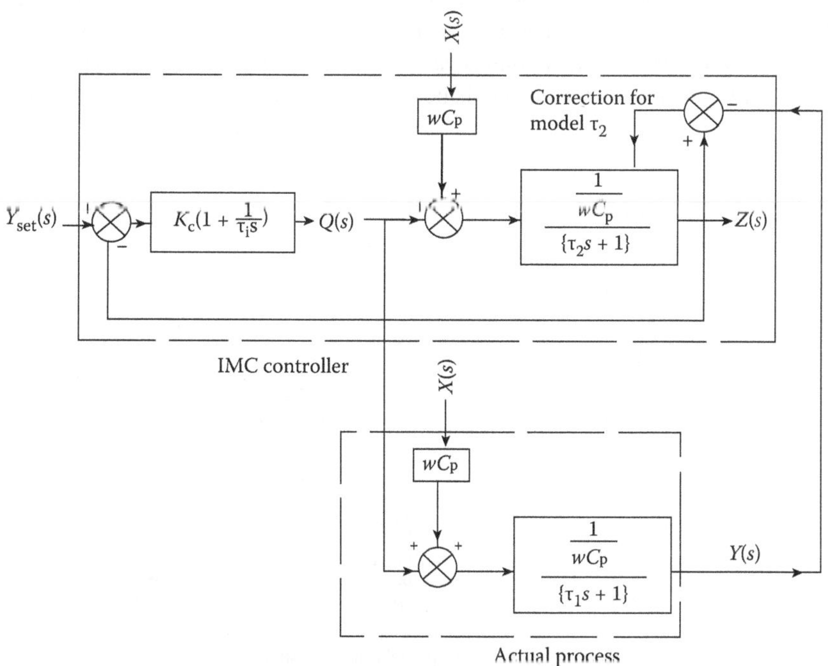

FIGURE 7.16 IMC control scheme for steam-heated tank problem.

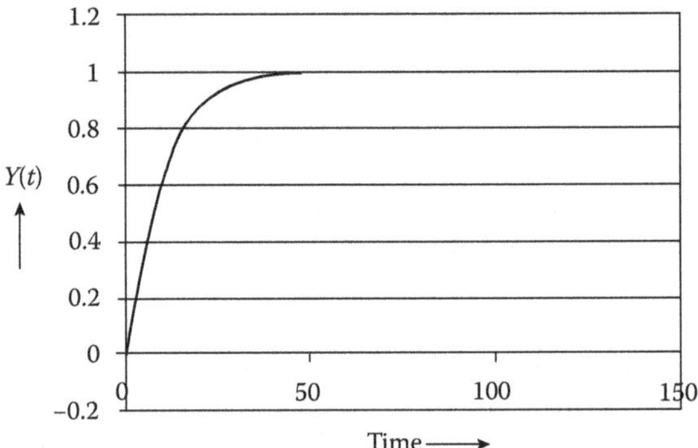

FIGURE 7.17 Response of the IMC control for large differences between model process and actual process time constants.

7.4 SPLIT RANGE CONTROL

A split range implies that more than one control action occurs from the output signal of a controller. Usually two control valves are actuated by a controller output while one operates in the opposite way with the other. This is explained in Figure 7.18. The control valve 1 is a normally closed valve (N/C), i.e., when the current signal is up to 12 mA, the valve is fully closed, starts opening while the current signal exceeds 12 mA, and is fully open when the current reaches 20 mA, the maximum output from the controller. The other control valve 2 is a normally open (N/O) type, i.e., the valve is fully open when the current is below 12 mA, starts closing when the current

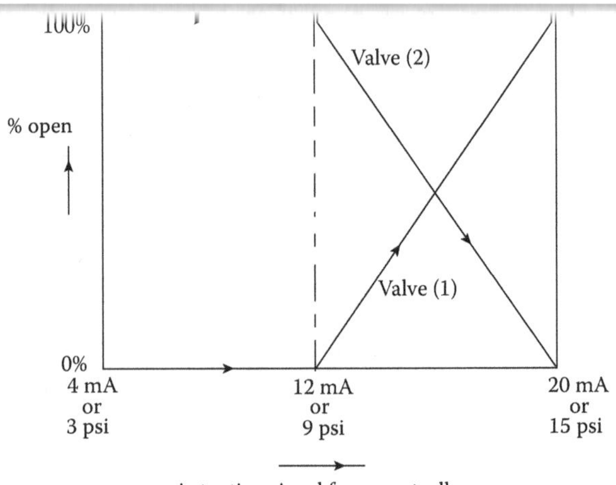

FIGURE 7.18 Control valves' opening as a function of controller signal.

exceeds 12 mA, and becomes fully closed when the current reaches its maximum of 20 mA. The valve openings and control signals are summarized below:

Control signal, mA	**4**	**4–12**	**12–20**	**20**
Valve opening	%	%	%	%
Valve 1, N/C	0	0	Partially open	100
Valve 2, N/O	100	100	Partially closed	0

For a pneumatic control valve, the pneumatic signal in the range of 3 to 15 psi will be applicable as the output signal from the current to pressure converter from the controller. Thus, the split of the signal will be equivalent to pneumatic signals of 3–9 psi for valve 1 to fully closed and which will open after 9 psi and be fully open as it reaches 15 psi. Control valve 2 will be fully open until 9 psi and starts closing above this and completely closes at 15 psi.

The valve opening and signal values are plotted in Figure 7.18 for linear control valves.

However, other configurations are also possible as shown in Figures 7.19 and 7.20.

As shown in Figure 7.19, control valve 2 is a N/O valve, which starts closing from 12 mA and is fully closed at 20 mA. Valve 1 is a N/C valve, which starts opening from 4 mA and is finally fully open at 12 mA. In this case, both control valves are fully open at 12 mA. However, in Figure 7.20, control valve 1 is a N/O type and closes at 12 mA; whereas, the control valve 2 is a N/C type and fully opens at 20 mA. Here, both the valves are fully closed at 12 mA.

Although the foregoing actions of split-range valves were shown around 12 mA, other asymmetrical split ranges are also applicable as per the requirement, i.e., split may be at 6 mA, 8 mA, 10 mA, etc., of course, process safety may dictates the type of control valve actions. Strategy, such as choosing both the valves partly open, one fully open and the other partially, both the valves fully open, or both valves

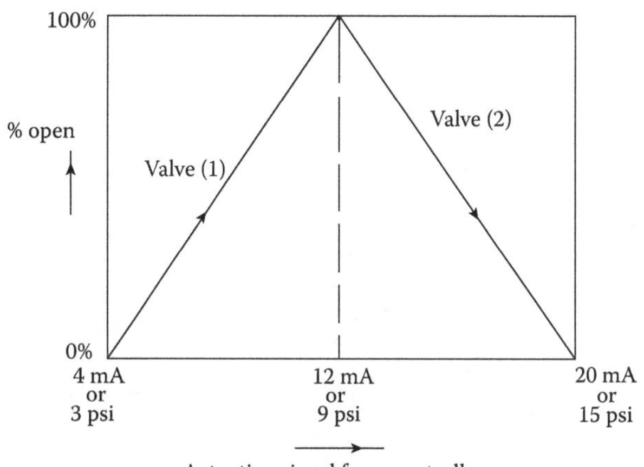

FIGURE 7.19 Split-range control valves. Both the valves are fully open at 12 mA.

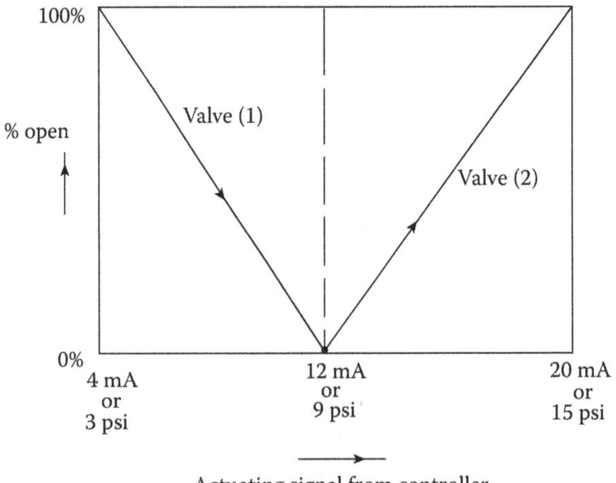

FIGURE 7.20 Split-range control valves. Both the valves are fully closed at 12 mA.

fully closed, etc., are decided by the control actions and the fail-safe conditions as well. A split-range control system for a temperature-control system is presented in Figure 7.21.

Consider temperature control in a tank problem where hot and cold liquid streams are introduced such that while the temperature is low, the flow of the hot stream will increase, and the cold stream will be decreased.

If the process fluid enters at a constant rate of w mass/time with a constant C_p and density as before, the rate of heat supplied by the hot stream is $Q_1(t)$ and the rate

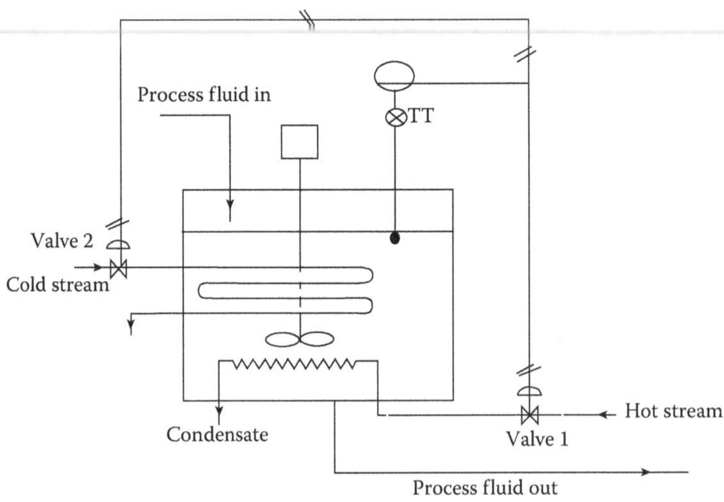

FIGURE 7.21 Split-range control scheme for temperature control.

of heat extraction by the cooling coil is $Q_2(t)$, then the heat balance equation of the process is given as

$$X(t) + [Q_1(t) - Q_2(t)]/wC_p = \tau\, dY(t)/dt + Y(t) \tag{7.18}$$

$X(t)$ is the load, $Q_1(t)$, $Q_2(t)$, and $Y(t)$ are the deviation variables with respect to the initial steady values. Heat addition and extraction rates are determined by the controller output and the split actions of the control valves. A schematic block diagram of such a split-range control is presented in Figure 7.22. A PI controller has been used to monitor both the hot and cold streams through valves 1 and 2.

Consider the above temperature control problem for which the following equations are valid for valves 1 and 2, respectively, as

$$Q_1(t) = K_{v1} Y_c(t) \tag{7.19}$$

and

$$Q_2(t) = K_{v2} Y_c(t) \tag{7.20}$$

K_{v1} and K_{v2} are the valve constants for the split actions as per Figure 7.19, and $Y_c(t)$ is the output signal of the controller. Figure 7.23 shows the response of the

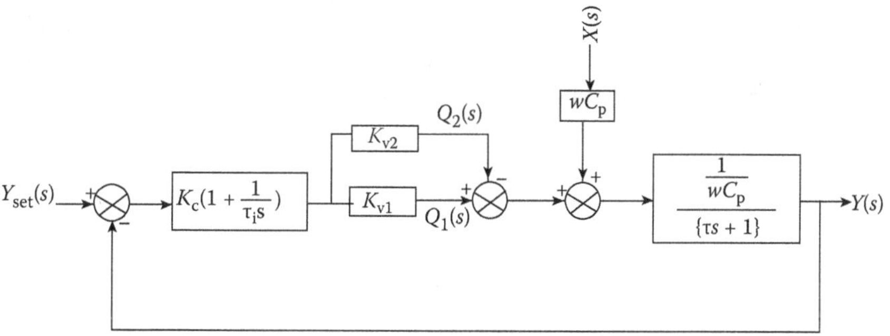

FIGURE 7.22 A split-range control block diagram of the temperature control in a tank.

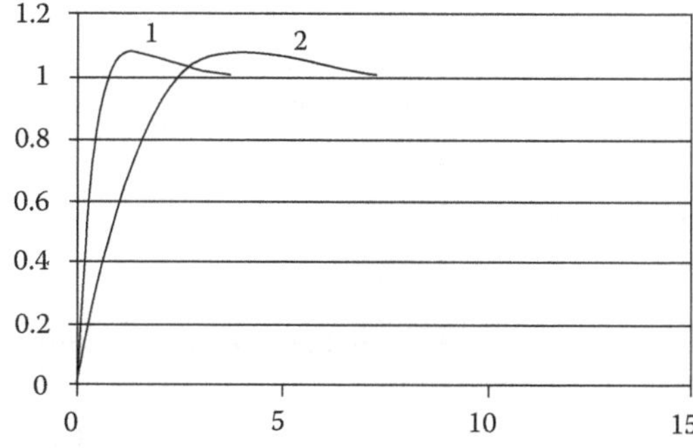

FIGURE 7.23 Split-range control of temperature while set point is disturbed by a unit step; curve 1 for $K_{v1} = 62.5$ and $K_{v2} = 12.5$; curve 2 for $K_{v1} = 25$ and $K_{v2} = 12.5$.

control system while K_{v1} and K_{v2} are varied with the PI controller having $K_c = 1$ and $\tau_i = 1$.

7.5 RATIO CONTROL

The ratios of air to fuel in a furnace, solvent to feed in an extraction plant, acid to alkali in a neutralization plant, etc., are common examples of where a ratio controller is very much required. Consider two different streams A and B have to be mixed in a certain ratio K_r. The flow rate of stream B has to be manipulated as a function of A in a ratio. Thus,

$$\frac{\text{flow rate of B}}{\text{flow rate of A}} = K_r. \tag{7.21}$$

A control scheme of such a ratio control system is presented in Figure 7.24.

The set point for the controller of stream B is determined by the product of the measured flow rate of A (Q_a) and K_r, i.e., $Q_{b\text{ set}} = Q_a K_r$. The equivalent Laplacian block diagram with respect to the deviation variable is presented in Figure 7.25.

G_{ma} and G_{mb} are the transfer functions of the flow-measuring transducers of streams A and B, respectively. Stream A is considered to be the wild stream as its flow is not dependent on the flow rate of stream B. G_c and G_{mb} are the transfer functions of the controller and the control valve, respectively, in stream B. In fact, there are two controllers; one is the ratio controller with the transfer function, a constant K_r, which can

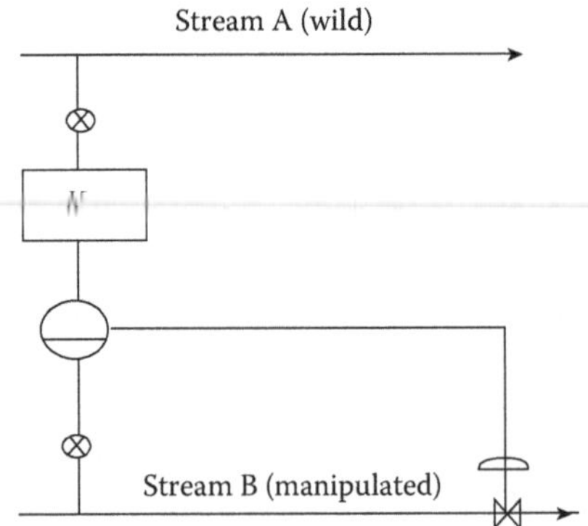

FIGURE 7.24 Ratio control scheme.

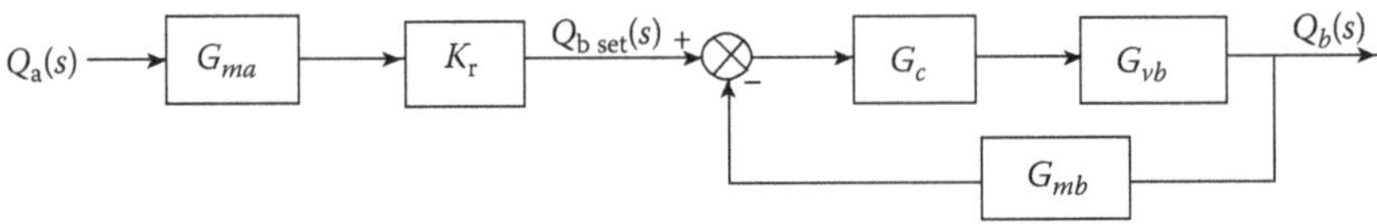

FIGURE 7.25 Laplacian block diagram of the ratio control system of two streams.

be changed at any desired value for the ratio, and the other is the flow controller of the manipulated stream whose set point is determined by the ratio controller ahead of it. The overall transfer function of the ratio of the flow rates in deviation form is given as

$$\frac{Q_b(s)}{Q_a(s)} = \frac{G_c G_{vb} G_{ma} K_r}{(1 + G_c G_{vb} G_{mb})}. \tag{7.22}$$

If the dynamic lags of the transducers and the control valves are neglected such that $G_{ma} = G_{mb} = G_{vb} = 1$, and the controller is a proportional controller with gain K_c, the variation of the ratio is determined as a function of time while the flow rate of stream A is increased by a unit step, i.e., $Q_{b\,set}(s) = K_r/s$

$$Q_b(s) = \frac{K_c K_r}{s(1 + K_c)} \tag{7.23}$$

or

$$Q_b(t) = \frac{K_c K_r}{(1 + K_c)}. \tag{7.24}$$

Thus, the flow of the manipulated stream will never achieve the value K_r unless K_c is very high. The instantaneous offset will be present theoretically until K_c is infinitely large. This is shown in Figure 7.26 where offset = $K_r/(1 + K_c)$.

If a PI controller is used with a gain K_c and integration time τ_i, the response is obtained as

$$Q_b(s) = \frac{K_c\left(1 + \frac{1}{\tau_i s}\right)K_r}{s\left\{1 + K_c\left(1 + \frac{1}{\tau_i s}\right)\right\}}. \tag{7.25}$$

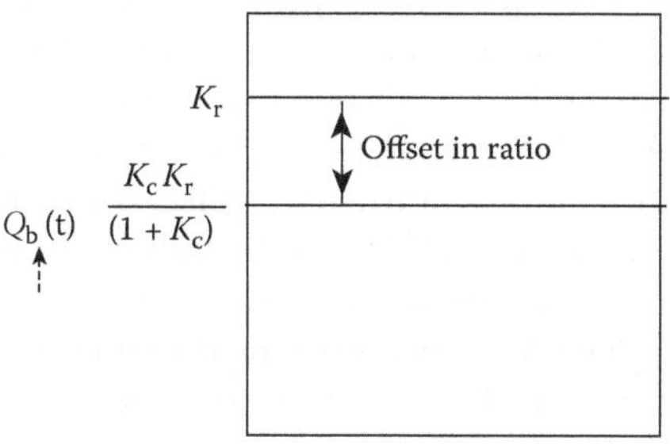

FIGURE 7.26 Response of proportional control in ratio control.

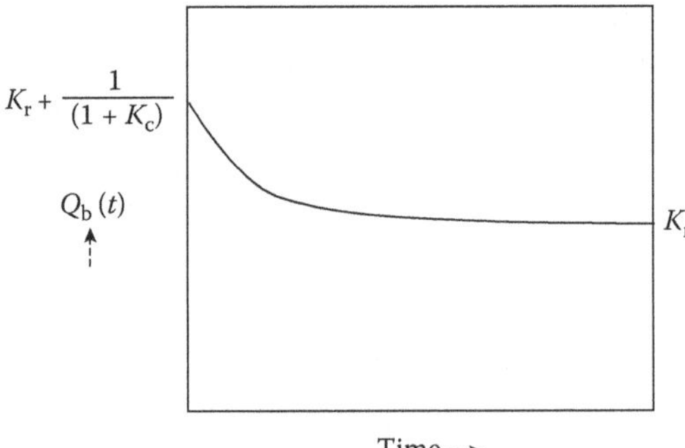

FIGURE 7.27 Response of proportional–integral control in ratio control.

The instantaneous response $Q_b(t)$ is given as

$$Q_b(t) = \left[K_r + \frac{1}{(1+K_c)} e^{-\frac{K_{c}t}{(1+K_c)\tau_i}} \right]. \tag{7.26}$$

The instantaneous value of $Q_b(t)$ as a function of time indicates that initially there is a wide difference between the desired ratio K_r and $Q_b(t)$, which can be reduced by using a small integration time, and finally, the ultimate value reaches K_r and offset becomes zero. Therefore, a PI controller is suitable to be used along with a ratio controller. The response is shown in Figure 7.27.

7.6 INFERENTIAL CONTROL

Inferential control is a strategy of control of an immeasurable process variable by measuring other continuously measurable variables that can be correlated with the immeasurable process variable. The unique example of such a control variable that cannot be measured continuously is the composition or property of a flowing fluid. Composition of an effluent from process equipment may not always be measurable instantaneously by a transducer, such as pressure, temperature, flow rate, and level. Rather, composition measurement is a time-consuming process and depends on the type and number of species present in the stream. Accurate measurement of composition requires rigorous laboratory analysis, which may involve the use of sophisticated analytical instruments, such as GC, HPLC, UV, AAS, FTIR, etc. The studies of all these instruments fall into the category of the science of analytical instruments, which is different from process instrument or transducers. Process instruments are basically transducers capable of quick and continuous measurement for the purpose of continuous control. Consider a stirred tank reactor, where a reactant A is converted to product B by a chemical reaction. If the concentration of A remaining in the product stream is measured as C_A and is given as

$$FC_{A0} - FC_A - KC_AV = V\frac{dC}{dt}A \tag{7.27}$$

where V is the volume of the reactor, F and C_{A0} are the flow rate and feed composition of the reactant, respectively, and K is the reaction rate constant, which is also a function of temperature T in the reactor as

$$K = K_0 e^{-E/RT} \tag{7.28}$$

K_0, E, and R are the constants.

Thus, it is understood from these relations that the conversion X_A is a function of both flow rate F and rate constant K or temperature T. So the conversion can be inferred by measuring flow rate F and temperature T. Because the relations are nonlinear and F and T are also time variants like C_A, it will not be possible to present a Laplacian block diagram. However, a control scheme can be presented in Figure 7.28 where independent temperature and flow control loops are employed, and the set points of F and T are manipulated by the computed C_A.

Flow and temperature corrector blocks estimate the corrected set points of flow and temperature control loops. G_{c1}, G_{c2}, G_{m1}, G_{m2}, G_{v1}, and G_{v2} are not necessarily transfer functions; they are only indicative of the controllers, transducers, and control valves of the respective loops. Noted that a process block containing F, K, and V in a differential equation form cannot be represented by a Laplace transformation as all the entities in the equation are nonlinear time variants. Corrector blocks shown are a simplified form of an error minimization program. Consider a stirred tank flow reactor as shown in Figure 7.29 is to be controlled for certain fractional conversions of the reactant A in the reaction. X_{Aset}, the desired conversion of A, is entered into the inferential controller, which generates the estimated set point of temperature T_{set} to the temperature controller. Usually, the throughput or the feed rate F is maintained as a constant while the heating rate Q is varied to adjust the rate of reaction. With the following reactor data, response of such an inferential control system can be studied.

$$K_0 = 7.8 \times 10^{10} \text{ and } E = 30{,}000,\ R = 1.98,\ V = 40,\ \rho = 1,\ C_p = 1, \text{ and } F = 100$$

where K_0 and E are the constants of reaction, R is the ideal gas constant, V is the volume of the reactor (taking constant liquid levels in the reactor), ρ and C_p are the average density and specific heat of the liquid feed and product mixture, and F is

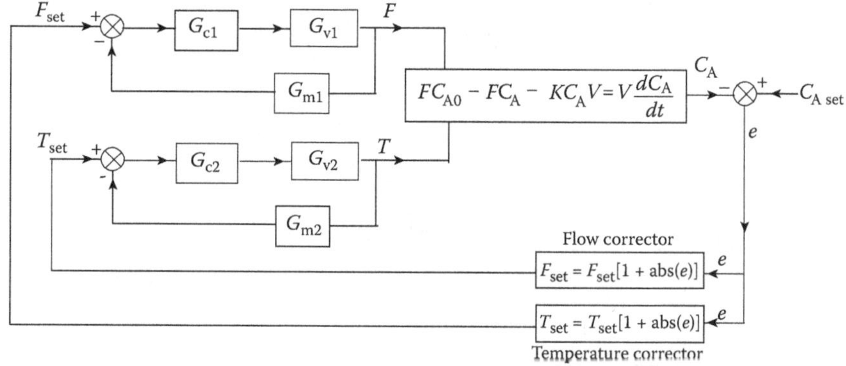

FIGURE 7.28 An inferential control scheme for composition control in a typical reactor.

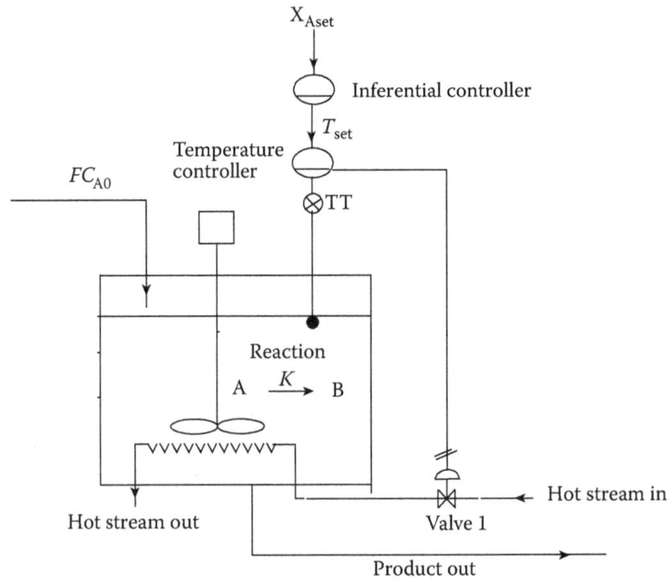

FIGURE 7.29 An example of an inferential control of a reactor.

the constant feed and product rate (independent of time). A schematic representation of the control system is presented in Figure 7.30 (which is strictly not a Laplacian control loop) with a PI temperature control.

A time domain analysis of the response has been carried out for changes in the desired conversion set point. The time variant relations for the above control system are presented as

$$T_{set} = 700 - K_c\varepsilon \tag{7.29}$$

where,

$$\varepsilon = X_{Aset} - X_A \tag{7.30}$$

$$Q(t) = 30{,}000 + K_{c1}\varepsilon_1 + \frac{K_{c1}}{\tau_i}\int \varepsilon_1 \, dt \tag{7.31}$$

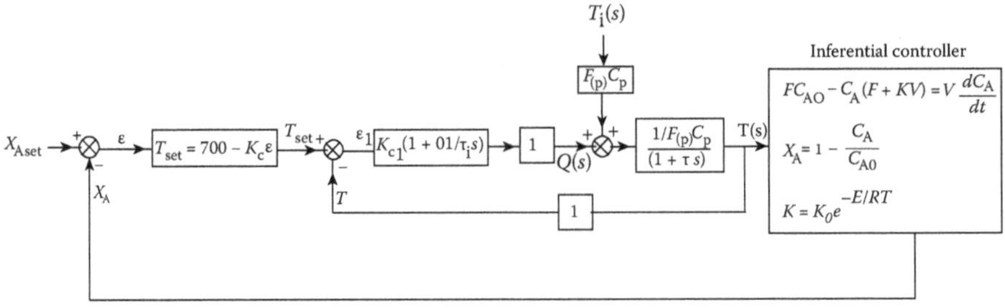

FIGURE 7.30 A schematic loop for reactor control using inferential control.

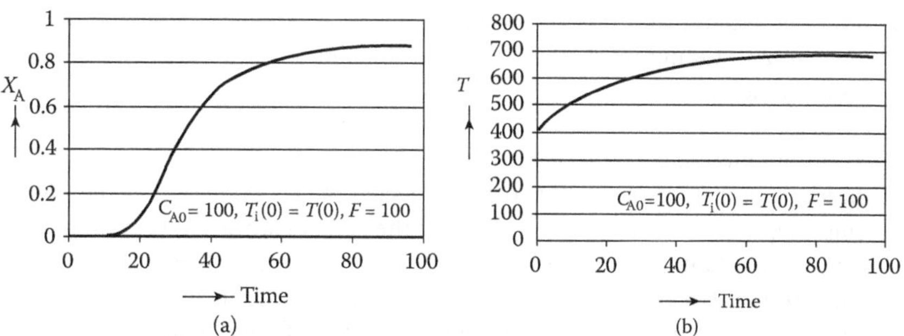

FIGURE 7.31 Response curves for inferential control: (a) conversion attaining 90% (b) corresponding temperature rise with zero offsets in both responses (obtained by computer solution).

where,

$$\varepsilon_1 = T_{set} - T \tag{7.32}$$

$$\frac{dT}{dt} = [(T_i - T) + Q/(F\rho\, C_p)]/v \tag{7.33}$$

$$\frac{dC_A}{dt} = [FC_{A0} - (F + KV)C_A]/V \tag{7.34}$$

with $K_c = 1$, $K_{c1} = 50$, and $\tau_i = 90$, the control actions are shown in Figure 7.31.

7.7 ADAPTIVE CONTROL

Adaptive control is a strategy to adjust the controller parameters whenever there is a change in the parameters of the process, control valve, or other elements involved other than the controller. When the controller is a PID controller, the parameters K_c, τ_i, and τ_d are automatically retuned by the controller itself if the process or control valve parameters change. In a cascade-control system, either the slave or master controller or both the controllers have the capability to adjust the respective controller parameters for any change in the process or control valve parameters. In the case of feed-forward or IMC control or inferential-control strategies, adaptation of the manipulated values is taken care of with the changing process or control valve parameters. In short, adaptive control means automatic adaptation of the controller for changes affecting the control system. Of course, an adaptive control system is successful only when the process model is either accurately known or determined online. Modern digital controllers have the provision of adaptability by an auto-tuning facility. In this auto-tuning mode, the controller changes its set point randomly, records the process variable under control, and determines the model of the controlled process and then tunes the controller parameters according to any standard method described elsewhere in this book. A complete program for model development and tuning must be available in the auto-tuning mode. Some of these are discussed next.

Self-tuning method:

Consider a process using a PID controller where the set point is disturbed by a unit step and the response is recorded. If the output happens to be a sigmoidal curve as determined by Cohen–Coon, then the process transfer function is determined as a first-order system coupled with an exponential lag. The tuning is straightaway obtained by the Cohen–Coon chart for the tuned parameters.

Gain scheduling method:

In this method, only the gain of the controller K_c is made adaptable for any change of the parameters of other elements, including the process. If a feedback control system contains a PID controller, a process, a transducer, and a control valve, having their respective gains as K_c, K_p, K_m, and K_v, then the product of these gains is determined initially as the overall gain. Thus,

$$K_{overall} = K_c K_p K_m K_v \tag{7.35}$$

If any or all of the gains K_p, K_m, or K_v are changed, K_c is altered automatically using the previously recorded value of $K_{overall}$ as,

$$K_c = K_{overall}/K_p K_m K_v \tag{7.36}$$

However, this technique cannot change the integral or derivative time constants of the controller nor does it take account of the changes of the time constants of the process or control valve.

Digital program method:

In order to avoid the limitations of the Laplace transformation technique for nonlinear processes, time domain analysis is the most attractive option for the availability of digital computers. This method has already been applied in the foregoing sections in IMC, ratio control, and inferential control.

7.8 OVERRIDE CONTROL

An override control is a strategy involved in the control system where more than one controller are used to control one or more control variables, where one controller overrides the other controllers' activities, i.e., the action of this controller will only be effective in a certain situation or moment while the other controllers will not be functioning. This can be understood from the following example. Consider the tank-temperature control system where steam is used to heat up the tank contents. A temperature controller (TC) and a pressure drop controller (PDC) are used to monitor the steam flow rate as shown in Figure 7.32. If the pressure drop across the control valve has a limitation to use such that when this reaches its maximum the TC will cease to act while the PDC will only take the action to maintain the steam flow rate.

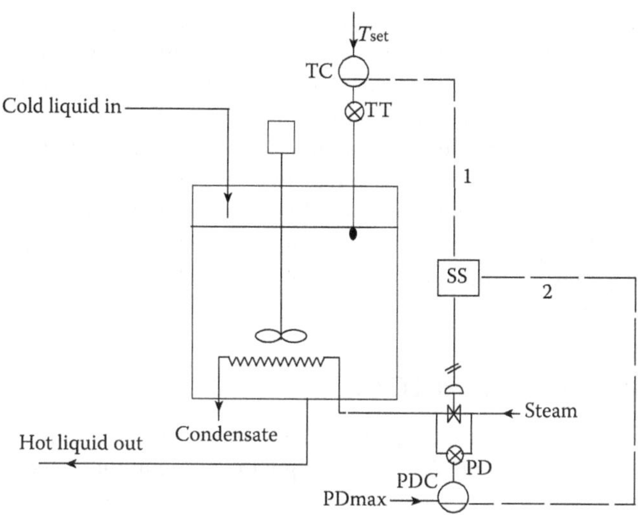

FIGURE 7.32 Override control in steam-heated tank.

The signal from both the controllers will be routed through a selector switch (SS), which will route only one signal to the controller either in loop 1, while the pressure drop is less than the maximum allowable pressure drop across the control valve, or loop 2, when the pressure drop exceeds or nears its maximum value. While the PDC action takes place, it will maintain the flow rate such that the pressure drop falls below the maximum value. This type of control system is helpful for protecting the control valve from damage resulting from excessive erosion because of friction at a high pressure drop. Another example of override control of the tank temperature where a level controller is used along with a TC is shown in Figure 7.33.

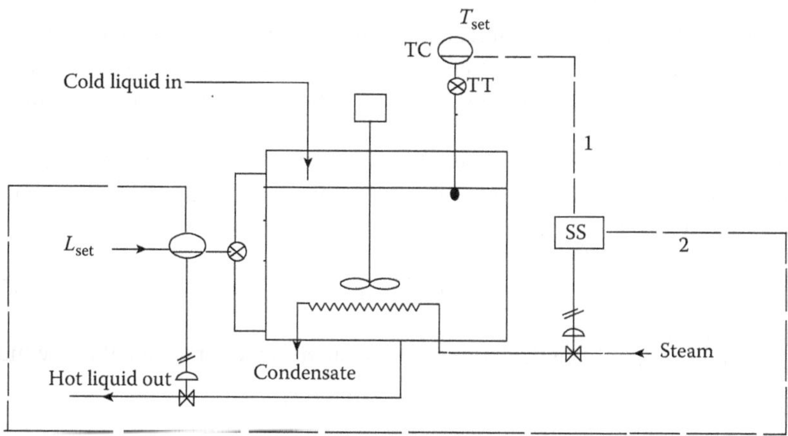

FIGURE 7.33 Override control of temperature and level.

In this control, the level and temperature will act separately as long as the level does not fall below a minimum. In case the level falls below the minimum, the steam control will cease to act, and the valve is closed. A normally closed (N/C) valve must be used for this purpose. The level control will alone continue until the level is built up in the tank. When the level becomes more than the minimum, the temperature control will resume. It is to be noted that two separate control valves have been employed for simultaneous control of level and temperature. This type of override control will protect the steam coil from damage resulting from overheating in a low-level or empty tank situation and also avoid vapor locking of the liquid pump (not shown in the figure).

7.9 ARTIFICIAL NEURAL LOGIC

Logic is the science of reasoning, which the human operator applies for achieving a certain goal. In a control system, the goal is to maintain the control variable at its set point at any disturbance during changing of set point or load or both. Human operator's logic can be expressed either in a mathematical relationship, such as the on/off, proportional, proportional–integral, or proportional–integral–derivative relationships; the use of these have been discussed in previous sections. But such logics are not successful in many chemical process control operations, which are mainly nonlinear and have sluggish dynamics. However, the limitations of mathematical logics have been overcome by the help of digital programmed logic. Artificial intelligence mimicking the human brain can be approximated by the neural-network method (or artificial neural-network method). In this logic, as with the biological system, the signal from the sensory glands or membrane (input nodes) propagates to the next gland or membrane (hidden nodes) through connecting nerve cells, which generate the output signal to the next set of nodes and finally comes out from the last node (output node), which connects the brain cells for corrective action. A neural-network model is thereafter developed based on the similarity with a biological [illegible] network. The network consists of three types of nodes, namely input nodes, hidden nodes, and output nodes. The information from the input nodes is communicated through connecting lines (like the nerve) with certain weights. Input vectors to the input layer are multiplied by weights and trigger excitation when entering the hidden nodes. Topology of the neural network is shown in Figure 7.34.

As shown in Figure 7.34, i-number of input nodes may be present. When the input signal (x_i) reaches the hidden nodes, it fires out an excitation, which is imagined as a sigmoidal function:

$$y_j = 1/\{1 + \exp(-a_j)\} \tag{7.37}$$

where y_j is the output from the hidden node j and a_j is the input signal to the hidden node and is given as

$$a_j = \sum w_{ij}\, x_i \tag{7.38}$$

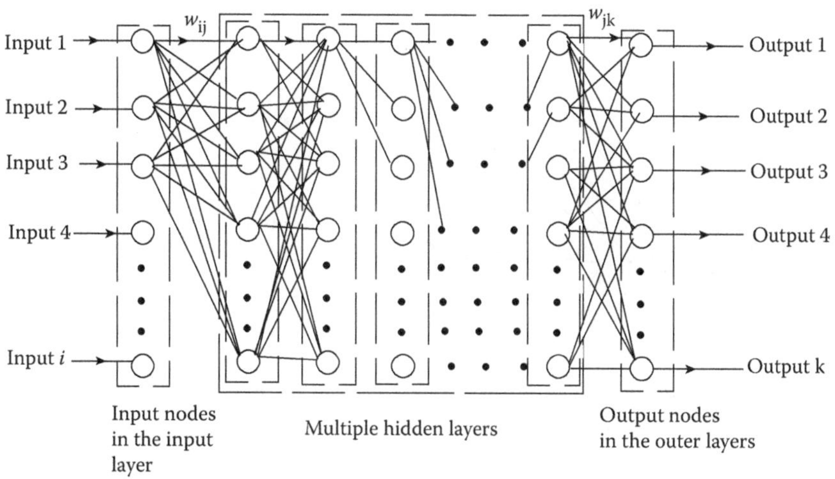

FIGURE 7.34 Anticipated topology of biological–neurological response system.

where w_{ij} are the weights connecting the input node i to the hidden node j, x_i is the input value (usually normalized to unity) to input node i. Signals from the hidden nodes then propagate to the output layer and generate an output signal similar to how the input did. Finally, the output signals from the outermost layer (output layer) is then delivered. The method of propagation of the signals through the network of neural nodes may be used to control logic development. For simplicity, a network containing one hidden layer of nodes may be thought of for the delivery of the manipulated variable. The input layer may consist of two input nodes where the input signals are the set point and the current process measurement (variable). The number of inner layers (hidden layers) is thought to be made of j numbers of nodes, and the output node is a single node delivering the manipulated variable. The process control scheme is shown in Figure 7.35.

In this controller, the number of nodes in the hidden layer can be varied to obtain optimum performance. W_{ij} are the weights connecting the input i nodes to the hidden j nodes, and W_{jk} are the weights connecting the hidden nodes to output nodes k. For this controller, the number of input nodes is 2 (i = 2), the number of hidden nodes is 2 (j = 2), and there is one output node (k = 1). The output signal is delivered to the process through the control valve. The normalized set point and process values are entered into the input nodes, and output values from each of the hidden nodes are evaluated using the initially assumed weights (W_{ij}). These outputs then enter the output node using the initially assumed weights (W_{jk}). This output is then converted from the normalized value to the controller. The resulting process variable C is then compared with the set point C_{set} and the error, $C_{set} - C$, is determined. An optimization program using a gradient reduction or generalized delta rule (GDR) method is then used to update the weights (W_{ij} and W_{jk}) and vary the manipulated variable continuously.

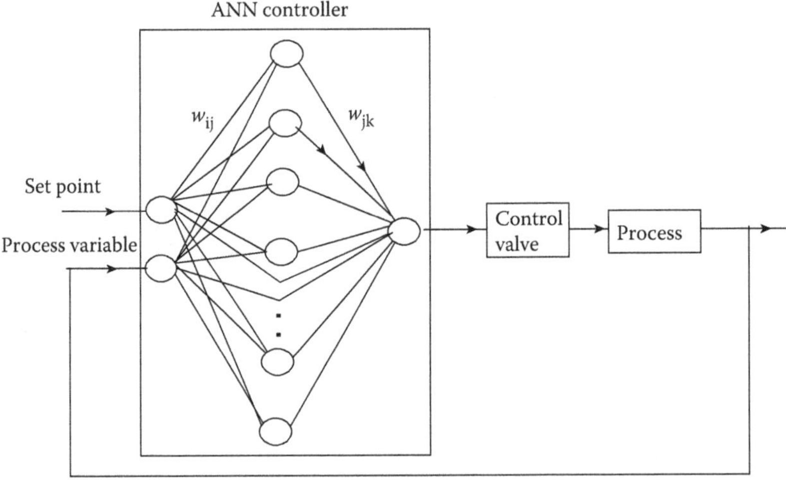

FIGURE 7.35 A simplified neural-network control scheme with one input layer, one hidden layer, and one output layer.

The control logic is explained stepwise below:

Let suffixes *i*, *j*, and *k* denote input, hidden, and output layers, respectively, *w* denotes weights, *b* is the bias, and *x* denotes the input to a node. *Oh* denotes outputs from the hidden nodes, and *OO* denotes output from the outer nodes. The algorithm for making a computer program to determine the output signal is listed below:

1. Guess w_{ij} and w_{jk}
2. Assign and normalize inputs x_i
3. Evaluate the input to hidden node *j*

$$s_j = \sum w_{ij} x_i + b_j. \tag{7.39}$$

4. Evaluate Oh_j as the output from hidden nodes

$$Oh_j = 1/\{1 + \exp(-s_j)\}. \tag{7.40}$$

5. Evaluate s_k as the input to the output nodes

$$s_k = \sum w_{jk}\, Oh_j + b_k. \tag{7.41}$$

6. Evaluate OO_k as the output from output nodes

$$OO_k = 1/\{1 + \exp(-s_k)\}. \tag{7.42}$$

7. Deliver the output to the process
8. Measure the resulting control variable *C*

9. Evaluate error

$$E = (C_{set} - C). \tag{7.43}$$

10. If the error $E <$ convergence criteria, then skip to step 14; otherwise go to the next step
11. Determine the derivative

$$fkk = dE/dO_k = (1 - OO_k)*E*OO_k. \tag{7.44}$$

12. Find the error signal to be propagated back to the hidden nodes as

$$dkj = E*fkk. \tag{7.45}$$

13. Determine change in weights w_{jk} as

$$\Delta w_{jk}^{(p+1)} = \beta\, dkj\, OO_k + \alpha\, \Delta w_{jk}^{(p)}. \tag{7.46}$$

where p is the pattern index, β is the learning rate, and α is the momentum term
14. Determine the error signal propagating back from the hidden to the preceding nodes (input nodes in this case)

$$dji = Oh_j*(1 - Oh_j)*\sum w_{jk} *dkj. \tag{7.47}$$

15. Determine change in w_{ij} as done in step 12
16. Change weights w_{ij} and w_{jk} as

$$w^{(p+1)} = \Delta w^{(p+1)} + w^{(p)}. \tag{7.48}$$

17. Go to step 3

Example 7.4

Consider the temperature control system of a steam-heated tank discussed earlier.

If the flow rate W of the process liquid is held constant with time at 10 units/time, the volume of the liquid in the tank unchanged with time (i.e., level is unchanged) is 50 units, and the average density and specific heat of the liquid are taken as unity, then compare the response systems of the controller while the logic in the controllers are (a) proportional–integral with $K_c = 2$ and $\tau_i = 1$ (b) neural-network logic with two hidden nodes, learning factor ($\beta = 0.35$), and momentum factor ($\alpha = 0.90$) given that the inlet liquid temperature ($T_i = 30°$).
Solution:

(a) The process model:

$$T(t) + \tau\, dT(t)/dt = Ti + Q/WC_p \tag{7.49}$$

where $\tau = V/W = 50/10 = 5$ time units.

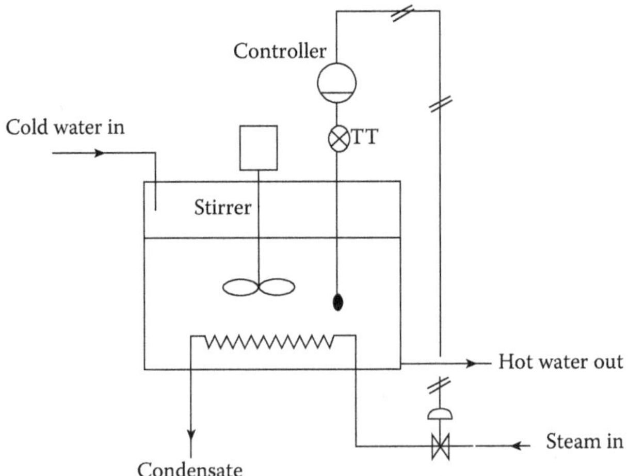

FIGURE 7.36 Temperature control scheme.

The controller logic is a PI relationship

$$Y_c(t) = Y_s + K_c\varepsilon + \frac{K_c}{\tau_i}\int \varepsilon\, dt \tag{7.50}$$

where Y_s is the bias value of the controller when steady state is reached, i.e., $\varepsilon = 0$.

If the initial value of the tank temperature T was 30°, the response of the controller is determined for a step change in set point from this initial temperature to 100°C. This is shown in Figure 7.37a. It is to be noted that the set point (100°C) is reached only after 10 seconds, and there is an overshoot up to 120°C.

(b) In this case, the process model is the same as that given in Equation 7.49, but the control logic is a neural-net logic as described earlier. The responses of the control system are obtained for a different number of hidden nodes as shown in Figure 7.37b, c, and d.

Note that, in all three artificial neural network (ANN) control responses, the fluctuation increases, but the set point is reached in all three cases near 5 seconds, which is much shorter as compared to the PI control response. However, the initial jump of the temperature is found in all the cases but later reduced drastically. In fact, this sudden wide overshoot sustains for a very short while at the start of control and later reduces as the net trains itself and modifies the weight. This is more clear if the performance of the controller is judged after a subsequent change in set point. For this, the set point is changed from 30°C to 50°C and then after sufficient time elapses, the set point is further raised to 100°C in steps. This is compared with the PI controller as shown in Figure 7.38a and b. The above responses were obtained by a computer-programmed solution in the time domain using a numerical method as it was difficult to use Laplacian mathematics or any other analytical method of solution.

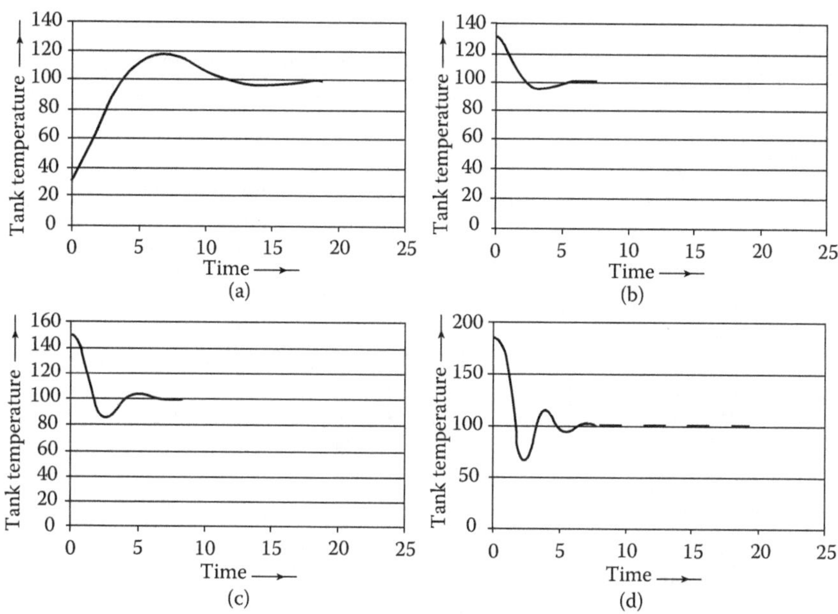

FIGURE 7.37 (a) Response of the PI controller; responses of the ANN controller with (b) one hidden node, (c) two hidden nodes, and (d) three hidden nodes.

It is seen from Figure 7.38 that the ANN controller is faster than the PI controller in achieving the set points, and the overshoot in the ANN controller after the second change in set point is almost negligible as that was in the beginning. Hence, it confirms that the ANN controller has quickly trained itself and smoothened the response.

The steps for the above ANN controller logic are combined with the process where the output from the process T was compared with the target T_{set}. The algorithm for making a computer program for evaluating the time-temperature relationship is given in Figure 7.39.

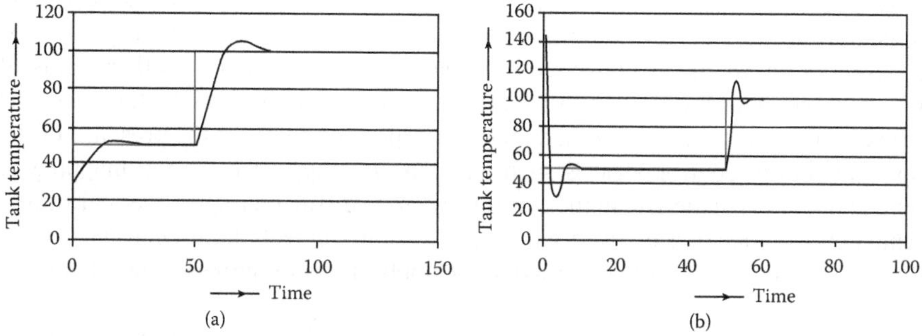

FIGURE 7.38 (a) Response controller for set point disturbed from 30°C to 50°C and again from 50°C to 100°C (a) for a PI controller (b) for an ANN controller.

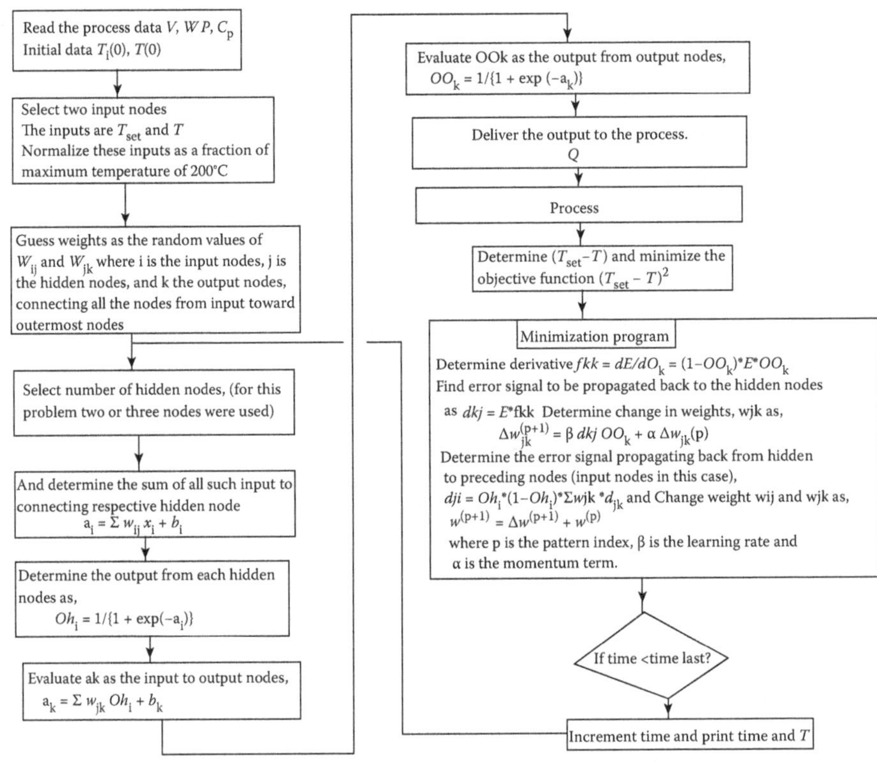

FIGURE 7.39 Artificial neural network controller for temperature control in a tank.

7.10 FUZZY LOGIC CONTROL

The word "fuzzy" means something indistinct; for example, the phrase "very high" may mean 100 feet or 1000 feet, or it may be 10,000 feet or 50,000 feet, etc. We, the human operators, use many such fuzzy words, which are intertwined in our language. In order to decode these fuzzy words in terms of numerical values, attempts have been made by mathematicians using statistical approaches. Because human logic used by an expert operator is sometimes better than PID control logic, if such an expert's knowledge is converted to a mathematical entity or, in other words, if the mapping of such expertise in a mathematical formula is possible, then, like the PID equation, this can be recorded and used repeatedly without the presence of the operator. Fuzzy control logic is such an innovative approach for the mapping of human operators' expertise in the control system to distinct (crisp) numerical values. Although the mapping of fuzzy entities in various subjects have been developed, discussions about this will be limited to the applications for process control only.

Let us study a simple example:

The temperature of a human body greater than 100°F is reported as high and 105°F as a very high temperature, respectively, by a group of physicians. Similarly, another group of physicians may consider 99°F as the high and 104°F as the very high temperatures, respectively. Thus, if the opinions of the many physicians are collected, then high (H) and very high (VH) temperatures can be obtained as a distribution of many numerical figures. If the number of such a group of physicians is 100, then a certain number of members (membership number) of this group or club will represent H or VH fuzzy words. The distribution may be a straight line (or bell shaped) curve as shown in Figure 7.40.

The membership values are depicted as

(i) At 99°F, the membership value of high temperature is 0.62.
(ii) At 100°F, the membership value is 0.18 for the very high temperature and 0.70 for the high temperature.

Thus, the crisp values of temperatures of 99°F can be expressed as fuzzy word "high" (H) and 100°F as both "high" (H) and "very high" (VH) fuzzy words.

Alternatively, if the membership value of a fuzzy word such as H is known (say 0.62), then the crisp value will be 99°F. Similarly, if the VH has a membership value of 0.18, then the crisp value is 100°F, or if the membership of the H is 0.70, the crisp value will be 100°F. Thus, membership distribution curves can be used for both conversion of crisp values into fuzzy entities or to determine crisp values from the fuzzy entities. In the control system, fuzzy entities are the process variables, set point, and manipulated variables. The fuzzy set is a discourse of fuzzy entities taken together at a particular time. Expertise or expert rules in the textual forms are translated using fuzzy words or the fuzzy rules. Let us take a simple example of a process of temperature control in a tank.

A human operator's expertise in his or her language is expressed in the following rules of control:

Rule 1: If the temperature is very low but the set point desired is very high, a very high rate of heating will be required.

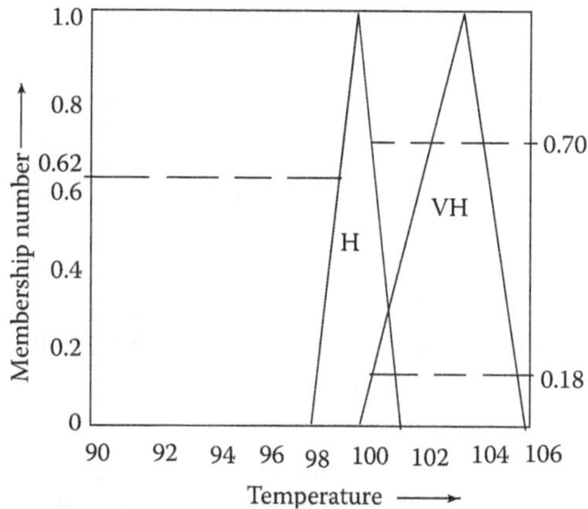

FIGURE 7.40 Membership distribution of fuzzy words of temperatures H and VH.

Rule 2: If the temperature is low, but the set point is very high, then a high rate of heating will be required.
Rule 3: If the temperature is high, but the set point is very high, then a low rate of heating will be required.

There may be a number of such instructions or rules available from the human operator. The above rules can be modified into fuzzy entities as listed below:

Rule 1: If temperature = VL, and set point = VH, then the rate of heating = VH.
Rule 2: If temperature = L, and set point = VH, then the rate of heating = H.
Rule 3: If temperature = H, and set point = VH, then the rate of heating = L.

The fuzzy words VL, L, H, and VH are applicable for very low, low, high, and very high, respectively. Thus, the linguistic rules are reduced to fuzzy rules involving only if, and, then. Now, these rules can be programmed in high-level languages, such as Basic, Fortran, C, etc.

Example 7.5

Let us take the membership distribution of the temperature, set point, and heating rate as shown in Figure 7.41. The control scheme is the same as that shown in Figure 7.38. If the initial inlet T_i and tank temperatures T are each 40°C, and the time constant of the tank is 2 seconds, determine the response of the control system using the following rules:

Rule 1: If T = VL, and T_{set} = VH, then Q = VH
Rule 2: If T = L, and T_{set} = VH, then Q = H

while the set point is raised from 40°C to 92°C.

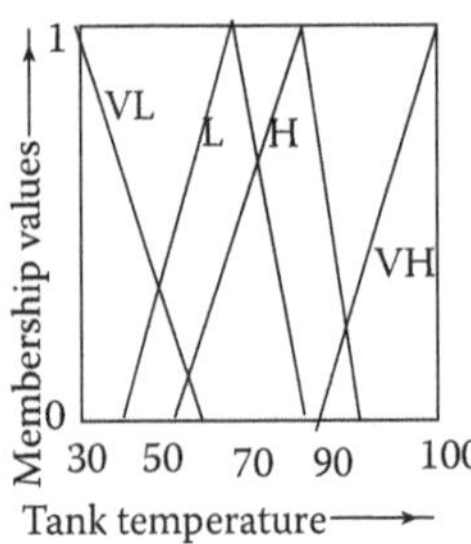

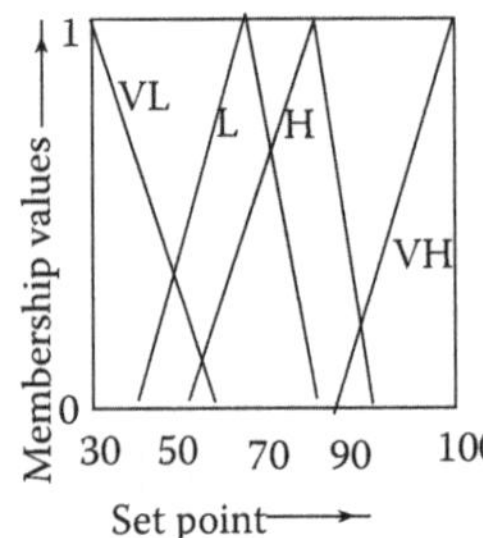

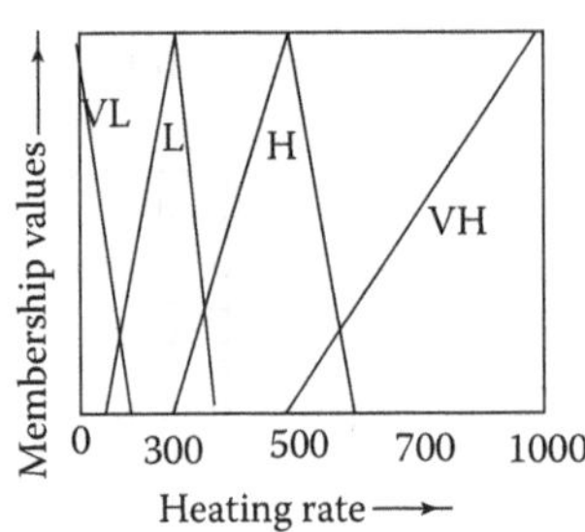

FIGURE 7.41 Membership distribution of fuzzy entities VL, L, H, and VH of the temperature T, set point T_{set}, and the heating rate Q. The maximum temperature and set point are each 100°C; whereas, the maximum heating rate possible is 1000 kcal/hr.

Steps of solution:

1. Fuzzification of the temperature and set point values at 40°, T = VL and L, and at 60°, it is VL, L, and H.
2. Determination of membership values of the fuzzified variables.

   The corresponding membership values (μ) are noted both for T and SP.

Crisp Value	**40**	**50**	**60**	**100**
Membership	**T**	**T**	**T**	**T**
VL	0.80	0.20	0.0	–
L	0.30	0.50	0.8	–
H	–	0.0	0.4	–
VH	–	–	–	1.0

   The above values are obtained from the distribution curves from Figure 7.41 at discrete points only. In a time-varying condition, these values have to be determined using the relationships (equations) of the distribution curves.
3. Logical operation: Logic gate AND is operated to choose a single value of the membership of T and T_{set}. This follows from the rules of "AND," "OR," etc., for combining two fuzzy entities. The membership value is selected by choosing the minimum of the membership values during "AND" operation, and the maximum between the two entities for the "OR" operation, etc.

   For example, from rule 1: if T = VL and T_{set} = VH, Q = VH.

   T = 40, and T_{set} = 92, hence, μ_{VL} = 0.8 for T and μ_{VH} = 0.2 for T_{set}, hence, the minimum value is chosen, i.e., μ = 0.2 for the heating rate, the output of rule 1.

   Similarly, for rule 2: if T = L and T_{set} = VH, then Q = H at T = 40, μ_L = 0.30, and for T_{set} = 92, μ_{VH} = 0.2, the minimum is 0.2.
4. Defuzzification of output to crisp value: The output of each rule is then determined with the help of the membership value chosen in step 2. For this, the area of the fuzzy region as per the rules is located as shown in Figure 7.42.

The crisp values of the output from the areas are determined by the center of gravity rule based on the area average method as shown in Figure 7.43.

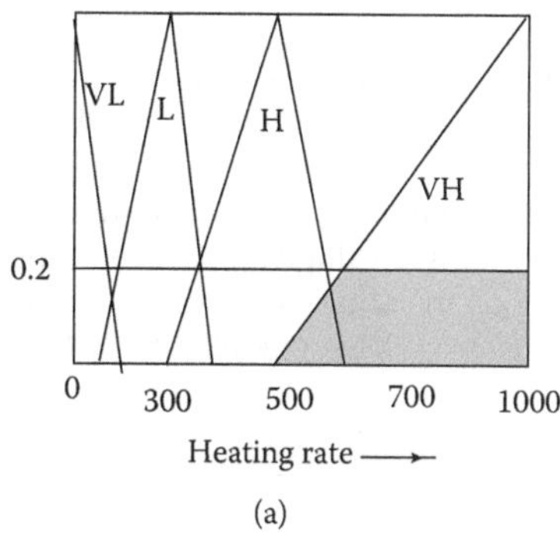

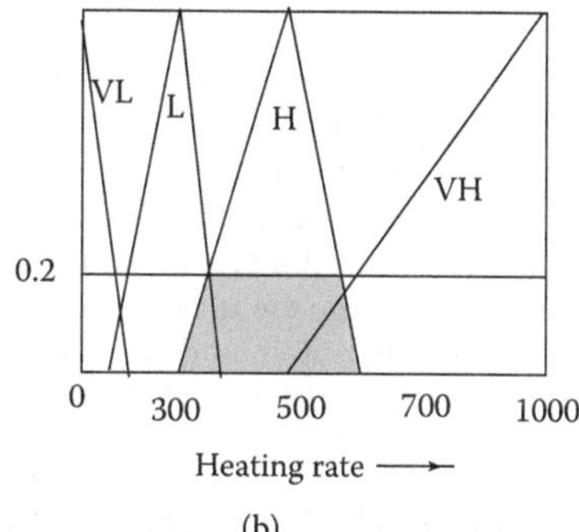

FIGURE 7.42 Membership values of the outputs from (a) rule 1 and (b) rule 2.

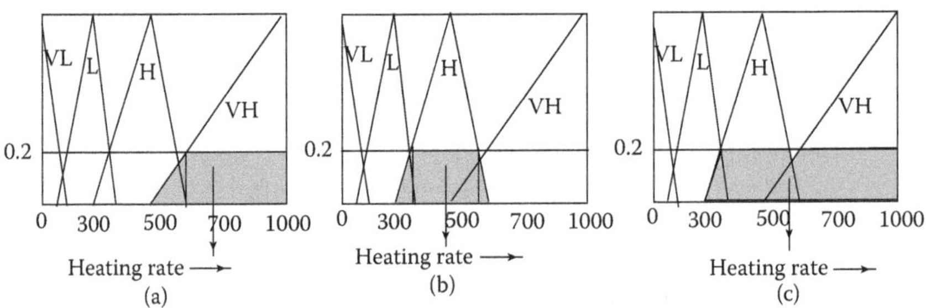

FIGURE 7.43 Determination of the crisp value of the output, the heating rate (a) from rule 1, (b) from rule 2, and (c) the final output from the two rules; the final weighted average center of gravity is 580 kcal/hr.

Note that the output of rule 1 has a membership value of 0.2, and the heating rate is the fuzzy entity VH as shown in the shaded area in Figure 7.43a. The defuzzification to crisp value is then determined as the center of gravity of the area, which is the weighted average of the area as

$$CG_1 = \frac{\int x\,dA}{\int dA} \tag{7.51}$$

The output, $CG_1 = 700$ kcal/hr.

Similarly, the output of rule 2 has a membership value of 0.2, and the heating rate is the fuzzy entity H as shown in the shaded area in Figure 7.43b. The defuzzification to a crisp value by the above center of gravity method is that CG_2 is found to be 480 kcal/hr.

Finally, the output from the combination of rules 1 and 2 is determined as the CG of CG_1 and CG_2 as

$$CG = \frac{CG_1A_1 + CG_2A_2}{A_1 + A_2} \tag{7.52}$$

where A_1 and A_2 are the areas of the output fields of rule 1 and 2, respectively. This value of the output CG of the controller is determined to be 580 kcal/hr as shown in Figure 7.43c.

The above example explained how the fuzzy controller will produce the output, which must be crisp and delivered to the process. As the temperature of the process changes, the fuzzy controller will repeat all the above steps to determine the necessary output and will continue with time as a tireless operator mimicking a human operator's expertise.

7.11 QUESTIONS AND ANSWERS

EXERCISE 7.1

What is a cascade control and where is it used?

Answer:
Cascade control is explained in Section 7.1. This is used when the response of the control variable is comparatively slower than that of the manipulated variable. For example, in a jacketed heater, the temperature in the jacket rises faster than the temperature of the liquid in the tank. Hence, a jacket temperature controller is used as the slave controller to increase the speed of control of the tank temperature. Another example is the control of distillation control where the composition of the vapor and liquid change slower than the change in reflux and reboiling rates.

EXERCISE 7.2

Determine the response of a jacketed tank temperature controlled by a cascade temperature controller with the jacket flow controller as shown in Figure 7.44. Also, compare it with a conventional control using a temperature controller alone.
Given the process data:

$V = 10$, $W = 10$, $\rho = 1$, $C_p = 1$, and the initial temperatures are $T_i = 30$, $T = 30$. The maximum heating rate is $Q = 900$ heat units/time.

Determine (a) the open-loop response (b) conventional control as shown in Figure 7.44a, and (c) cascade control as shown in Figure 7.44b.
Answer:

(a) The tank model is given at a constant flow rate of cold fluid in

$$T_i(t) + Q(t)/WC_p = T(t) + (V\rho/W)\, dT/dt \tag{7.53}$$

where $Q(t) = 900$, the open-loop tank temperature response is obtained by integrating the above differential equation with the given initial values of temperatures. This is shown in Figure 7.45, and the maximum temperature of the tank rose to 120°C.

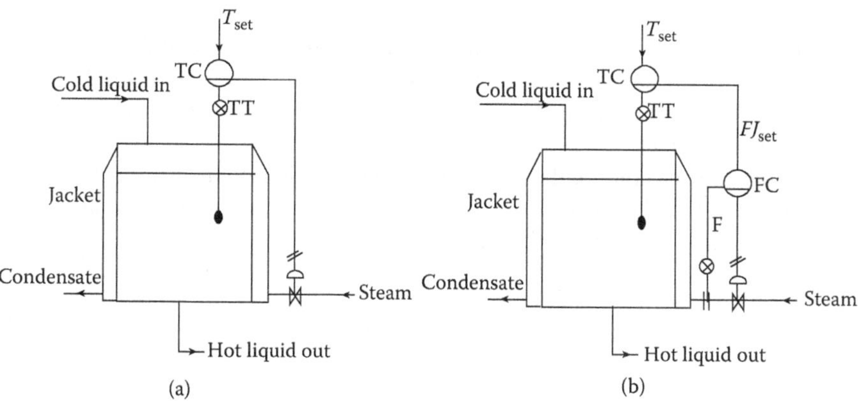

FIGURE 7.44 Temperature control in a jacketed tank: (a) conventional control and (b) cascade control.

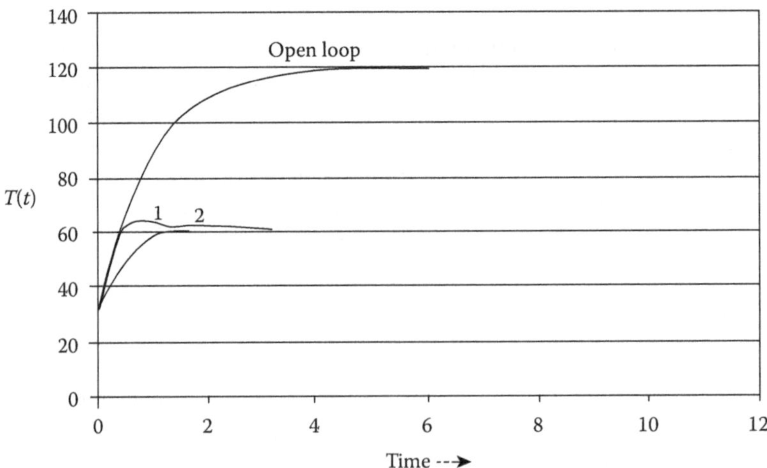

FIGURE 7.45 Comparative responses of tank temperature control problem open loop, (1) conventional control with $K_c = 40$, $\tau_i = 0.5$ and (2) cascade control with $K_c = 40$, $\tau_i = 0.5$, and $K_{c1} = 0.5$.

(b) The conventional control is carried out by using a PI controller with $K_c = 40$ and $\tau_i = 0.5$. Considering the valve has no lag, the output of the controller is taken as the heating rate as

$$Q = K_c \varepsilon + K_c \int \varepsilon \, dt/\tau_i \tag{7.54}$$

and

$$\varepsilon = T_{set} - T \tag{7.55}$$

Equations 7.53, 7.54, and 7.55 are solved, and the response of this is determined for the new set point as $T_{set} = 60°C$ as shown in the curve 1 in Figure 7.45. It was observed that little overshoot above set point occurred in this control scheme.

(c) In the cascade control, the set point of the jacket flow controller is set by the output of the PI controller, and the output of the jacket controller is taken as the heating rate to the tank. Thus,

$$F_{set} = K_c \varepsilon + K_c \int \varepsilon \, dt/\tau_i \tag{7.56}$$

and

$$\varepsilon = T_{set} - T \tag{7.57}$$

and

$$Q = K_{c1}\,(F_{set} - F) \tag{7.58}$$

where $F = Q$ for no valve lag.
Equations 7.53, 7.56, 7.57, and 7.58 are solved for new set point $T_{set} = 60°C$. The response is presented in curve 2 in Figure 7.45, which indicates that the overshoot is reduced as compared to conventional control.

Note that steam condensing in a jacket in this example is the source of the heating rate to the tank content. Hence, the temperature in the jacket will be constant at the condensing temperature. If a temperature controller was used (in the place of the flow controller) in the jacket, cascade control will not be feasible. However, jacket temperature will vary if the heating fluid is a hot gas, superheated steam (not condensing) or liquid.

EXERCISE 7.3

Consider a pressure control system in a vessel as shown in Figure 7.46. A high-pressure source and a low-pressure exit system are provided through two different control valves. When the pressure in the vessel P is lower than the desired pressure P_{set}, gas enters from the high-pressure source at P_1 until this reaches the set point. If the pressure increases over the set point, gas leaves through the low-pressure exit line at P_2. The high-pressure line valve is a normally closed (N/C) type, and the low-pressure exit line is a normally open (N/O)-type control valve.

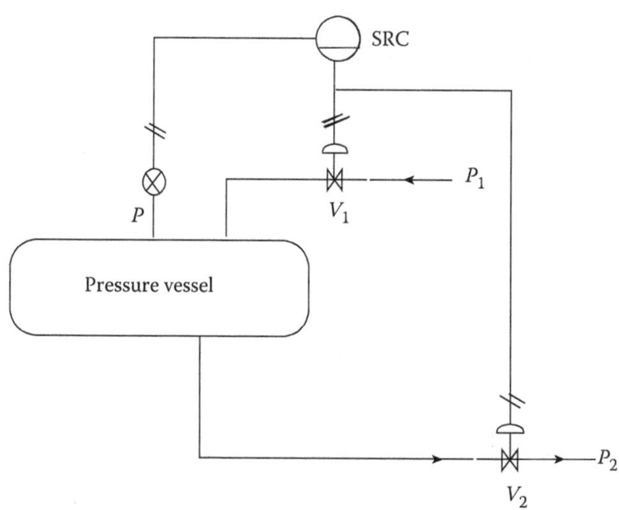

FIGURE 7.46 Split range control of pressure.

Solution:

The material balance in the vessel is obtained as

$$W_1 - W_2 = VM/RT \frac{dP}{dt} \tag{7.59}$$

where W_1 and W_2 are the mass flow rates through high-pressure and low-pressure lines, respectively. The controller output is given as

$$W_1 \text{ or } W_2 = K_c\,(P_{set} - P). \tag{7.60}$$

The values of gas flow from the high-pressure source are given as

$$W_1 \neq 0 \text{ while } P < P_{set} < P_1$$

$$W_1 = 0 \text{ and } W_2 = 0 \text{ while } P_{set} = P$$

whereas $W_2 \neq 0$ while $P_2 < P > P_{set}$.

Given that maximum values of W_1 and W_2 are each 100 units/time, pressure P rises to a maximum of 20 units, and the vessel volume V is 1 liter at a temperature of 30°C. Such that $VM/RT = 1.4$. Determine the responses for the following cases:

(a) the open-loop response when $W_1 = 100$ and W_2 is suddenly decreased from 100 to 90 units (b) a proportional controller with $K_c = 0.4$ is used such that the controller output is $O_c = 9 + k_c * e$ and varies from 3 to 15 psi where e = set point pressure – instantaneous pressure = $SP - P$. The controller is used to manipulate the exit flow W_2 while $W_1 = 100$ units fixed (c) the same proportional controller is used as a split-range controller to drive two control valves; one drives the inlet flow $W_1 = -150 + 100\,O_c/6$, and the other $W_2 = 250 - 100\,O_c/6$ such that the split range represents the following valve actions as shown in Figure 7.47.

(a) The integration of the

$$dp/dt = (W_1 - W_2)\,RT/(VM) \tag{7.61}$$

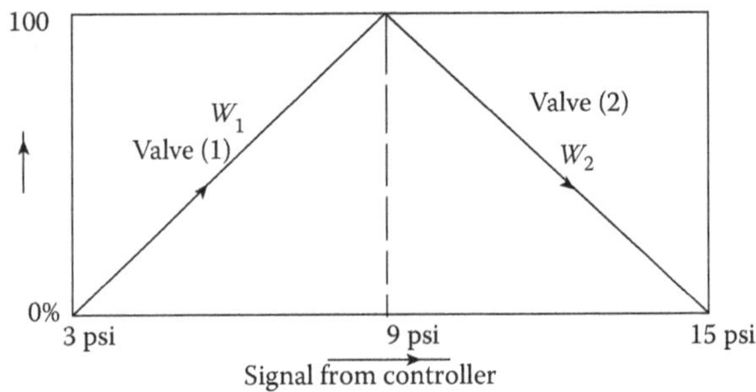

FIGURE 7.47 Split range actions of control valves.

is obtained for changing set point pressure from initial 3 psi and is plotted as a function of time as presented in Figure 7.48 for the open-loop system.

(b) Simultaneous integration of equation (7.61) while $W_1 = 100$ and W_2 is manipulated by the controller. The response of the controlled system at a set point of 10 units of pressure is shown in Figure 7.49.

(c) Simultaneous integration of Equation 7.61 while W_1 and W_2 are manipulated by the controller as the split-range controller. The response of the controlled system at a set point of 10 units of pressure is shown in Figure 7.50.

EXERCISE 7.4

Consider an override control arrangement as shown in Figure 7.51 where a pressure drop across the control valve is controlled by a DPC. A temperature controller is also in action driving the same control valve through an SS, which diverts the control signal of the DPC to the control valve, keeping the temperature controller's signal on hold when the pressure drop across the valve reaches the maximum value (DP max). The process has a time constant of 2 seconds, the maximum pressure drop across the valve allowable is 0.5 psi, and the initial inlet temperature T_i and the tank

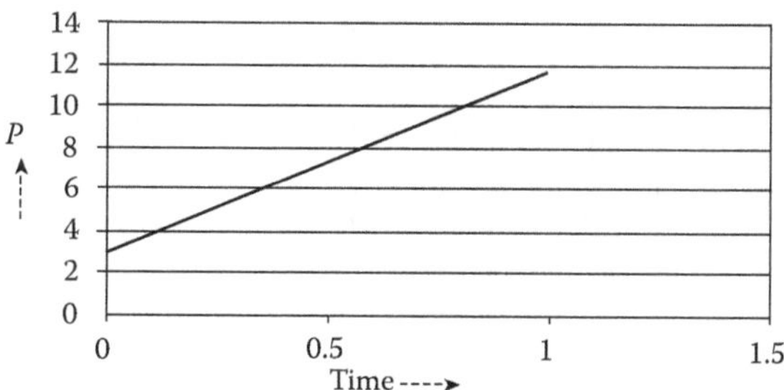

FIGURE 7.48 Open-loop response for sudden reduction of exit flow.

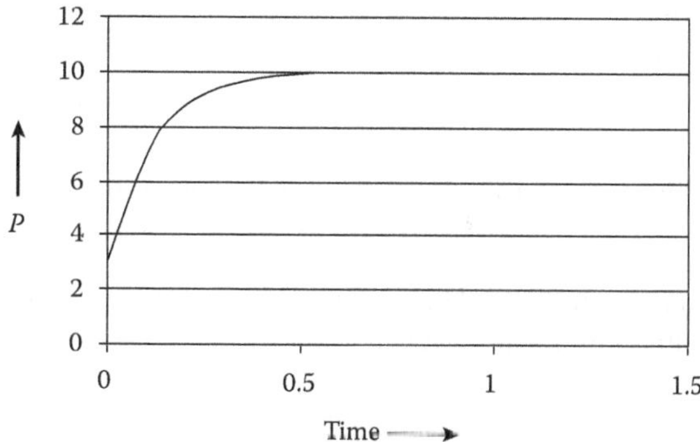

FIGURE 7.49 Conventional control of pressure using a control valve at the exit.

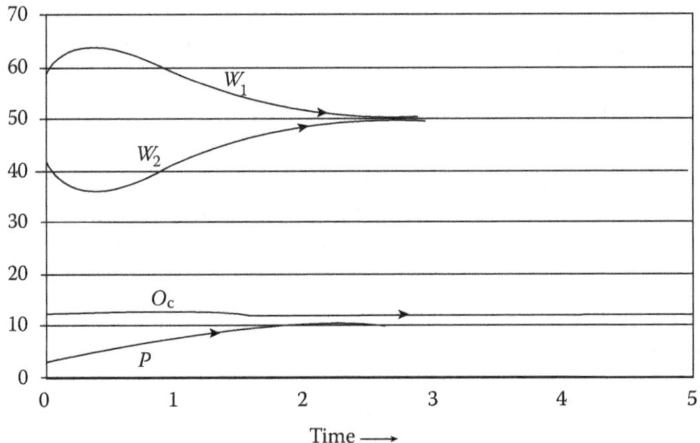

FIGURE 7.50 Split-range control of pressure at a set point of 10 units.

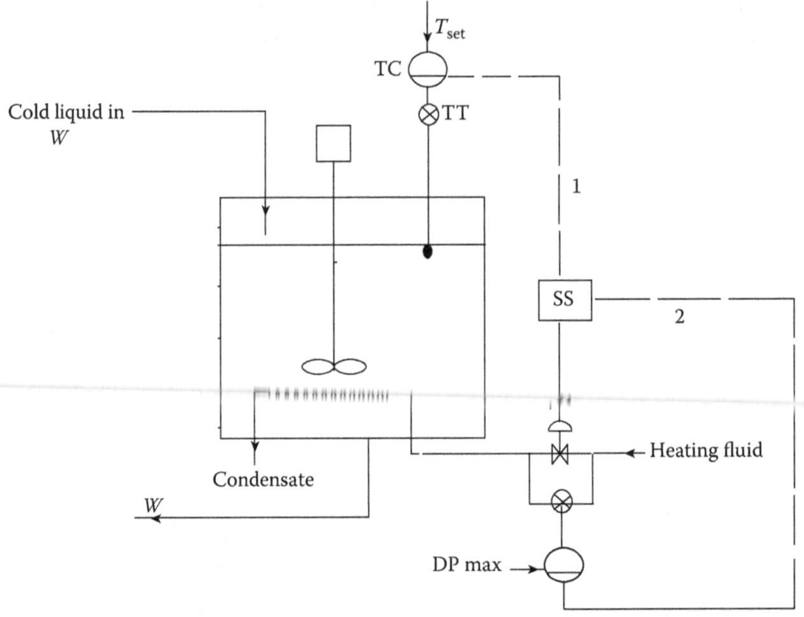

FIGURE 7.51 Override control arrangement for a temperature control in a tank.

temperature T were each 30°C. The process data, temperature controller, and the pressure drop controller data are given as

Process data:

> W: Constant flow rate of the process liquid, 100 liter/sec
> C_p: Average specific heat of the process liquid, 1 Kj/kg K

τ: First-order time constant of the process, 2 sec
λ: Latent heat of vaporization of heating fluid, 10,000 kcal/kg

Temperature controller (PI) data:

Controller gain: $K_{c1} = 0.05$ psi/K
Integration time, $\tau_i = 1$ sec

Differential pressure controller (P) data:

Controller gain: $K_{c2} = 1$ psi/psi

Control valve data:

$Q = \lambda(O_c - 3)/12$, kcal/sec

where O_c is the output of the respective controllers, psi.
Answer:
The process model:

$$T(t) + \tau\, dT(t)/dt = T_i + Q/WC_p \quad (7.62)$$

Temperature controller:

$$O_{c1} = K_{c1}\varepsilon_1 + K_{c1}/\tau_i \int \varepsilon_1\, dt \quad (7.63)$$

where

$$\varepsilon_1 = T_{set} - T \quad (7.64)$$

Pressure drop controller:

$$O_{c2} = K_{c2}\, \varepsilon_2 \quad (7.65)$$

where

$$\varepsilon_2 = \Delta P_{max} - \Delta P \quad (7.66)$$

$$Q = \lambda(O_c - 3)/12 \quad (7.67)$$

Override logic:
If the pressure drop ΔP across the control valve is less than ΔP_{max} (0.5 psi), $O_c = O_{c1}$, and if the pressure drop ΔP across the control valve is more than ΔP_{max} (0.5 psi), $O_c = O_{c2}$.

Considering a linear control valve with the assumption that flow rate F through the valve is proportional to the controller output O_c, i.e., $F = O_c$ and pressure drop,

$$\Delta P = F^2. \tag{7.68}$$

The above equations are solved numerically with the help of a computer program, and the responses are presented in Figures 7.52 and 7.53, respectively, for $T_{set} = 100°C$ and $T_{set} = 200°C$.

EXERCISE 7.5

Consider a furnace using propane as the fuel to heat a charge stock where the combustion air is preheated and the stack gas leaving the combustion zone. It was

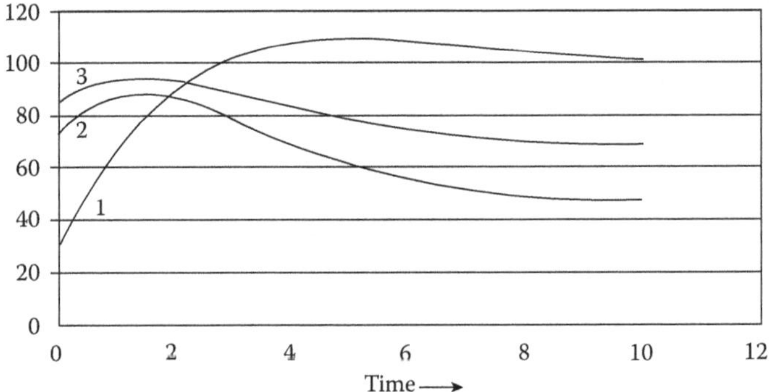

FIGURE 7.52 Override control in a heated tank where (1) temperature T, (2) heating rate $Q/100$, and (3) pressure drop ($DP \times 100$) while the pressure drop fluctuates above or below 0.5 psi and the temperature set point is 100°C.

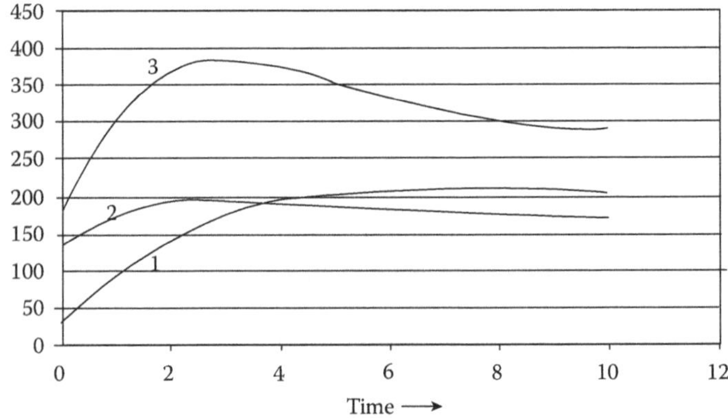

FIGURE 7.53 Override control in a heated tank where (1) temperature T, (2) heating rate $Q/100$, and (3) pressure drop ($DP \times 100$) while the pressure drop is above 0.5 psi and the pressure drop controller overrides the temperature controller at a set point of 200°C.

found that the lower the air-to-fuel ratio, the higher the efficiency of the furnace and vice versa. Similarly, the efficiency increases with the rise in preheated air temperature. A ratio controller is required to monitor the combustion air rate and the air temperature for optimum efficiency. Present an appropriate control scheme and analyze the response of the system for variation of air-to-fuel ratio while the temperature of combustion air varies as a sinusoidal function above and below 460°C, given as

$$T_a = 460 + 10\sin(2t) \tag{7.69}$$

where time t is in minutes.

The air-to-fuel ratio (A/F) in kg/kg, as a function of the air temperature for optimum efficiency is approximated as

$$A/F = 30 - 0.5(T_a - 459.8). \tag{7.70}$$

If the initial steady-state fuel firing rate was 10 kg/min, determine the response of the air flow controller if the fueling rate suddenly increases to 20 kg/min. The proportional flow controller for air has a $K_c = 0.5$ and the control valve is a linear valve and has no dynamic lag such that the output of the controller is taken as the flow rate of air.

Answer:

The control scheme is presented in Figure 7.54.

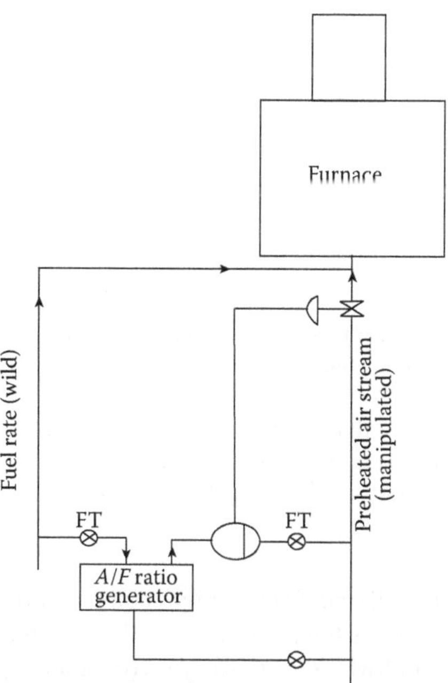

FIGURE 7.54 Control scheme for air-to-fuel ratio of a typical furnace.

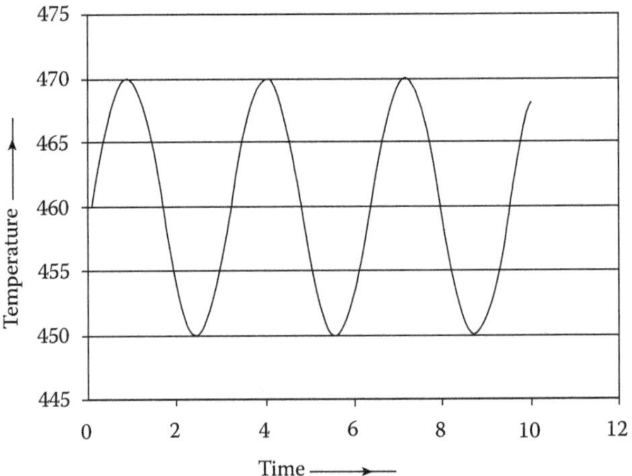

FIGURE 7.55 Fluctuation of temperature of combustion air.

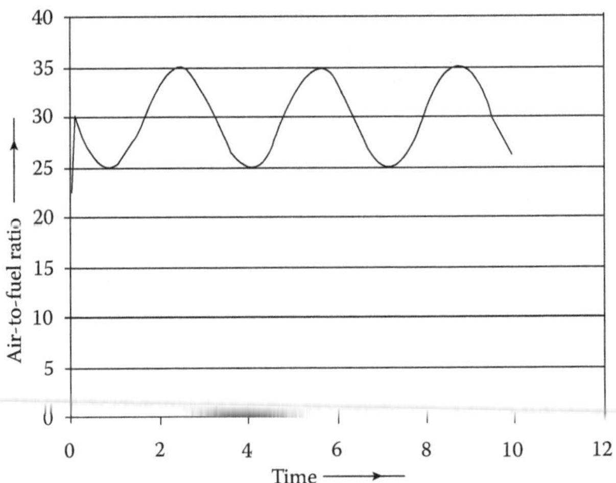

FIGURE 7.56 Response of the ratio controller changing the air-to-fuel ratio for change in fueling rate and temperature fluctuation.

The figures also indicate that while the air temperature rises, the air-to-fuel ratio is reduced and vice versa to maintain optimum efficiency.

EXERCISE 7.6

Determine the response of the fuzzy temperature controller with seven fuzzy entities for set point, process temperature, and the heating rate as negative large (NL), negative medium (NM), zero (ZR), positive small (PS), positive (P), positive medium (PM), and positive large (PL). The membership distribution of these entities are taken as linear for simplicity as shown in Figure 7.57. However, any

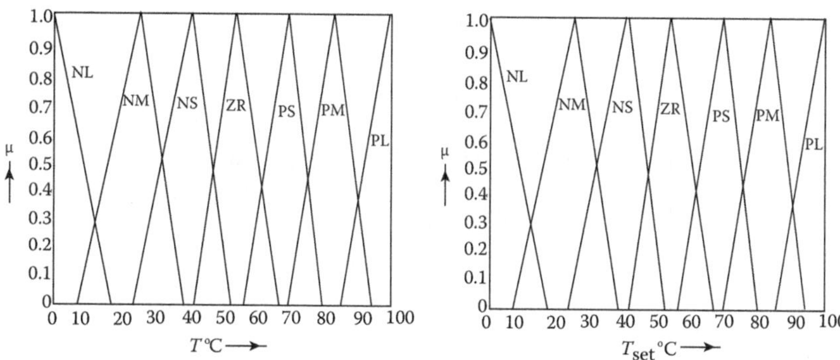

FIGURE 7.57 Distribution of fuzzy entities of process temperature T and set point T_{set} as inputs to the fuzzy controller.

other abbreviation could be used in place of NL, NM, NS, ZR, PS, PM, and PL indicating very large to very little difference of T_{set} and T from NL to PL in this order.

The expert rules are listed in Table 7.1.

The process parameters for the temperature control of the tank are given as

Liquid flow rate, $W = 10$ kg/hr
Specific heat of the liquid, $C_p = 1$ kcal/kmole K
Density of the liquid, $\rho = 1$ kg/m^3
Volume of tank (liquid), $V = 20$ m^3
Maximum heating rate, $Q = 850$ kcal/hr

The output heating rate is shown in Figure 7.58.

The responses of the fuzzy controllers are obtained by using the above parameters, the fuzzy membership distribution values, and all 49 rules as listed in the table. These responses are shown in Figure 7.59.

TABLE 7.1
Expert Rules in Fuzzy Entities

	T_{set} →						
T ↓	**NL**	**NM**	**NS**	**ZR**	**PS**	**PM**	**PL**
NL	NL	NL	NL	NL	NL	NL	NL
NM	NM	NM	NM	NM	NM	NM	NM
NS	NS	NS	NS	NS	NS	NS	NS
ZR	ZR	ZR	ZR	ZR	ZR	ZR	ZR
PS	PS	PS	PS	PS	PS	PS	PS
PM	PM	PM	PM	PM	PM	PM	PM
PL	PL	PL	PL	PL	PL	PL	PL

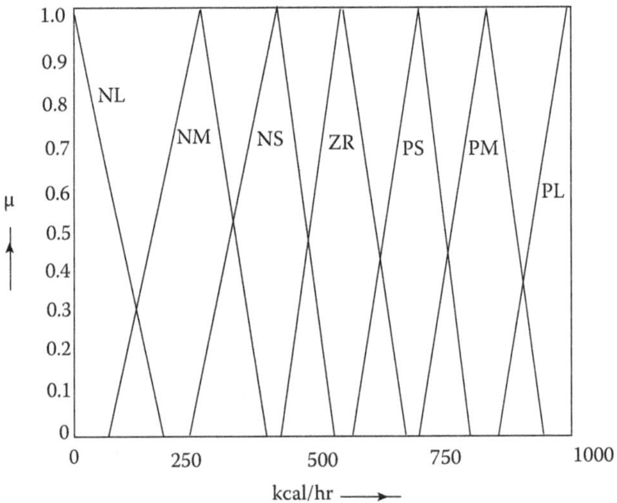

FIGURE 7.58 The output of the controller as the heating rate is available in seven fuzzy entities varying from 100 to 1000 kcal/hr.

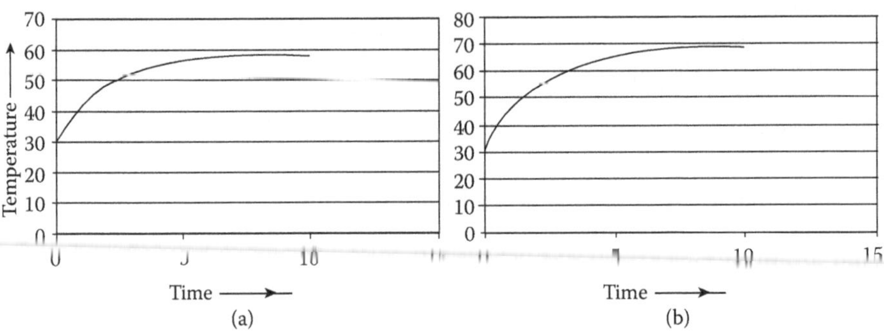

FIGURE 7.59 Responses of the fuzzy logic controller (a) for set point changed from 30°C to 60°C (b) set point changed from 30°C to 70°C.

The response of the fuzzy controller is also shown for change of set points from 30°C to 70°C and 70°C to 100°C in Figure 7.60. All the responses indicate that the fuzzy controller is more a robust control logic without any fluctuation with a little acceptable offset. However, the controller performance is highly dependent on the rules. Hence, faulty rules may be responsible for poor control.

However, no overshoots were observed even though the speed of response is very fast.

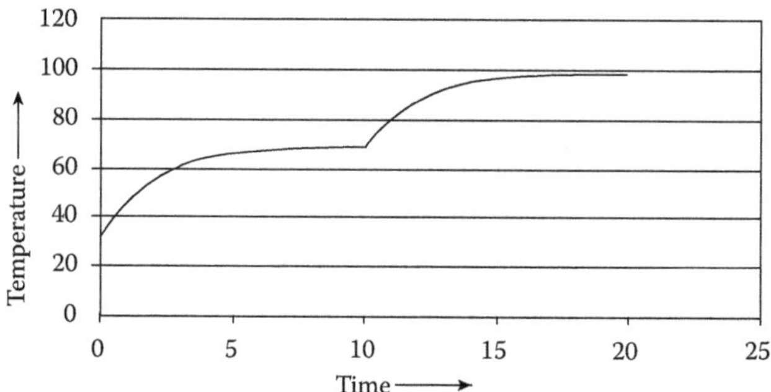

FIGURE 7.60 Response of fuzzy controller for set point changed from 30°C to 70°C and 70°C to 100°C.

8 Virtual Laboratory

8.1 GETTING STARTED IN THE VIRTUAL LABORATORY

Virtual Laboratory software, developed by the authors of this book, deals with the various problems of process dynamics and control systems. It is contained in the CD that accompanies this book. (The CD-ROM can be found at www.routledge.com/9781466514201.) Open-loop process dynamics have been prepared to study systems of a first order, two first orders in series, a general second order, and a first order coupled with a dead lag, involving common disturbances, such as step, ramp, sine, and impulse. In the closed-loop control systems analysis, processes of a first order, a second order, a third order, and a first order with a dead lag have been involved along with control elements consisting of a PID controller, a control valve, and a transducer. Studies can be made by selecting any of the four control options, such as P, PI, PD, or PID. Three types of control valves have been involved, having linear, equal percent, and hyperbolic characteristics and each with first-order dynamics. A transducer having first- and second-order dynamics has been used in the feedback control loop. In the tuning problems, the Ziegler–Nichols and Cohen–Coon methods have been presented, which can be experimented with a controller and a selected process. The user may select any of the process parameters and tune the controller, applying either of the two methods as mentioned above. In the advanced control systems, cascade, ANN, and fuzzy control systems have been presented. Process and controller parameters can be changed by the user for understanding the responses. In all the above four features, a two-dimensional graphical presentation has been provided. The start-up window of this virtual world is displayed as shown in Figure 8.1, indicating the above features.

8.2 STUDY OF OPEN-LOOP SYSTEMS

From the start-up window, select the "Open-Loop System" button and click it to get the next window, which shows additional buttons to select the desired "Process" and "Disturbance." The corresponding response will be observed by clicking the "Response" button. The window is shown in Figure 8.2.

In order to carry out the open-loop experiments, click the "Disturbance" button to select one type of disturbance available in the next window as shown in Figure 8.3.

Select the "Step Disturbance" button and enter the magnitude of the disturbance in the next window as shown in Figure 8.4.

The default value of the magnitude of the step can be changed by the up and down arrow scroll bar. After entering unity, "1," or any other value, click the "Back" button to return to the previous window. After returning to the window in Figure 8.2, select and click the "Process" button and select any one of the processes available in the window as shown in Figure 8.5.

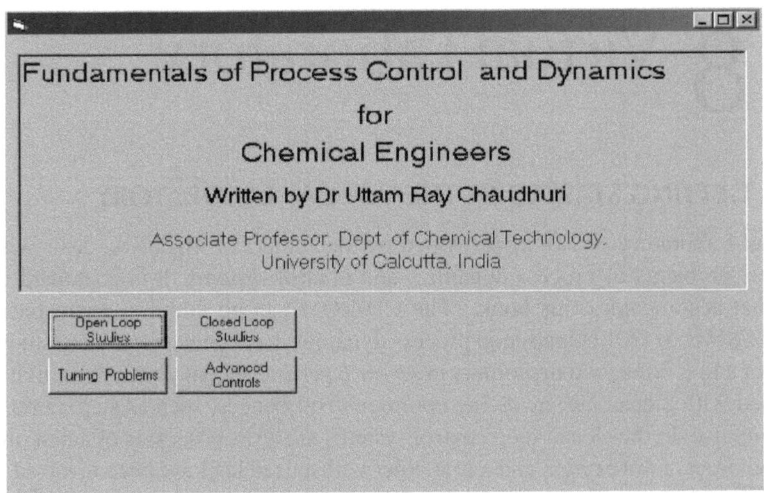

FIGURE 8.1 Start-up window of Virtual Laboratory.

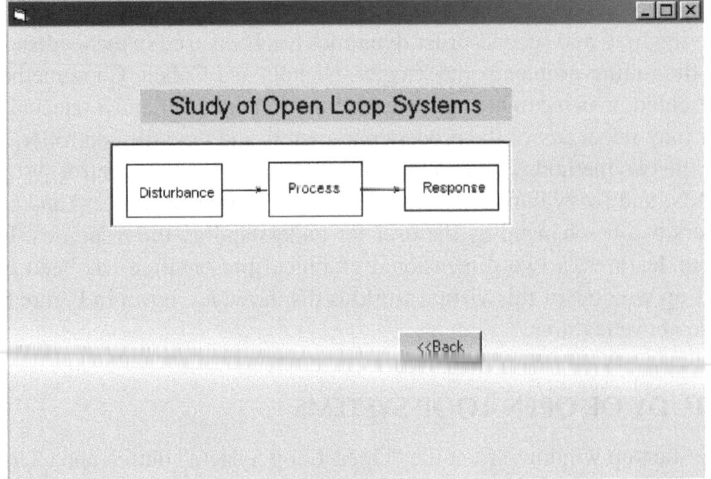

FIGURE 8.2 Open-loop systems experiments.

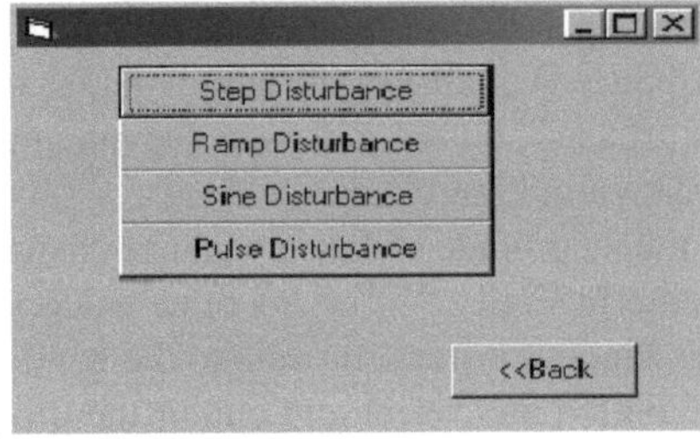

FIGURE 8.3 Window for selection of type of disturbance.

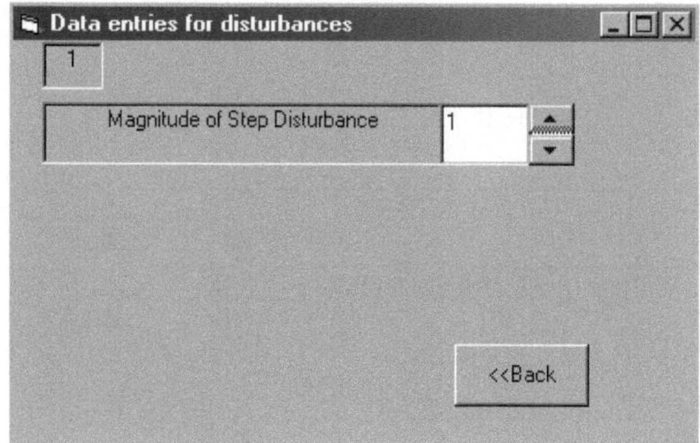

FIGURE 8.4 Window for entering magnitude of disturbance.

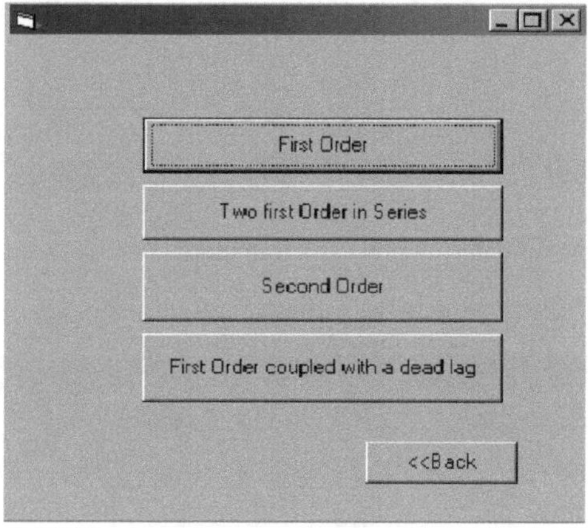

FIGURE 8.5 Window for selection of process.

8.2.1 Experiments with First-Order Systems

Select and click on the "First Order" button and enter the process parameters in the window displayed in Figure 8.6.

Change the gain and time constant of the first-order process selected with the vertical scroll bar from the default values and click back to the previous window and finally back to the window shown in Figure 8.2. Thus, you have selected a first-order system with a gain of 10 units and a time constant of 5 units, which will be disturbed by a unity step disturbance. To get the response of the process by combining the selected process and the disturbance, click the "Response" button seen in Figure 8.2. At this point it may be necessary for you to note the type and magnitudes of the process and the disturbance selected before viewing the response, so that you can compare the response with what you studied earlier. The response window will appear as you click on the "Response" button and will display the response curve when you click on the "View Response" button as shown in Figure 8.7. Responses

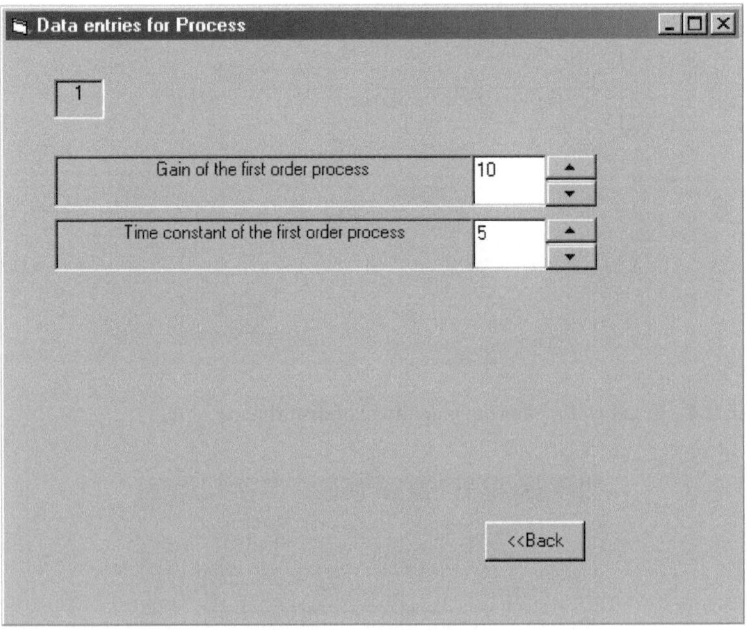

Data entries for Process

1

Gain of the first order process 10

Time constant of the first order process 5

<<Back

FIGURE 8.6 Window for entering process parameters.

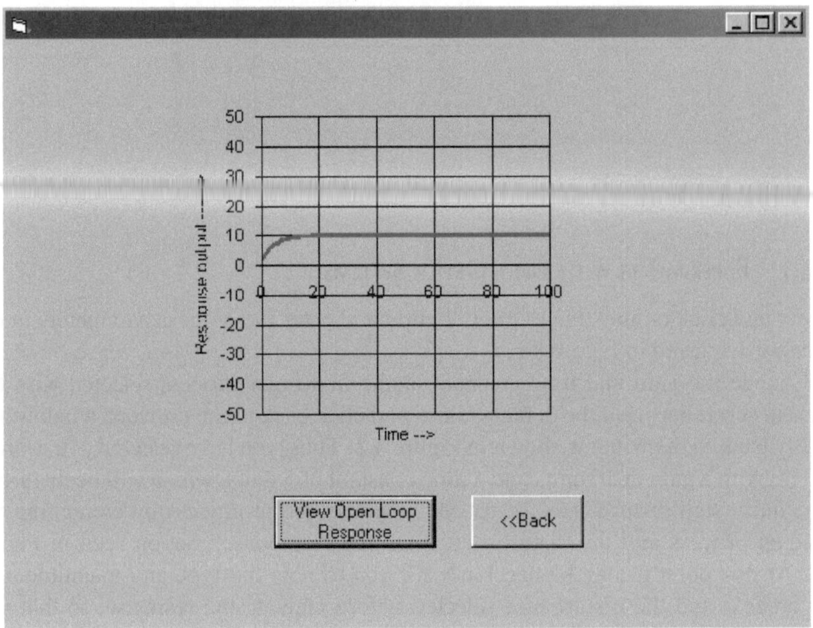

FIGURE 8.7 Response curve resulting from a unit step disturbance in a first-order process having a gain of 10 units and a time constant of 5 units.

for other values of the step disturbances and process parameters can also be viewed. For example, if the step magnitude is increased from unity to 5, the response curve will also be augmented as shown in Figure 8.8.

If a ramp disturbance of slope unity is chosen to disturb the process, the following steps should be followed. Click the "Disturbance" button from Figure 8.2, select the "Ramp" button window and enter the magnitude of the rate of ramp disturbance using the vertical scroll bar button as shown in Figure 8.9.

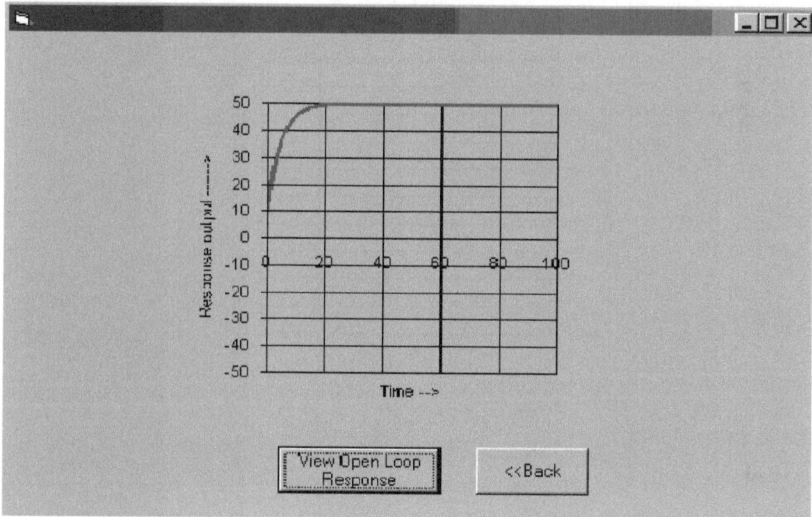

FIGURE 8.8 Response curve resulting from a step disturbance of 5 units in a first-order process having a gain of 10 units and a time constant of 5 units.

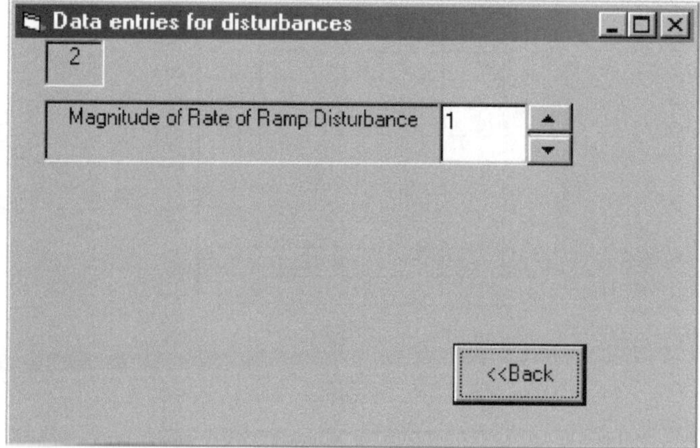

FIGURE 8.9 Selecting a unit ramp disturbance.

The response of the above first-order system (Figure 8.10) to a sine disturbance with an amplitude of 10 units and unit radian frequency is presented in Figure 8.11.

For an impulse disturbance with a magnitude of 10 units and unity radian frequency is also played with a first-order process having a gain of 10 units and a time constant of 5 units. The response is displayed in Figure 8.12.

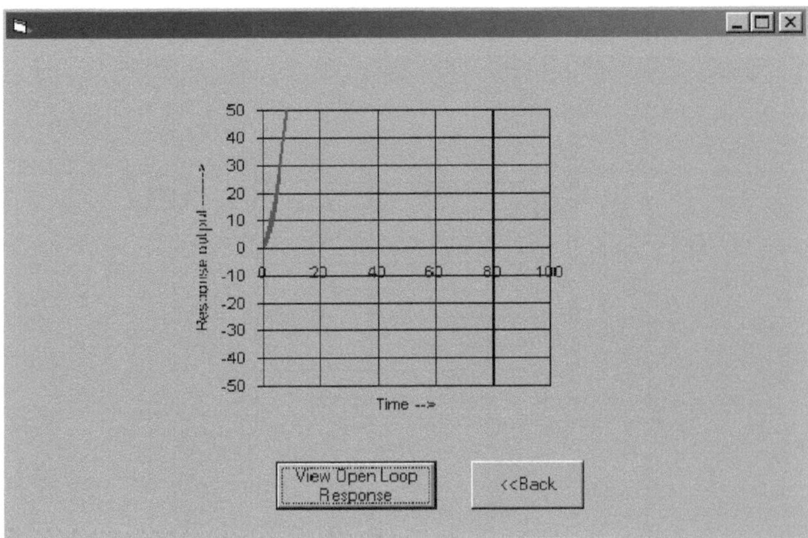

FIGURE 8.10 Response of the first-order process of gain 10 and time constant 5 units resulting from a ramp disturbance of unit rate.

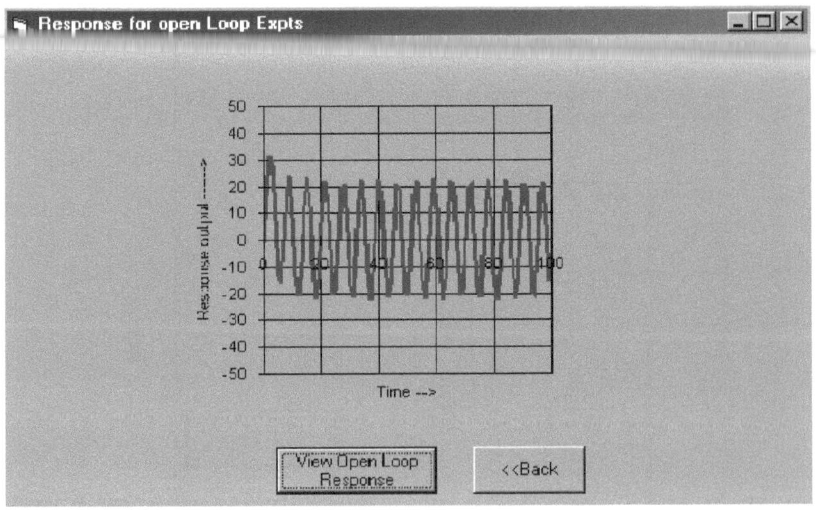

FIGURE 8.11 Response of a first-order system to a sine disturbance.

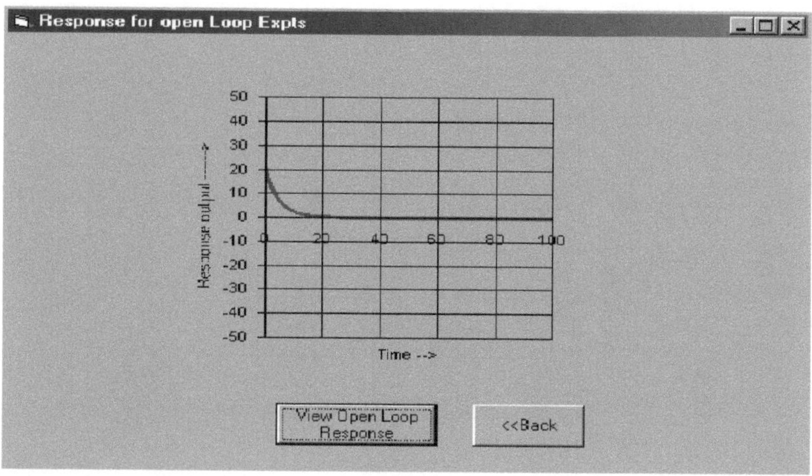

FIGURE 8.12 Response to an impulse disturbance of 10 units to a first-order process with a gain of 10 units and a time constant of 5 units.

8.2.2 Experiments with Two First-Order Systems in Series

Two first-order systems with gains of 1 and 2, respectively, and time constants of 2 and 3 units are experimented with a step disturbance of 10 units. The response is presented in Figure 8.13.

The response of a ramp disturbance with a slope of 1 unit over the two first-order systems in series as described above is obtained as shown in Figure 8.14.

The response to a sine disturbance with an amplitude of 10 units and unit frequency to the above two first-order systems in series. This is shown in Figure 8.15.

The response of an impulse disturbance of 10 units to the above two first order systems in series is shown in Figure 8.16.

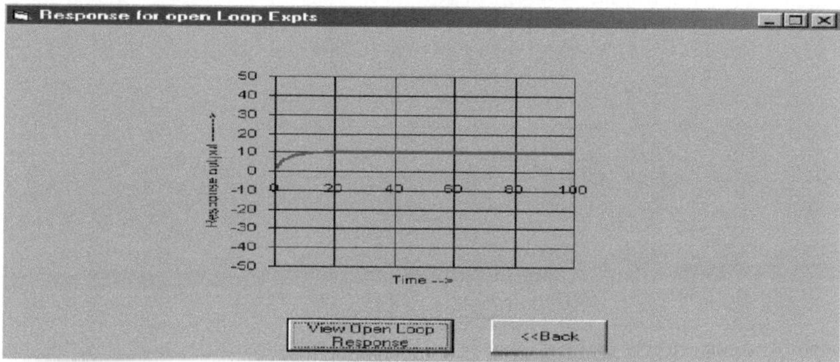

FIGURE 8.13 Overdamped response of two first-order systems in series.

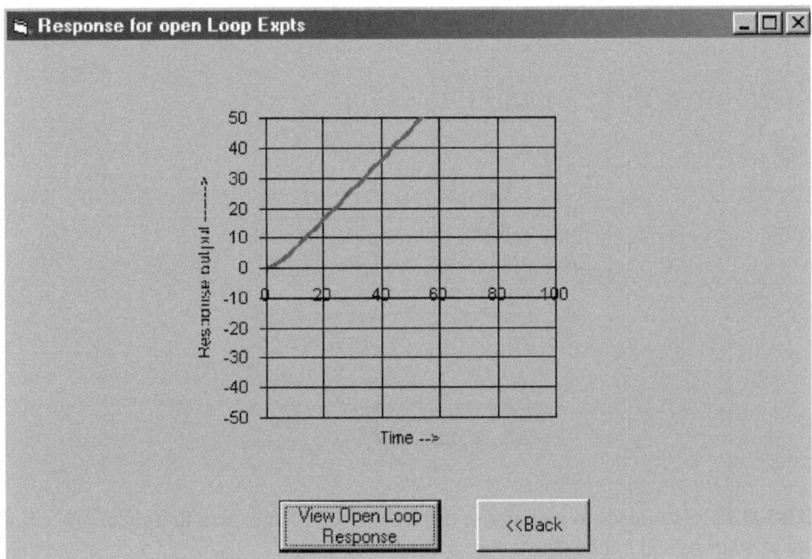

FIGURE 8.14 Response to a unit ramp disturbance to two first-order systems in series.

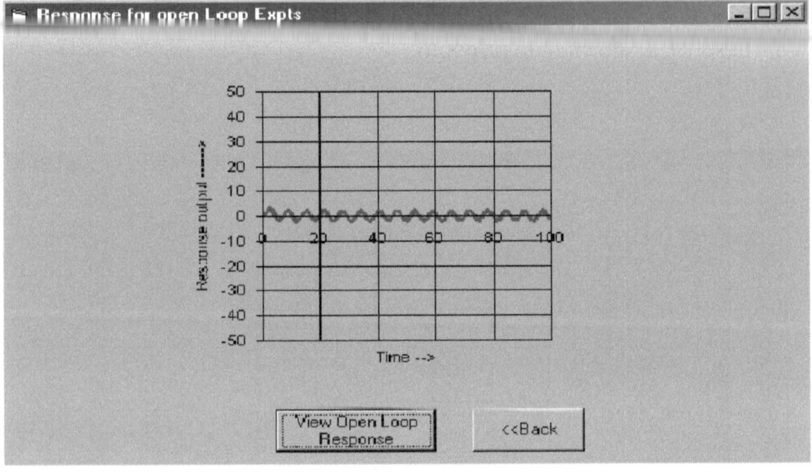

FIGURE 8.15 Response of a sine disturbance to two first-order systems in series.

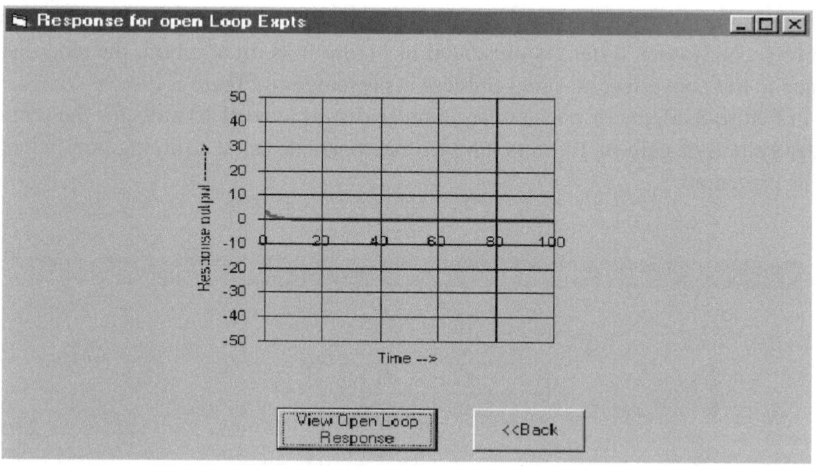

FIGURE 8.16 Pulse disturbance of 10 units to the above first-order systems in series.

8.2.3 Experiments with a Second-Order System

Similarly, a second-order process can be studied with different time constants and damping coefficients for any of the four types of disturbances. For example, select a step disturbance of 5 units and select a second order with a gain of 5 units and a damping coefficient of 0.1 unit; the response curve will be visualized as shown in Figure 8.17. It is observed that the response reaching toward an ultimate value of 25 units is theoretically desirable.

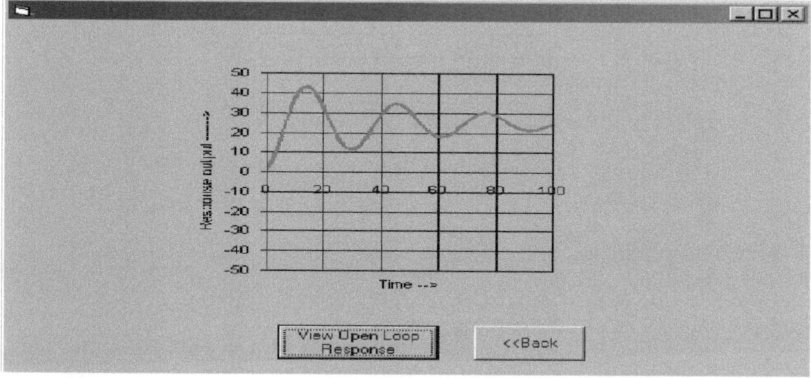

FIGURE 8.17 Response to a step disturbance of 5 units to a second-order process with a gain of 5 units, a unity time constant, and a damping coefficient of 0.1.

Similarly, the response with a ramp disturbance with a rate of 2 units for the above second-order system is presented in Figure 8.18. In addition, the sine disturbance to the above second-order process is presented in Figure 8.19.

In Figure 8.20, the response to an impulse disturbance of 10 units for the second-order system of gain of 10 units, unity time constant, and a damping coefficient of 0.1 is presented.

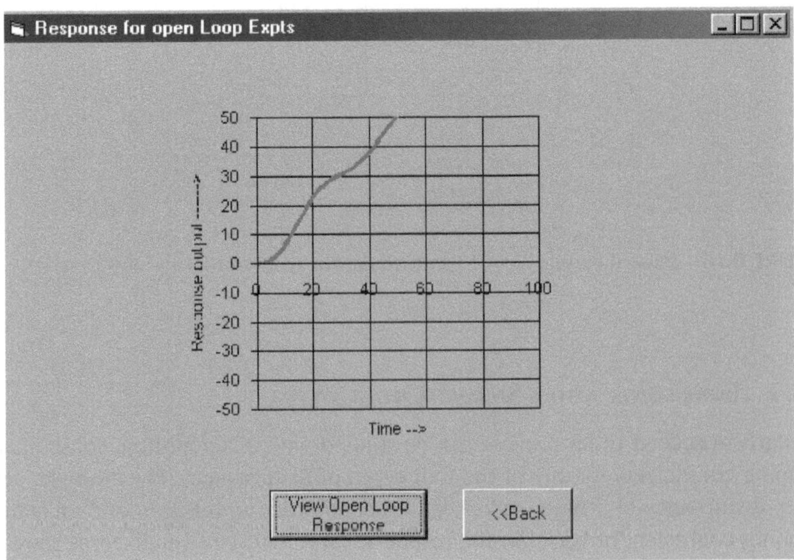

FIGURE 8.18 Response to a unit ramp disturbance to a second-order system with a unity gain, a time constant of 5 units, and a damping coefficient of 0.1.

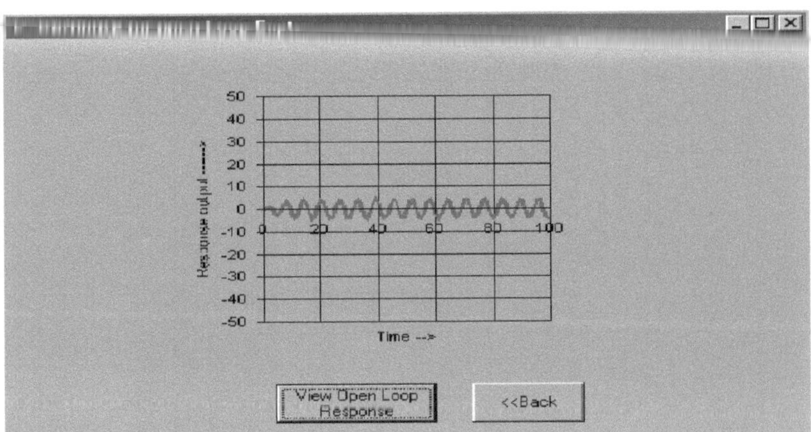

FIGURE 8.19 Response of a second-order system with unity gain and unity time constant and with a damping coefficient of 0.1 for a sine disturbance of unit magnitude and unit frequency.

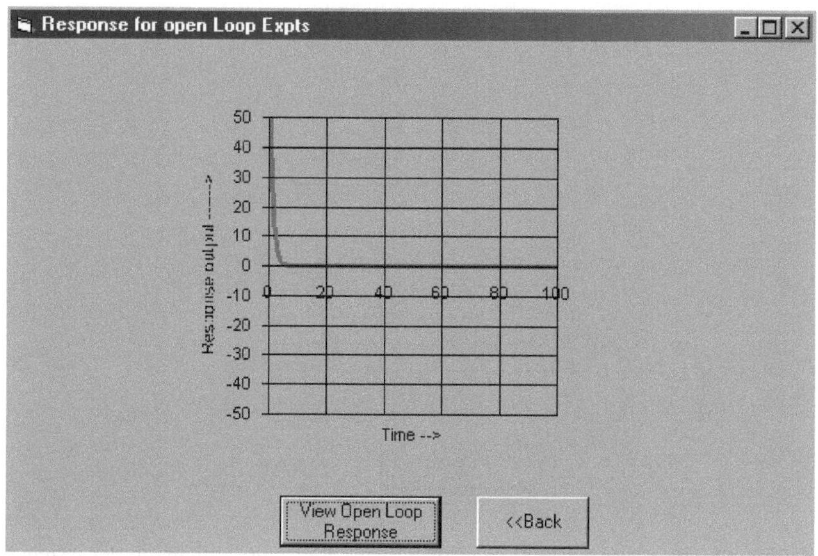

FIGURE 8.20 Response to an impulse disturbance to a second-order system.

8.2.4 First-Order System with a Dead Lag

The response of a first-order system to a gain of 10 units and time constant of 2 units with a dead lag time of 5 units for a unit disturbance is presented in Figure 8.21.

The response to a unity ramp for the same process as described above is presented in Figure 8.22.

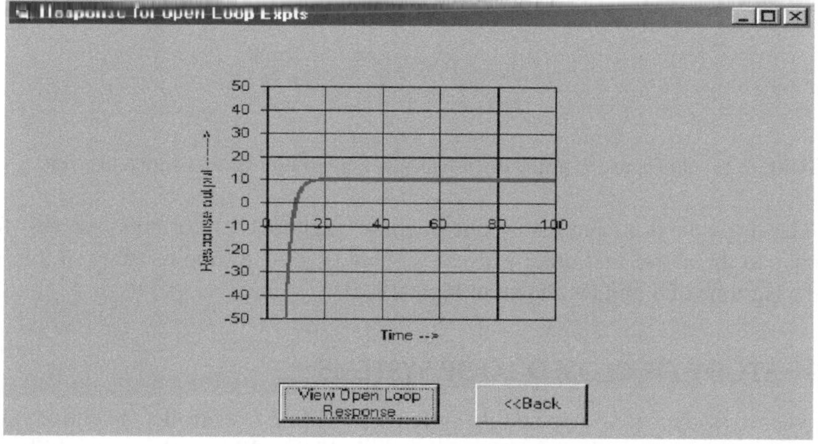

FIGURE 8.21 Response to a unit step disturbance of a first order of gain 10 units, time constant of 2 units, and lag time of 10 units.

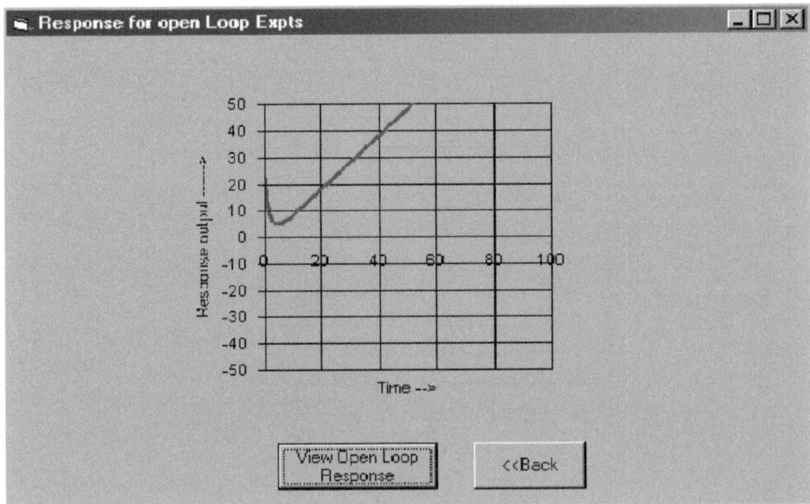

FIGURE 8.22 Response to a unity ramp disturbance for a unity gain first-order system with a time constant of 2 units and a lag time of 5 units.

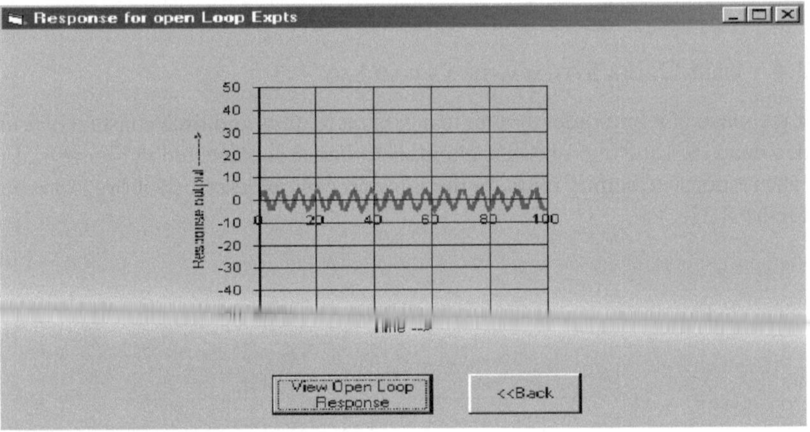

FIGURE 8.23 Response to a sine disturbance to a first-order system with a lag time.

The response of a sine disturbance with a magnitude of 10 units and unit frequency to the above first-order process with unity gain, a time constant of 2 units, and a lag time of 5 units is shown in Figure 8.23.

8.3 STUDY OF CLOSED-LOOP SYSTEMS

In order to study closed-loop control systems, including a controller, a control valve, and a transducer, go to the start-up window as in Figure 8.1 and choose the "Closed-Loop Study" button and click it. The closed-loop window will appear as shown in Figure 8.24.

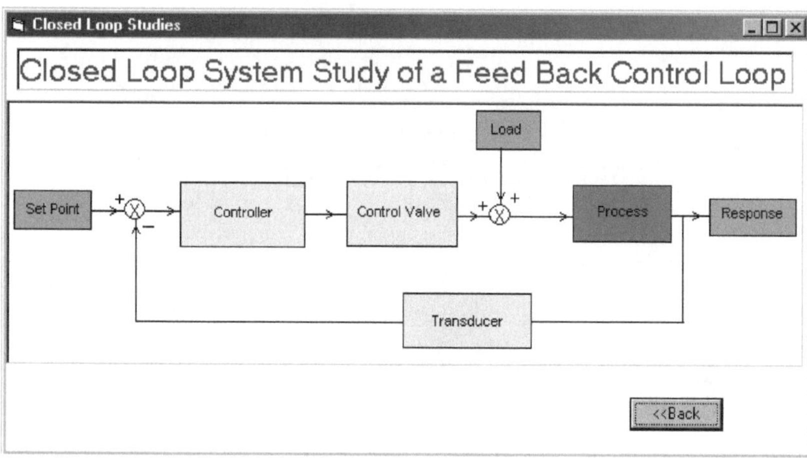

FIGURE 8.24 Window for closed-loop control system experiments.

8.3.1 First-Order Process with a Linear Control Valve

In order to study a first-order system, choose the "Process" button and select any of the four processes as shown in the window appearing in Figure 8.25.

Click on the "First Order" button for studies in the closed-loop system and enter the process data as shown in Figure 8.26.

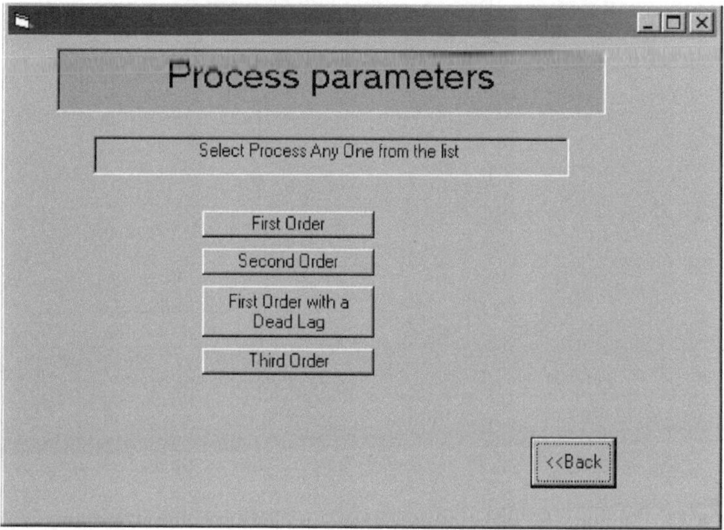

FIGURE 8.25 Selection of process for closed-loop system studies.

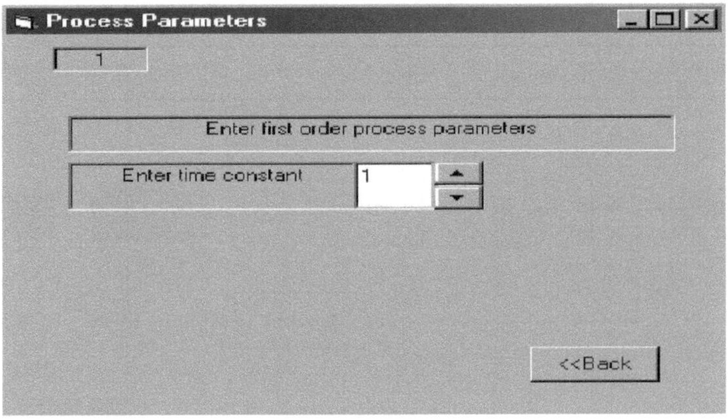

FIGURE 8.26 Window for entering time constant for the process.

Enter unity time constant for the process with an in-built gain of unity. Now return to the window as shown in Figure 8.24 and select "Transducer" from the window as shown in Figure 8.27.

Parameters for a first-order transducer are then entered by clicking the "First-Order Instrument" button and entering unity time constant as shown in Figure 8.28.

Now, return to the window as in Figure 8.24 and select the "Controller" button, and the window as shown in Figure 8.29 will pop up for controller data entry.

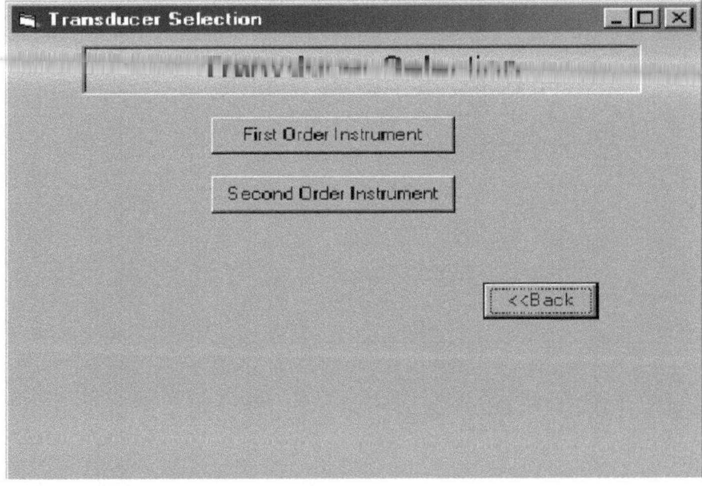

FIGURE 8.27 Window for selecting the transducer element in the closed loop.

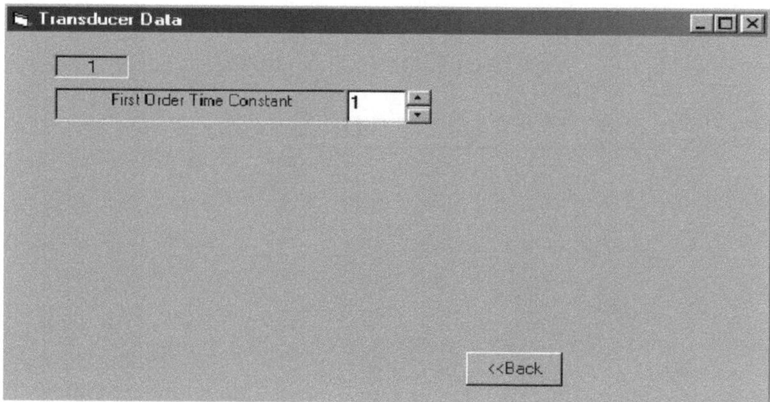

FIGURE 8.28 Window for data entry for the transducer.

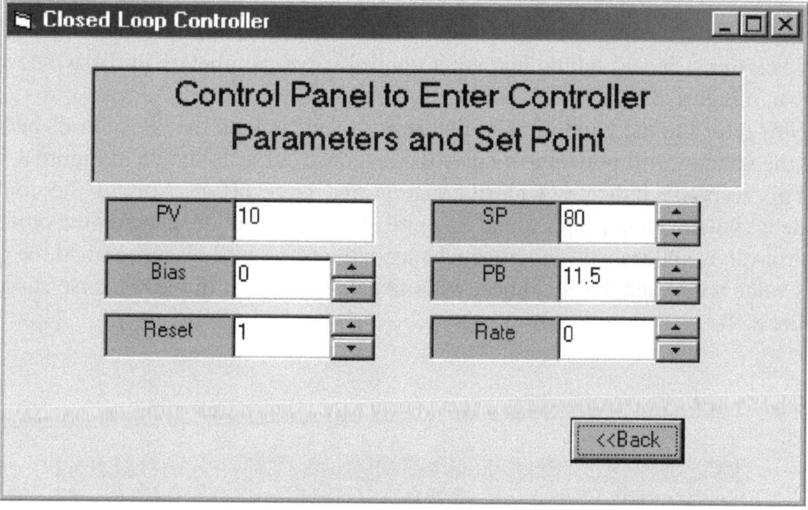

FIGURE 8.29 Data entry window for PID controller.

where process value (PV), set point (SP), and proportional band (PB) are entered as 10, 80, and 11.5, respectively. The bias value, reset (inverse of integration time constant), and rate (derivative time constant) of the controller are set as zeroes. This will make the controller the proportional (P) controller where the gain of the controller is $K_c = 100/\text{PB}$. After entering these data, return to the window as in Figure 8.24 and click the "Control Valve" button, which will display the window for selecting the control valves listed as shown in Figure 8.30.

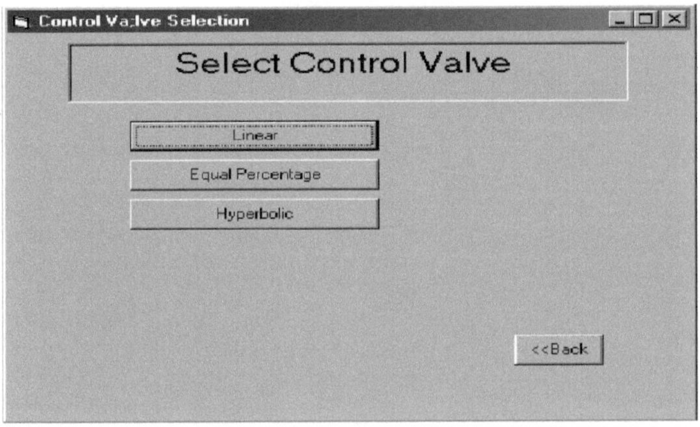

FIGURE 8.30 Window for selection of control valve for closed-loop studies.

Click the "Linear" button and enter control valve parameters in the window as shown in Figure 8.31.

Now return to the window as in Figure 8.24 and click on the "Response" button, and the window will pop up to display the response curve as shown in Figure 8.32.

This response indicates wide fluctuation and large offset. Now, if the control mode is switched to a proportional–integral (PI) control mode, keeping the process, control valve, and transducer data unchanged, the offset will disappear, and the process value will reach the SP almost without delay as shown in the response curve in Figure 8.33.

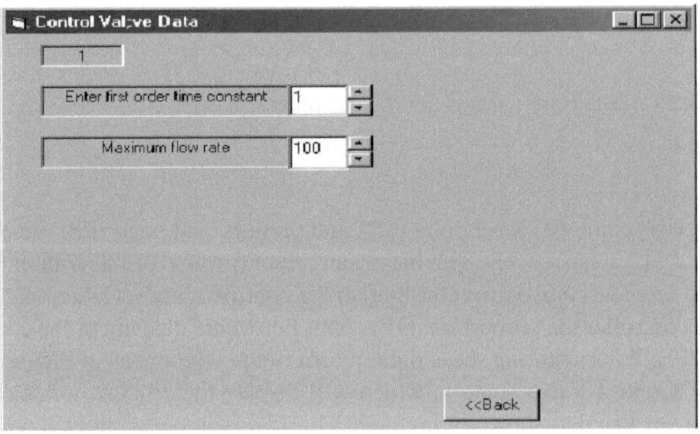

FIGURE 8.31 Window for entering control valve data.

(a)

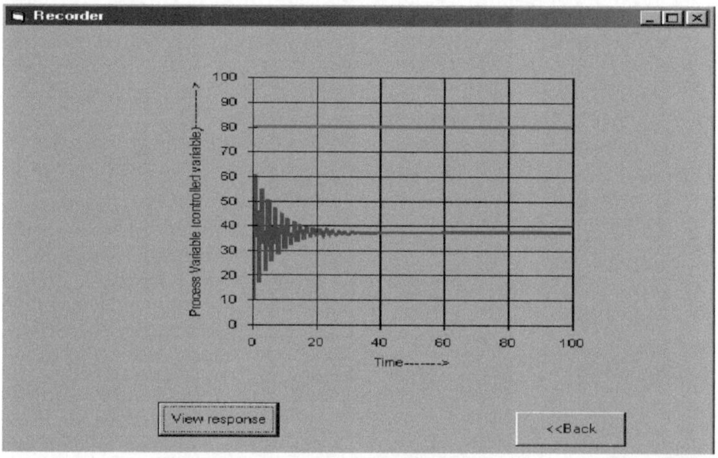

(b)

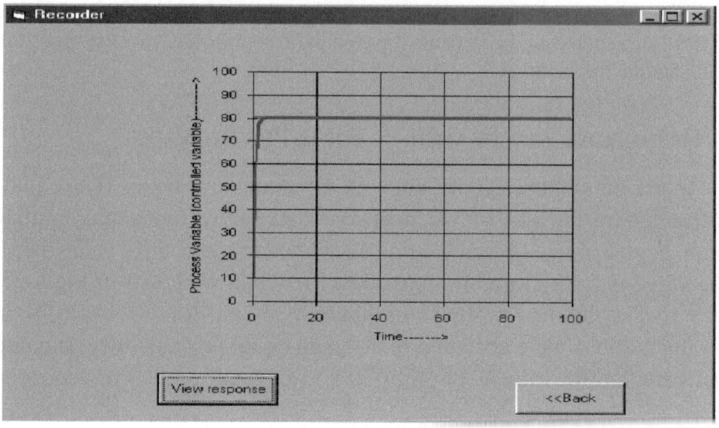

FIGURE 8.32 (a) Window showing the response of the proportional controlled system with a first-order process, first-order transducer, and a first-order linear control valve. (b) Window showing the response of the proportional–integral controlled system with a first-order process, first-order transducer, and a first-order linear control valve.

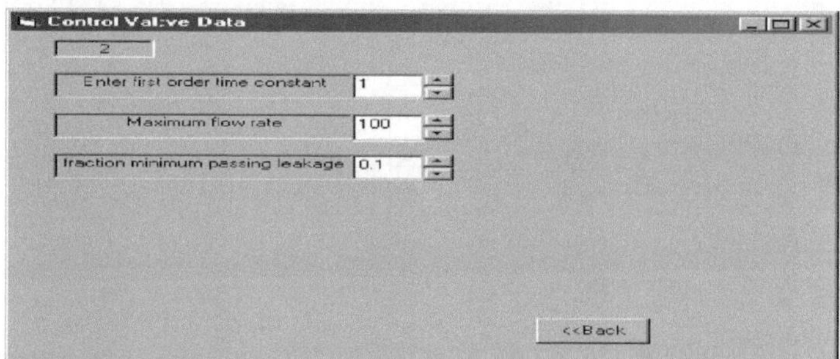

FIGURE 8.33 Window for data entered for the equal percent valve.

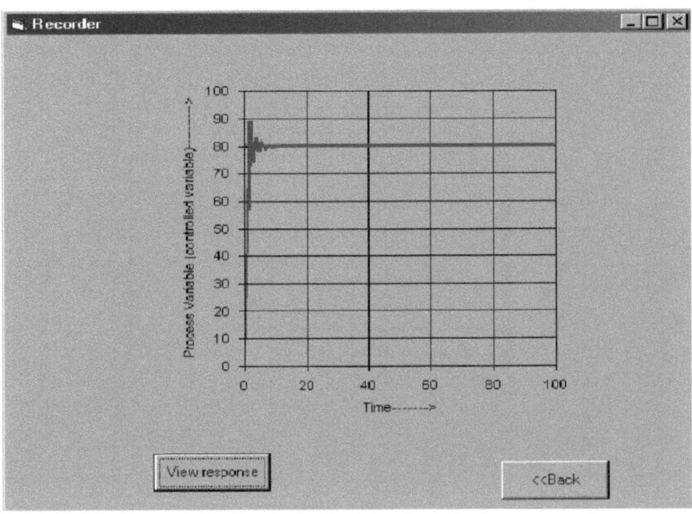

FIGURE 8.34 Response of the PI control system with a first-order process, first-order transducer, and an equal percent valve.

8.3.2 First-Order Process with an Equal Percent Valve

Further experiments can be carried out with different valve types. If an equal percent valve is selected having a first-order time constant of unity and a maximum flow rate of 100 units with leakage flow of 0.1 unit as entered in the control value data window when the valve is fully closed as entered in the window shown in Figure 8.33, the response curve is obtained as shown in Figure 8.34.

Slight fluctuations were observed in using an equal percent valve as compared to that of a linear-control valve.

8.3.3 First-Order Process with a Hyperbolic Valve

For the same closed-loop configuration, the response of a hyperbolic valve with the following data as shown in Figure 8.35 are obtained in Figure 8.36.

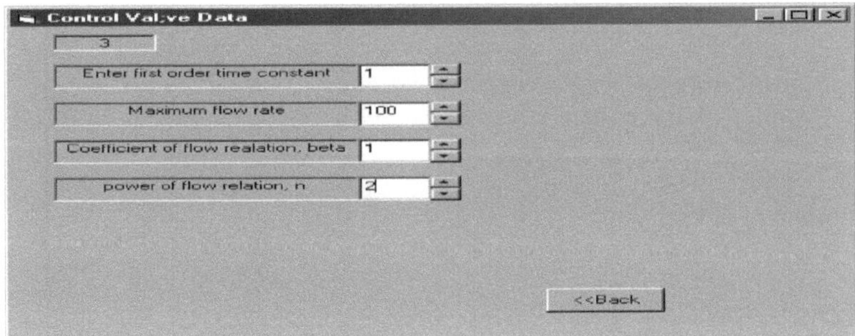

FIGURE 8.35 Window showing entries of hyperbolic control valve data.

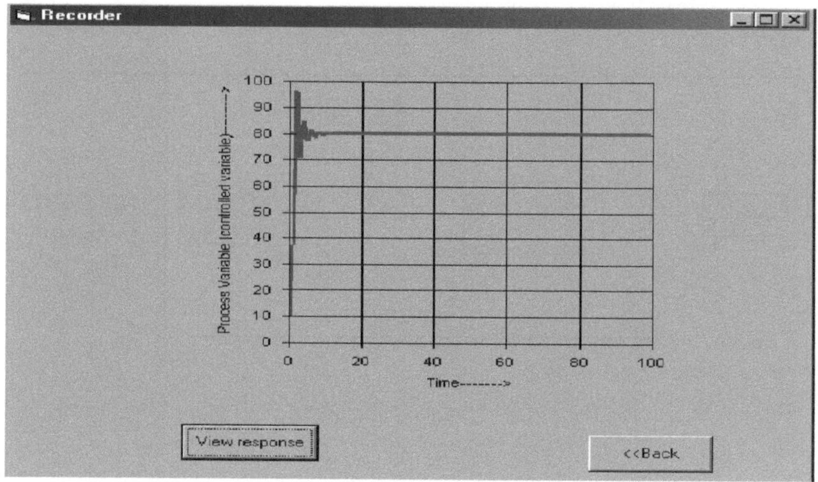

FIGURE 8.36 Response of the closed-loop control for the same control configuration but with a hyperbolic control valve.

This indicates slightly more fluctuation as compared to the equal percent valve observed in the performance of a hyperbolic valve. However, several responses can be studied using different control valve types and parameters.

8.4 TUNING PROBLEMS

8.4.1 Ziegler–Nichols Tuning

Now return to the start-up window as shown in Figure 8.1 and click the "Tuning Problems" button, which will pop up the selection window for either of the Ziegler–Nichols and Cohen–Coon methods as shown in Figure 8.37.

Click the "Ziegler–Nichols Method" button and obtain the pop up window as shown in Figure 8.38.

In this figure, the control valve and the transducer have been omitted to simplify the tuning procedure. Click either the "Process" or "Controller" button to enter the

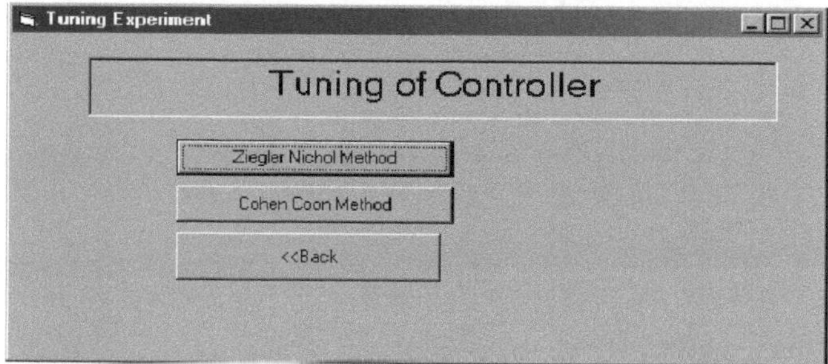

FIGURE 8.37 Window for selecting tuning methods.

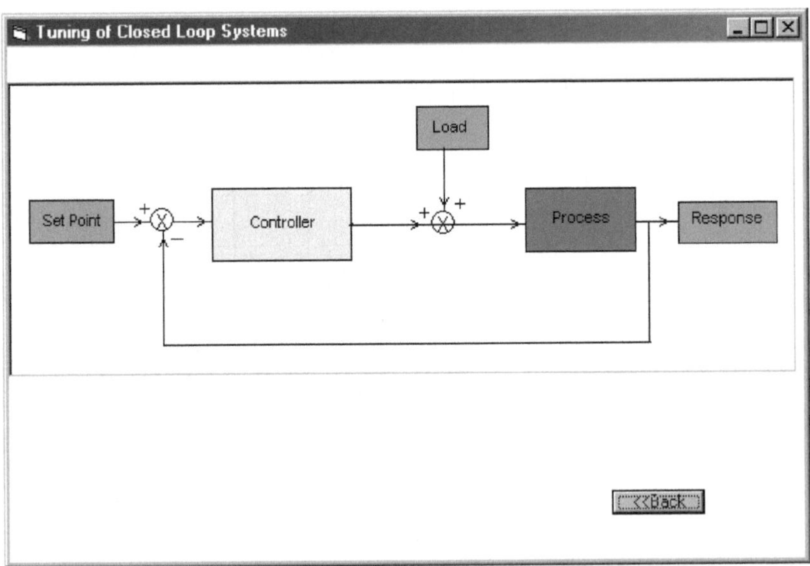

FIGURE 8.38 Window showing the closed-loop control system for Ziegler–Nichols tuning.

necessary process data and controller data in the proportional mode. This is shown in Figure 8.39 where a third-order process with an exponential lag has been taken for studying the tuning. The controller is in proportional control mode, and the corresponding response curve is obtained as in Figure 8.40.

The above response curve shows a stable performance like an underdamped response curve. The red horizontal line indicates the SP, and the blue fluctuating line shows the process value (controlled variable). However, according to the Ziegler–Nichols method of tuning, the proportional band of the controller has to be varied in the proportional-control mode to get a response curve like a sine function. Hence, let us reduce the PB value to 1.02 (so K_c will be high for fluctuation), and a response is observed as shown in Figure 8.41. Increasing above this PB value, the sine curve

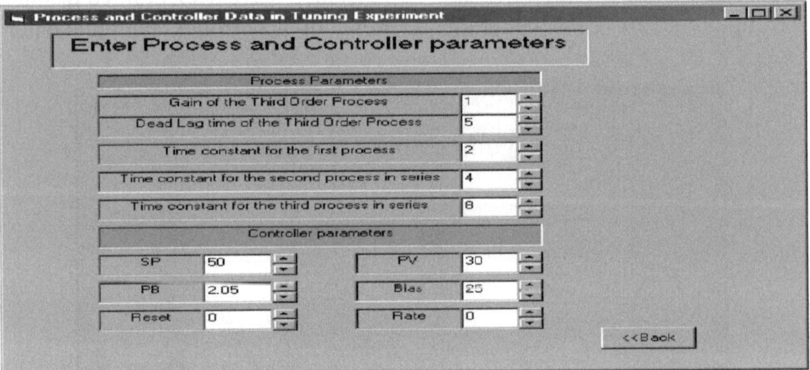

FIGURE 8.39 Process and controller data already in proportional-control mode.

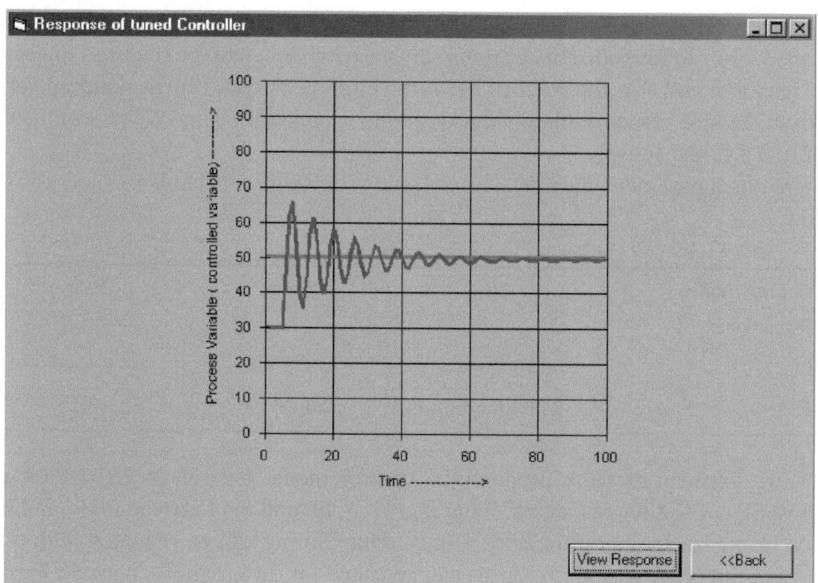

FIGURE 8.40 Closed-loop response curve in the proportional-control mode for the process and controller parameters shown in Figure 8.39.

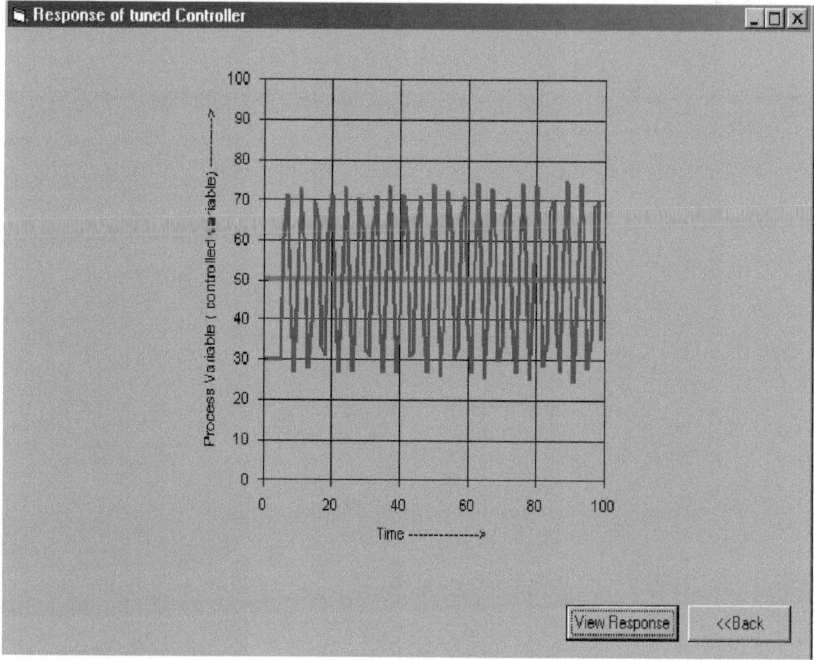

FIGURE 8.41 Response of the proportional controller at *PB* = 1.02 in proportional mode indicating undamped or sine oscillations.

may be distorted either as the unstable response with increasing amplitude, underdamped, or overdamped. This trial-and-error experiment may be required to get this sine response and note the value of PB as the ultimate PB_u and corresponding period P_u from the sine response obtained above. The values of PB_u and P_u are obtained as 1.02 and 0.1, respectively.

The tuned parameters are determined using the Ziegler–Nichols method as listed below:

Control mode	*PB*	Reset	Rate
Proportional	2 × 1.02 = 2.04	————	————
PI	1.02/0.45 = 2.27	1.2/0.1 = 12	————
PID	1.02/0.6 = 1.77	2/0.1 = 20	0.0125

Now, consider first the proportional-control mode and select $PB = 2.04$ and change SPs to test for responses. Figures 8.42, 8.43, and 8.44 are the P, PI, and PID mode responses based on the above tuned data. Tuning can be repeated with other values of the process, and responses can be tested for different SP changes (Figures 8.42 through 8.44).

Better control responses are, in fact, found to be at higher values of PB, reset, and derivative time obtained by tuning. For changes in process parameters, retuning has to be carried out. For more accurate tuning, many more trials may be made by this method. This is left as an exercise for the students.

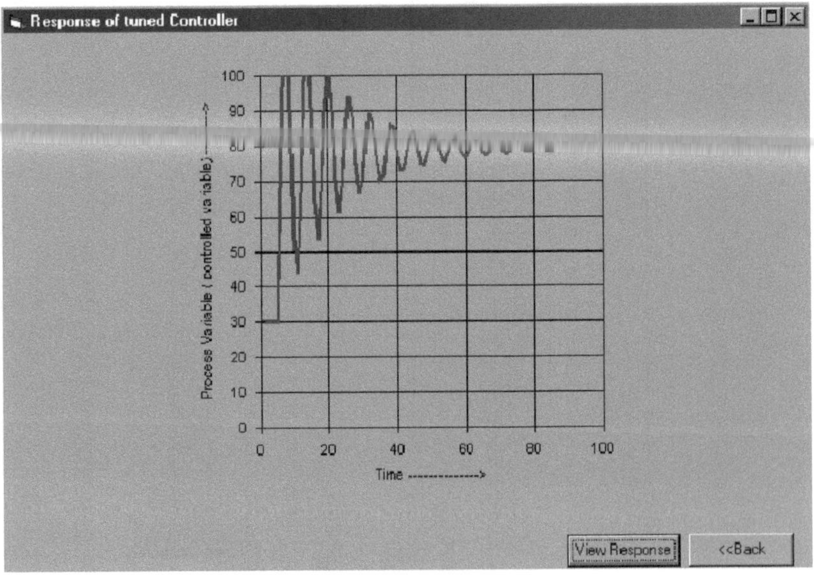

FIGURE 8.42 Proportional control mode while the set point is changed from 30 to 80 at $PB = 2.04$.

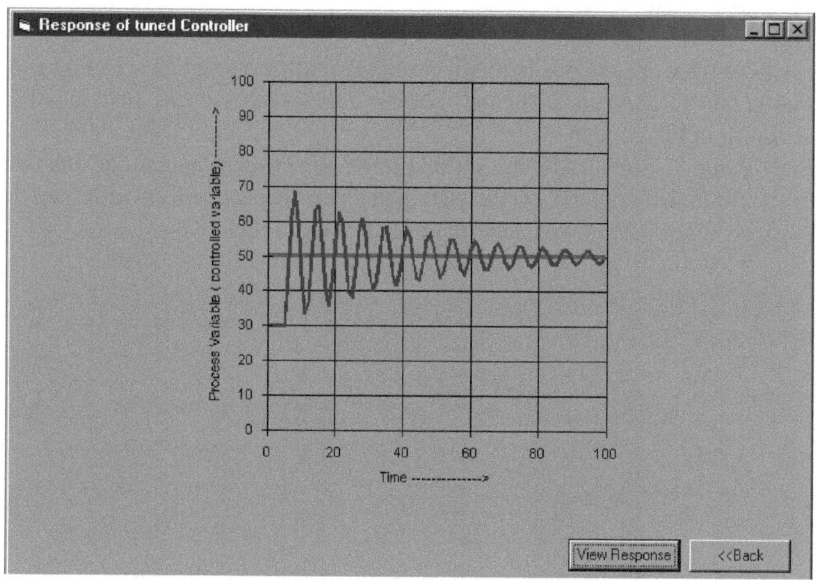

FIGURE 8.43 Proportional–integral–control mode while the set point is changed from 30 to 50 at PB = 2.27 and reset = 12.

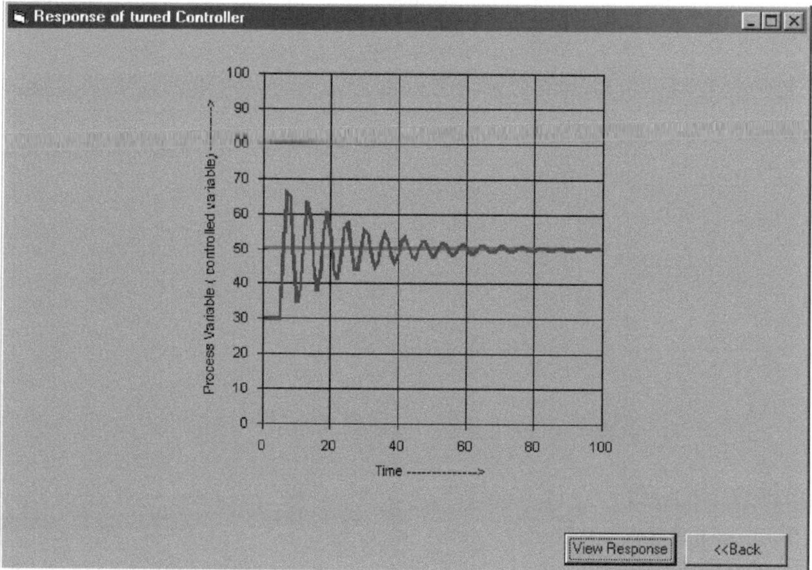

FIGURE 8.44 Proportional–integral–derivative control mode while the set point is changed from 30 to 50 at PB = 1.77, reset = 20, and rate = 0.0125.

8.4.2 Cohen–Coon Tuning

To carry out the Cohen–Coon method of tuning, return to the window in Figure 8.37 and select the "Cohen–Coon Method" button. The window for this method will pop up as shown in Figure 8.45.

Enter a step disturbance by the vertical scroll bar in small magnitude and obtain the response (reaction) curve on the right and determine the process gain, lag time, and process time constant and enter them as shown in Figure 8.46.

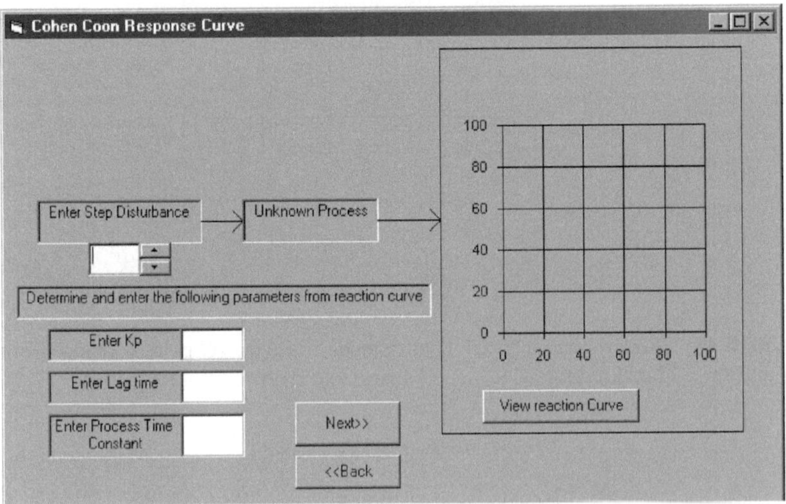

FIGURE 8.45 Window for the Cohen–Coon method of tuning experiment.

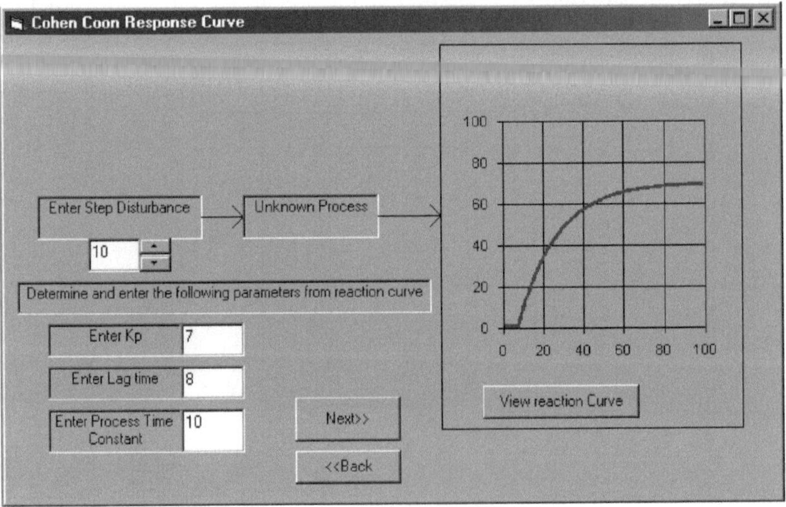

FIGURE 8.46 Window showing the reaction curve resulting from the step disturbance of 10 units entered.

From the observation of the response curve K_p, lag time and process time constants are determined as 7, 8, and 10, respectively. These are entered as shown and the "Next" button is pressed to see the tuned parameters according to the Cohen–Coon table based on the values of K_p, lag time, and time constant determined. This is displayed in the window shown in Figure 8.47.

The tuned parameters are next entered in the controller, and the response is visualized for PI and PID modes as shown in Figures 8.48 and 8.49, respectively.

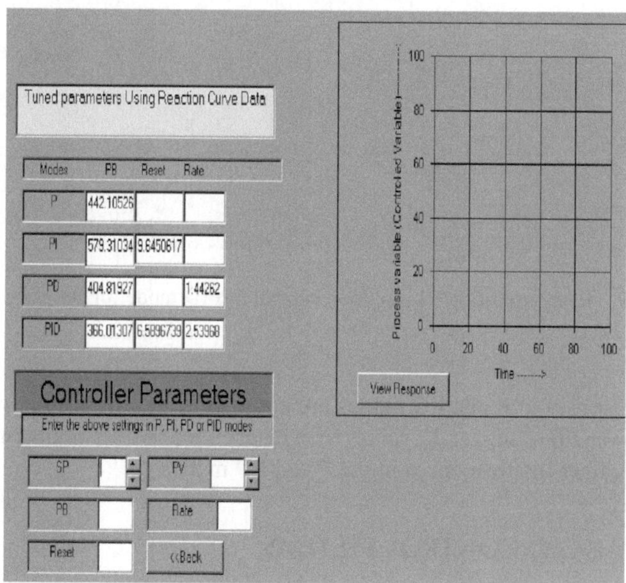

FIGURE 8.47 Window showing the Cohen–Coon tuned parameters.

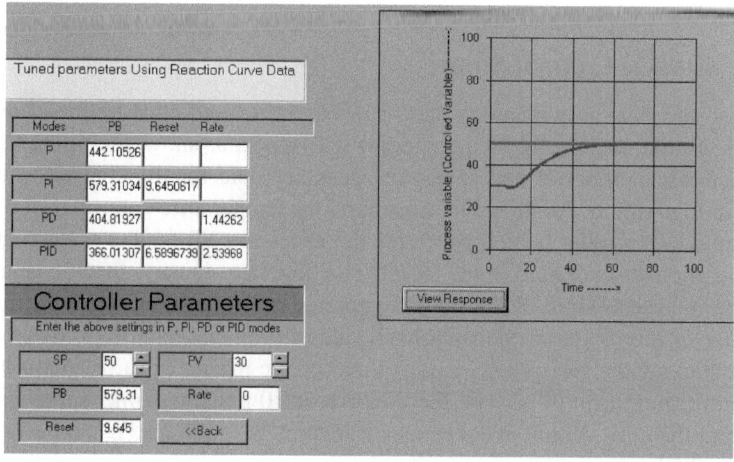

FIGURE 8.48 Response of the closed-loop control in the PI control mode for set point change from 30 to 50.

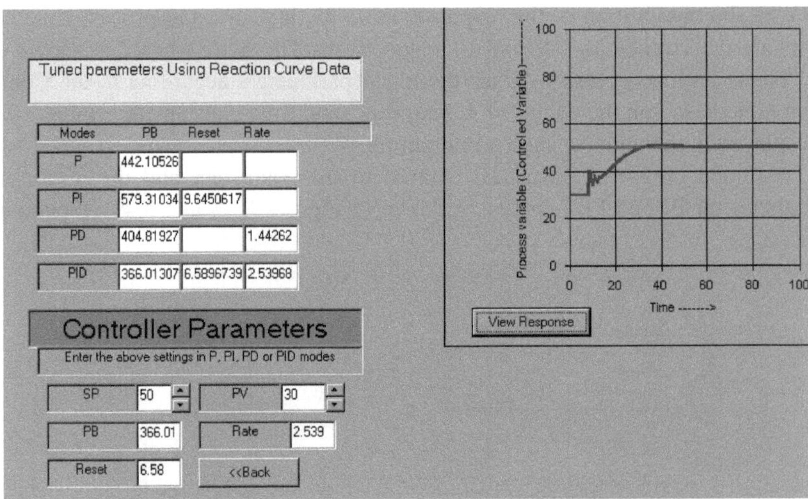

FIGURE 8.49 Response of the closed-loop control in PID mode for set point change from 30 to 50.

However, responses in P and PD modes are not satisfactory, indicating wide offsets. However, lower *PB* or higher K_c values than determined by this tuning method need to be set for offset-free responses in the P and PD modes.

8.5 ADVANCED CONTROL SYSTEMS

In this session of experiments, three advanced control strategies have been included for experimental studies. In these schemes, a temperature-control system in a tank is the process that has been chosen for control.

8.5.1 Cascade Control Study

Return to the start-up window as shown in Figure 8.1 and click on the "Advanced Control" button, and the following window will appear as shown in Figure 8.50.

Select cascade control and enter the necessary data for the control scheme as shown in Figure 8.51. As shown in this figure, the temperature controller of the tank is the master controller, which generates the necessary SP for the slave controller, which manipulates hot fluid in the jacket. The flow rate, density, specific heat of the influent, and the volume of the tank are entered on the left-hand side of the form and those of the jacket on the right-hand side of the tank. The response is visualized by clicking the "Response" button either on the master or slave controller. The inlet temperature is at 30°C, and the flow rate is 10 units per time, specific heat is unity, and the tank volume is taken as constant at 10 units. The response is shown in Figure 8.52.

The students can make changes in the parameters of the process and the controllers as well for experiments and visualize the responses.

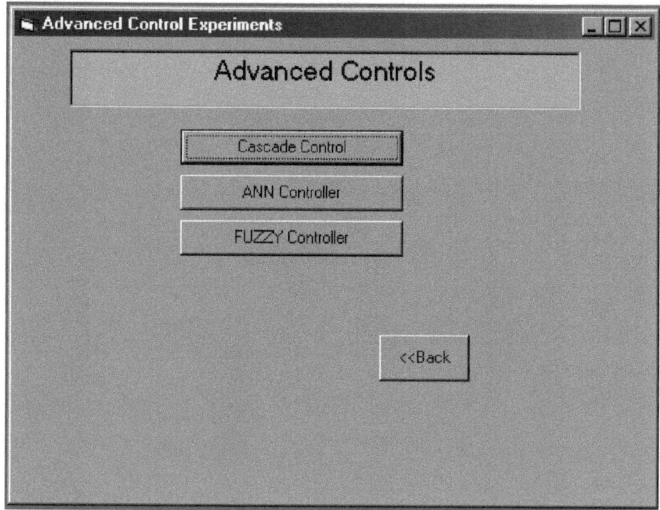

FIGURE 8.50 Window for selecting advanced control schemes.

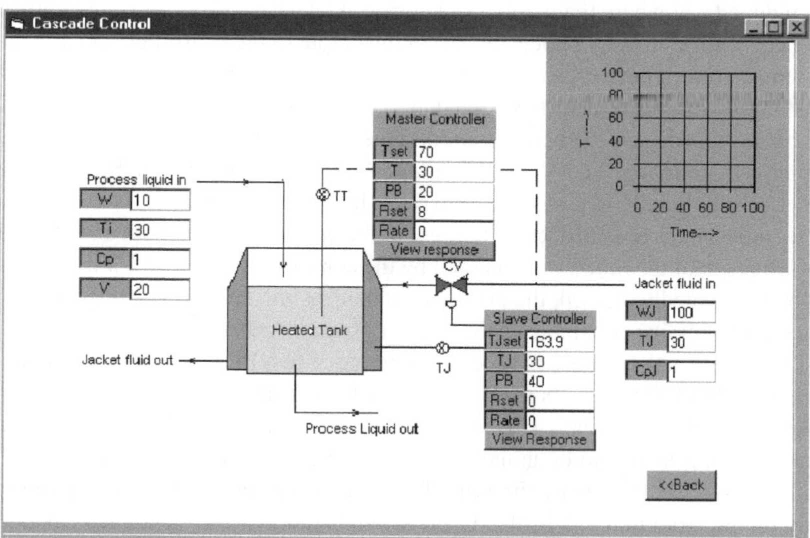

FIGURE 8.51 Cascade control of tank temperature entering the process and controllers parameters.

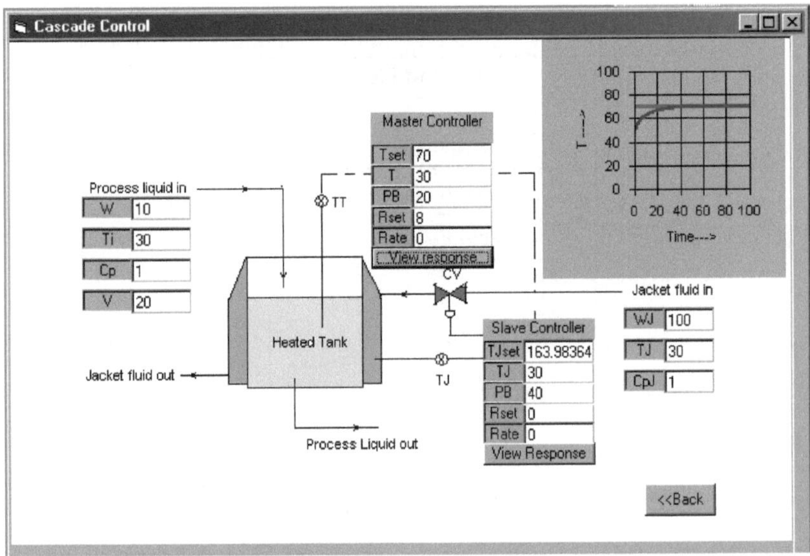

FIGURE 8.52 Response of the cascade control of a tank temperature.

8.5.2 Artificial Neural Network Control

To begin, let us return back to the window as shown in Figure 8.50 and click on the "ANN Controller" button. The previous tank temperature control is repeated with a ANN (artificial neural network) controller, and the response is viewed. The ANN controller has two input nodes, consisting of the SP (T_{set}) and the process temperature *T*, a hidden layer where the number of nodes can be varied up to seven, and an outer layer containing an output node delivering the output of the controller, which changes the temperature, which is then compared with the T_{set}, and the error is evaluated. The error is then minimized to make *T* equals to T_{set}, and the weights are changed in the back propagation method. This continues indefinitely. You can visualize the response by clicking the "View Response" button for a particular SP selected as shown in Figure 8.53. The response curve is shown in Figure 8.54.

The response is not satisfactory because of the presence of overshoot and offset. But this is a result of the initial weights used by the controller but as the "View Response" button is repeatedly pressed, the overshoot and offset will vanish. The response is displayed after pressing once more on the "View Response" button as shown in Figure 8.55.

Thus, the ANN response is superb as compared to a PID controller except at the beginning, but later it adapts itself quickly by back propagation, and as a result, the overshoot or offset vanishes. This is also true for other ranges of SP and can be experimented. But it is also to be noted that for high SPs, it is necessary to increase the heating power by increasing the **scale factor**, and for lower SPs by decreasing the **scale factor** from unity. It is also to be remembered that the scale factor is one of the process parameters and not the controller parameter. For example, if the SP is to be changed from 30 to 90, the scale factor is made 2 and ANN response will achieve the SP without overshoot and offset as shown in Figure 8.56.

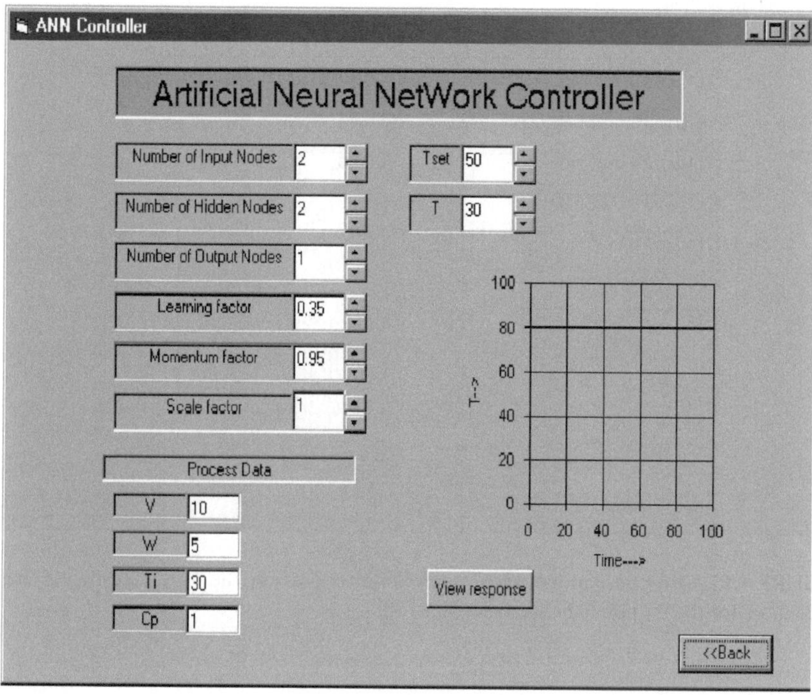

FIGURE 8.53 Window for data entry for the process and ANN controllers.

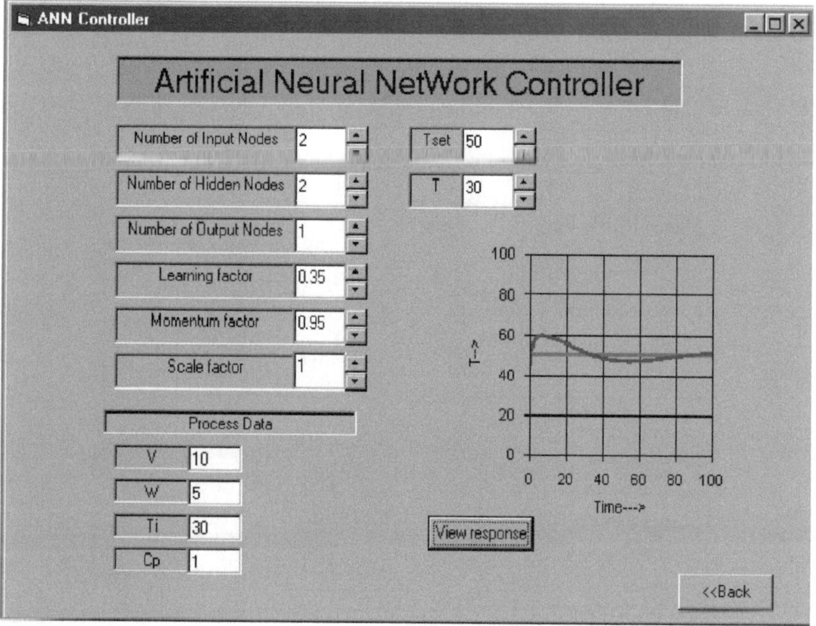

FIGURE 8.54 Window showing the initial view of the response of the ANN controller.

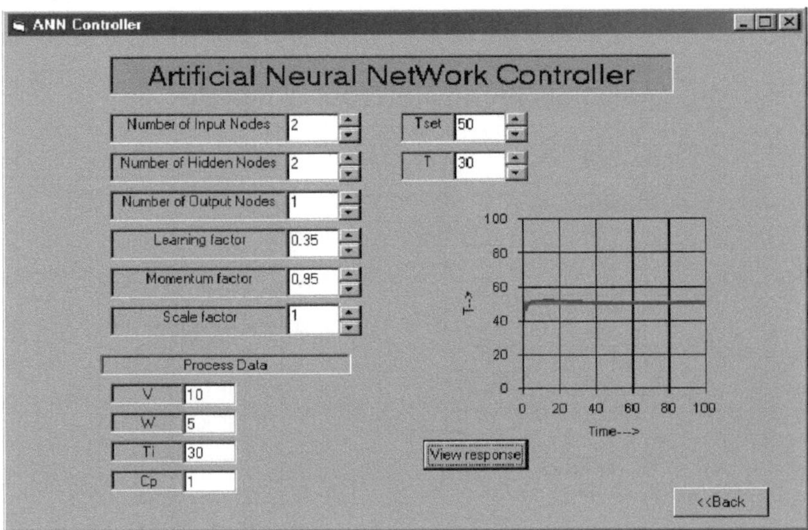

FIGURE 8.55 Response of the ANN controller after pressing the "View Response" button once more for the set point change from 30 to 50.

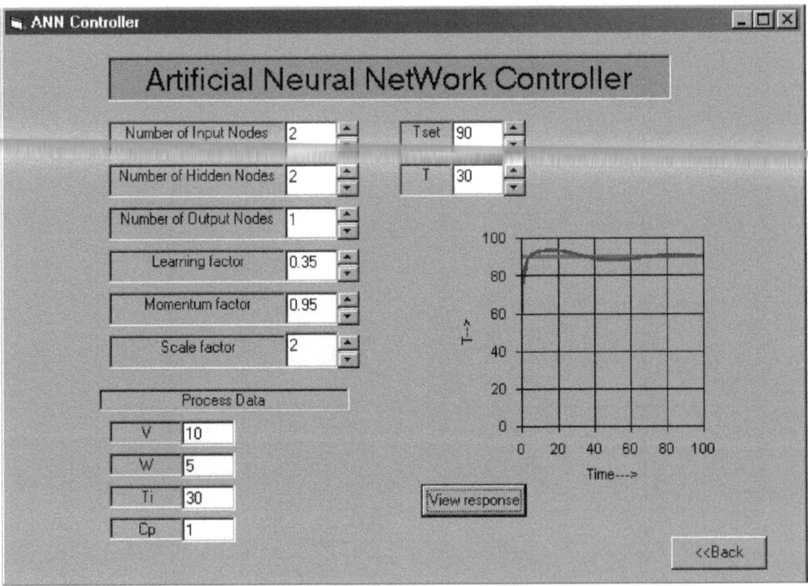

FIGURE 8.56 Response for the ANN controller for set point change from 30 to 90 where the scale factor for heating rate is doubled.

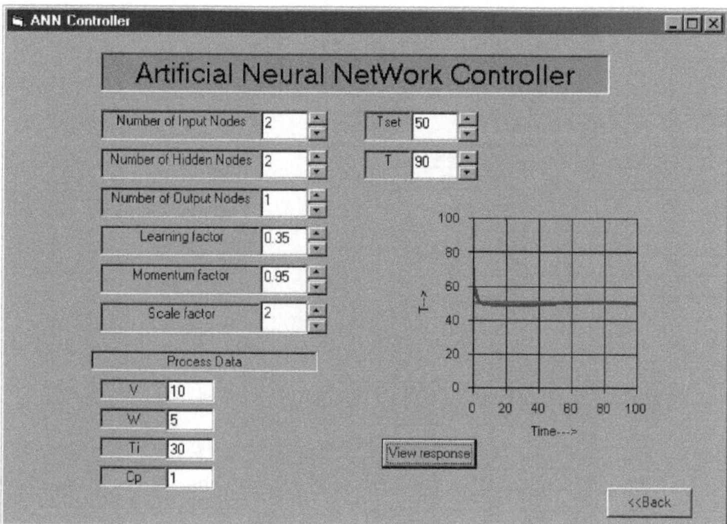

FIGURE 8.57 Response of ANN controller for change of set point from 90 to 50.

Similarly for the changing of SP from 90 to 50, the response is shown in Figure 8.57. Thus, at the same scale factor (2), the SP can be achieved from 30 to 100 or even from 100 down to 30. Hence, this also rectifies the design of the process parameters. Further experiments can be carried out by the students.

8.5.3 Fuzzy Logic Control

In order to study fuzzy logic, return to the window as shown in Figure 8.50 and click on the "Fuzzy Control" button. The window will pop up and will display the default rules as shown in Figure 8.58. The SP and PV can be changed; press the "View Response" button. The default 49 rules are fired for the SP 30 and the PV 30 as shown, where NL, NM, NS, ZR, PS, PM, PL are the abbreviated terms of fuzzy words negative large, negative medium, negative small, zero, positive small, positive medium, and positive large, respectively. The columns correspond to the error, and rows correspond to T_{set}. Thus, at the extreme top left of the rule sets, NL indicates that the rule as T_{set} is NL and error is NL, then the output is NL. Similarly, other rules are read. The user may change the rules by typing capitalized letters NL, NM, NS, ZR, PS, PM, or PL and observe the response. Instead of manually changing the rules, automatic change in the rules can be done by pressing the "Change Rules" button and viewing the corresponding response. The rules will change every time this button is pressed. Find the satisfactory rule set for the corresponding SP. When the SP is changed to 90 from its initial value of 30, the response is not satisfactory as shown in Figure 8.59. But it becomes satisfactory after changing the rules by pressing the "Change" button a few times. These are shown in Figures 8.60 and 8.61.

Students may try for other set point variations and change rules by either manually typing the two-lettered words of rules in the rule sets or pressing the change rule button any number of times to find the best rule sets.

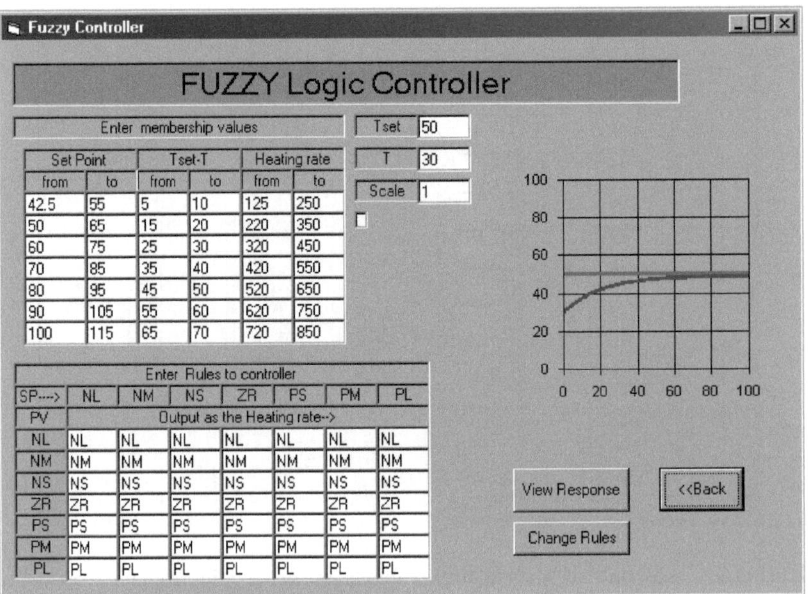

FIGURE 8.58 Window showing default rules and response.

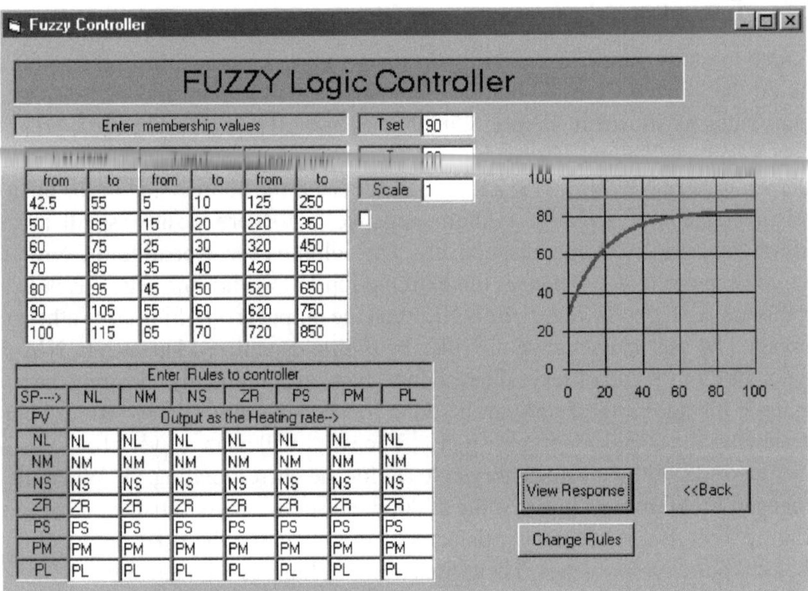

FIGURE 8.59 Window showing unsatisfactory response for change of set point from 30 to 90 using the default rules shown.

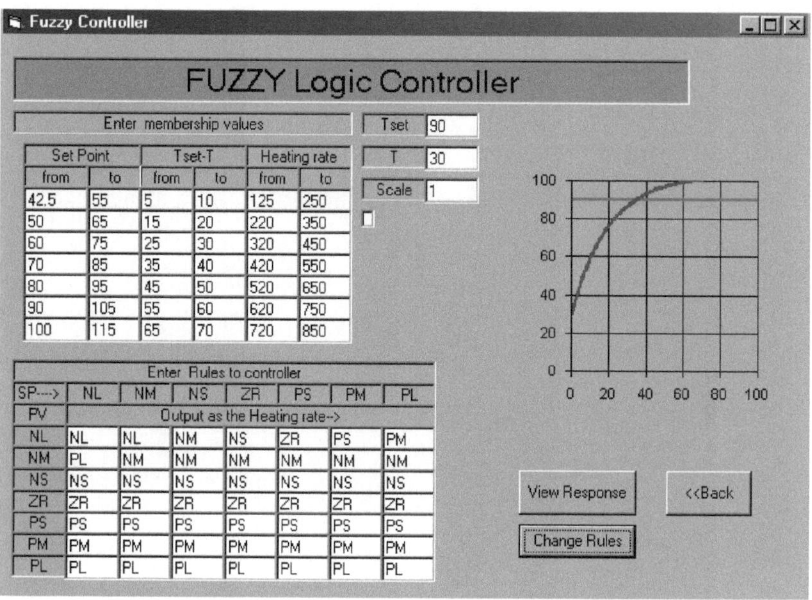

FIGURE 8.60 Rules changed for control, but response is still unsatisfactory for set point from 30 to 90.

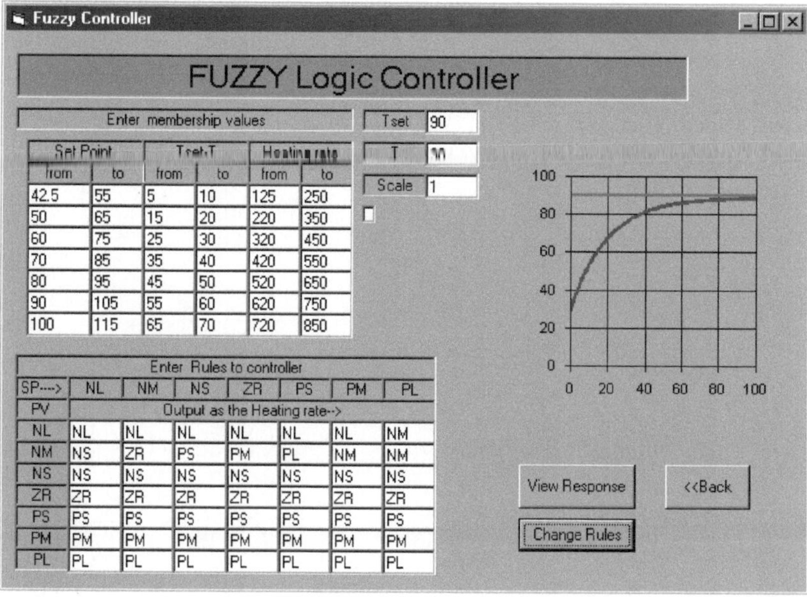

FIGURE 8.61 Rules are found for satisfactory response for change of set point from 30 to 90 by pressing "Change Rules" button a few times.

9 Computer Control

9.1 HARDWARE

A computer-controlled system consists of the process, the controlling elements, and the digital processing units (microprocessors or computers). The transducer, controller, and control valve are the controlling elements or instruments with appropriate communication capabilities. The transducer is loosely understood as the sensing element, which generates an output signal, which can be transmitted, processed, and recorded. Usually, the signal is generated (by the sensor itself or with the help of an additional instrument called the signal conditioner) as electrical voltage or current. This signal will be continuously produced by the transducer and will vary with the change in the process variable it measures. However, it is desirable that when the transducer is not in contact with the process it will send a bias signal indicating its active condition. Thus, the absence of the bias signal will indicate the failure of the instrument, which may be located in a remote control room where it communicates with the controller, and action can be taken. The controller in the computer-controlled system is either a microprocessor, a desktop or laptop computer, or a big work station consisting of a network of computers. Usually, the transducer generates an electrical signal in a continuous manner, which is therefore known as the analog instrument, while the microprocessor controller has to convert this electrical signal into a digital signal with the help of an in built analog-to-digital (A/D) converter. The control logic is programmed and resides in the memory of the controller. For a PID controller, parameters, such as the set point, bias value, proportional band, reset time, rate time, sampling time, modes of action (direct or reverse), etc., are accessed and can be changed as and when required. The program then computes the output signal, which is a digital signal and needs to be converted to an analog signal as the final control element, such as the motorized or solenoid control valve that can be operated only by electrical current or voltage of a desired level. In case the final control element is a pneumatic control valve, an additional instrument known as the electric-to-pressure or -pneumatic (E/P) converter or current-to-pressure (I/P) converter will be required to actuate the control valve. A microprocessor controlled loop is shown in Figure 9.1.

In many processing plants, explosive or flammable fluids are involved, and hence, pneumatic valves in place of electric motor-operated valves are used. So modern chemical plants involve electro–pneumatic–digital signal systems as shown in Figure 9.2.

9.2 SUPERVISORY CONTROL AND DATA ACQUISITION

Digital input and output signals to and from the controller can also be accessed by a computer (laptop, desktop, or a work station) through a two-wire communication

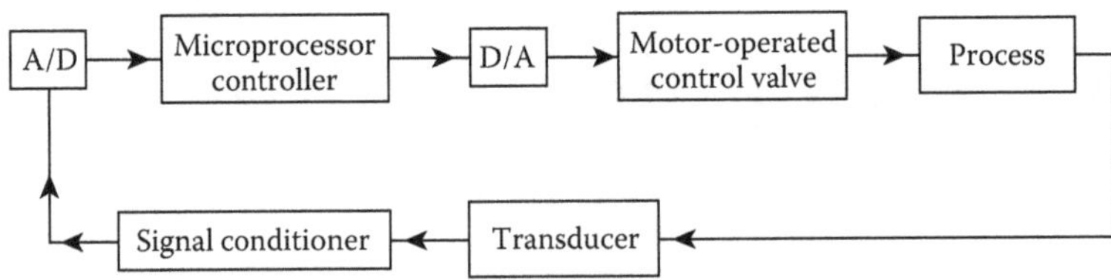

FIGURE 9.1 Microprocessor-controlled system in electro–digital environment.

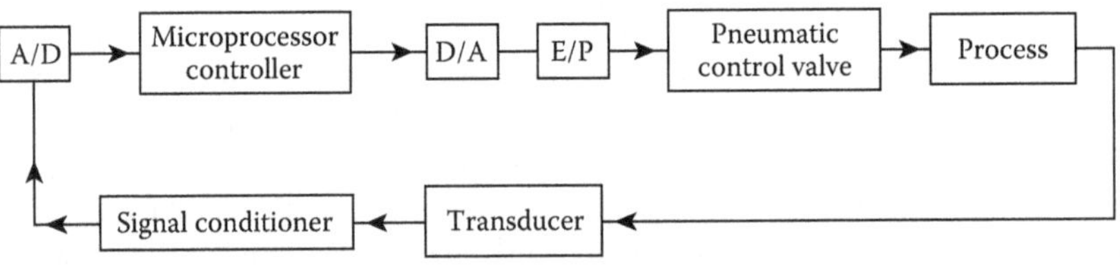

FIGURE 9.2 Microprocessor-controlled loop in electro–pneumatic–digital system.

cable with the help of an Rs485–Rs232 converter connecting the CPU of the computer. However, USB communication ports are the recent facility for communication of digital signals. A microprocessor-controlled system with the communication facility of an external computer is shown in Figure 9.3.

Although the control action can be sufficient with the microprocessor controller for a single-process variable (a single input–single output controller) or multiple process variables and multiple final control elements (multiple input–multiple output controller), the addition of a computer or computers connecting the microprocessor controllers makes large number of processing operations possible. In fact, external computers connected with the control loops can be monitored by a single operator supervising multiple operations. Such a system is elaborated in Figure 9.4.

As shown in Figure 9.4, the microprocessor controllers are usually located in a control room along with the computer as shown in a boundary. The transducers and control valves are located outside the control room, usually at a distance

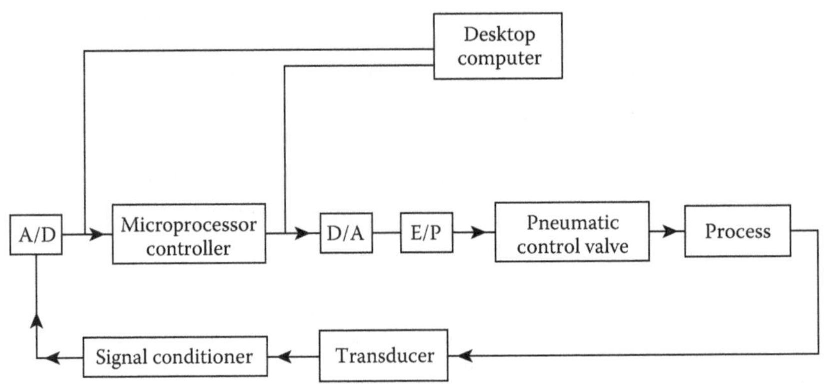

FIGURE 9.3 Microprocessor-controlled loop with communication to an external computer.

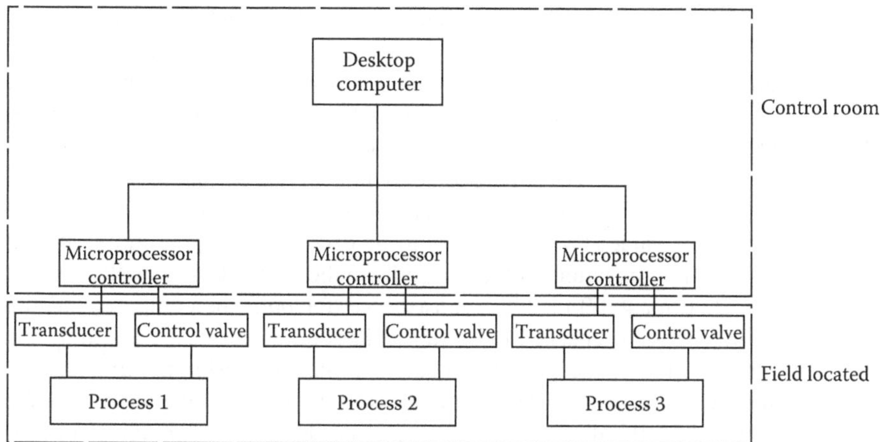

FIGURE 9.4 Multiple control operations with an external computer.

varying from a few meters to kilometers from the processes. Thus, the transducers and control valves are also known as the field instruments. Sometimes, microprocessor controllers may be placed near the processes and are also included as field instruments. The parameters of the microprocessor controllers can be changed from the computer by a single operator who will identify the process and its controller by unique addresses. Depending upon the number of processes and the convenience of operations and supervision, multiple computers may be connected for segregated groups of field controllers. In any modern plant today, a network of external computers located at various places outside a control room are used to gather information on the processes required for day-to-day operations. Redundant communication wires are also connected with this network without affecting the processing controllers. However, the information flow should be unidirectional from the process controllers to the network computers located outside the control room to safeguard the process operations from external interferences. Proper authority must be delegated to certain personnel for changing the operating parameters (set points) and controller parameters (e.g., PB, reset, and rate time, etc.). The control system is also known as the distributed control system (DCS). Usually, the control program applicable for such a distributed system is known as the supervisory control and data acquisition (SCADA). The field controllers will be identified by their unique addresses, which will be scanned by the computer and will decode the communicated signals on the computer screen. The operator will then send and record necessary information and actions for rectifying the process conditions.

9.3 PROTOCOL

When the operator activates the computer, it fetches the address of the field controller and identifies the process information. For this, the computer sends a signal through the communication port, usually by USB port or by RS232 pins via RS485 converter to the controller or through a dedicated communication port provided for this purpose. The controller will reciprocate this signal only when the set of information, such as the BAUD rate, parity (even or odd), specific number of data bits, specific number of stop bits, etc., sent by the computer that exactly matches that accepted by the controller and vice versa. In the RS485 serial communication mode, the controller will send the data signal to the computer. The data signals delivered are synchronized with time, one after another, with a gap of one sampling time usually on the order of 1 millisecond. This data signal containing certain information, such as the process value (PV), which has to be decoded by a program following certain rules or communication protocol specific to the controllers. Usually, a single Baud rate of 9600 (or 4800 or 2400) must be set for both the computer and the controller. Parity must be checked by setting either odd or even as required by the protocol to work. In a typical example of such a protocol transfer, the first character of the message signifies the start of the message followed by the address of the controller connected, the data character string, and the end-of-message character. This message string has to be decoded according to the direction of the protocol supplied by the vendor of the controllers. For a typical controller, an example is given. The message "L2M?*" sent by the computer means the computer is asking for the PV in the controller having address "2". Similarly, the controller sends the message "L2M#25561*", which means the PV value of the controller 2 has a value of +255.6 where the last digit "1" indicates data should have one decimal point. If the computer wants to set the value of the SP to the controller 2, it has to send two messages. At the beginning, the message is "L2S#23051*" where the set point is to be +230.5, and the next message is "L2SI*", which sets the set point to the controller as per the previous message. However, the protocol of such communications are proprietary to the controller manufacturers.

9.4 DIRECT DIGITAL CONTROL

Instead of using any microprocessor controller, a computer (desktop, laptop, or a work station) can be used as a controller by interfacing the transducer input via A/D converter to the CPU and sending the output signal via D/A converter to the control valve. This type of direct control was carried out in the past, but the control system gets badly affected when the computer is out of order or there is a power failure. In an indirect or DCS control, even though computers in the network may be out of service, the control system of the processes will be unaffected as long as the microprocessors are powered. Modern-day controllers are backed by charged batteries or small UPS units. However, small processes or batch processes can be controlled by a direct digital control system where the failure of the control system is not costly.

9.5 LABORATORY CONTROL SYSTEM

A laboratory computer-controlled system is shown in Figure 9.5 where the process is an acid-based neutralization system, and the pH of the effluent has to be controlled. Acid and alkali flow rates are controlled separately by two controllers. A pH transducer and a glass electrode transducer, is used for transducing the signal to the controllers simultaneously.

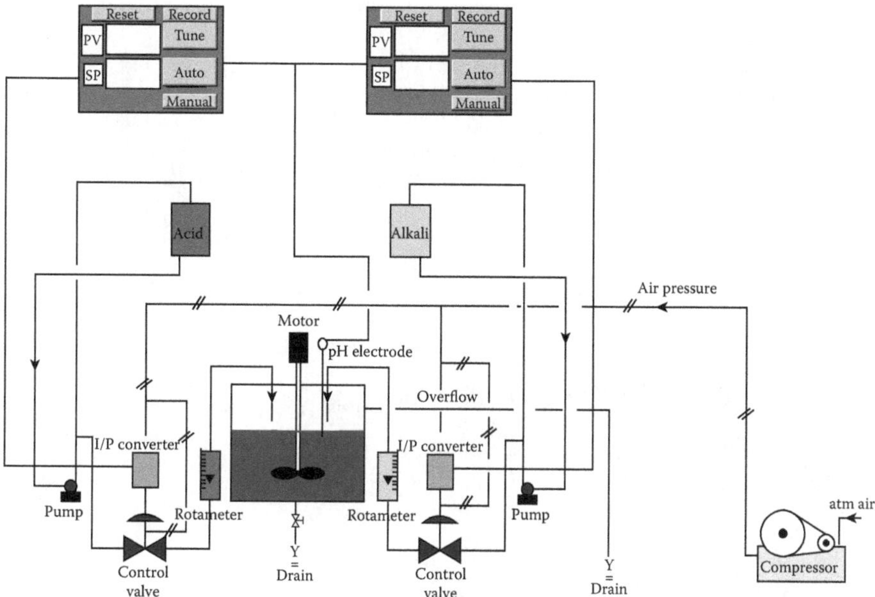

FIGURE 9.5 Laboratory computer-controlled system.

Controllers send output current signals (4–20 mA) to separate I/P converters to actuate two pneumatic control valves. Compressed air from a compressor is used to drive the pneumatic control valves, which are air-to-open type control valves. Variation of the pH causes the output of the controllers; hence, the output pressure signals from the I/P converters vary and actuate the control valves. The controllers can be operated in two modes, such as "auto" and "manual." In the auto mode, the PID program of the microprocessor controllers work according to the PB, bias, reset, and rate time values set through the controller front panels. In order to connect the system with an external desktop computer, a separate two-wire cable is connected with the controllers through the appropriate pins for transmission of data through a RS485 converter to the serial port (RS232) of the CPU of the computer. This type of connection may be useful for a long-distance transmission without any loss of data even up to a distance of about half a kilometer. A computer program has been written for accessing the data from the controllers, and set point and the controller parameters are changed right from the computer keyboard without the need to change from the controller front panel switches. The picture as shown in Figure 9.5 is drawn on the computer screen, which can be used by the operator to access the controllers' information. The actual field connected with the hardware in a typical laboratory is shown in Figure 9.6.

From the computer keyboard, auto and manual control modes can be selected by the operator. The "Tune" button is also available for changing the controller parameters for a tuning experiment. However, when auto is selected, the controllers

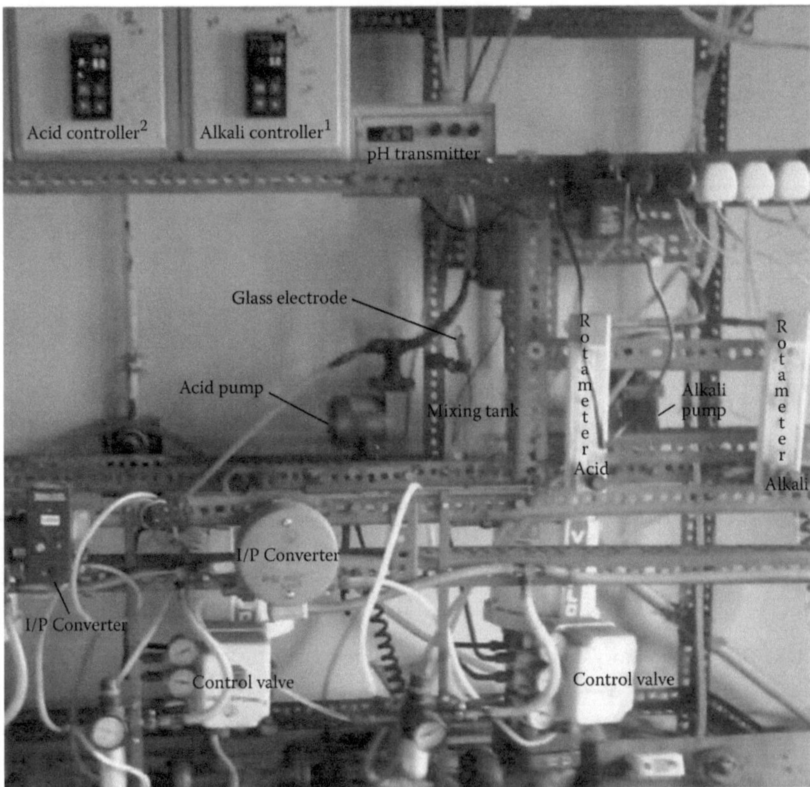

FIGURE 9.6 Laboratory pH control system using microprocessor controllers in author's laboratory.

perform according to the PID program residing in the controllers' memory and will continue the operation even when the computer is off. However, the set points and the controller parameters as last set will be maintained for the operation. This is like a supervisory control and data acquisition working system. In Figure 9.7, the video camera view of the field instruments, which are away from the control room, are shown above the computer monitor.

When manual mode is selected, the controller will bypass the PID program, and the output of it will be exactly the set point entered either from the controller front panel or from the computer keyboard. However, when the computer is on and communicating with the controllers, the parameters of the controllers cannot be changed from the controller front panels as the computer will override the controller's actions. While the "Auto" or "Manual" button is pressed, the current SP and PV values will be displayed on the window as shown in Figure 9.5. However, when the "Tune" button is pressed from the computer screen, the following window as shown in Figure 9.8 will be displayed.

(a)

(b)

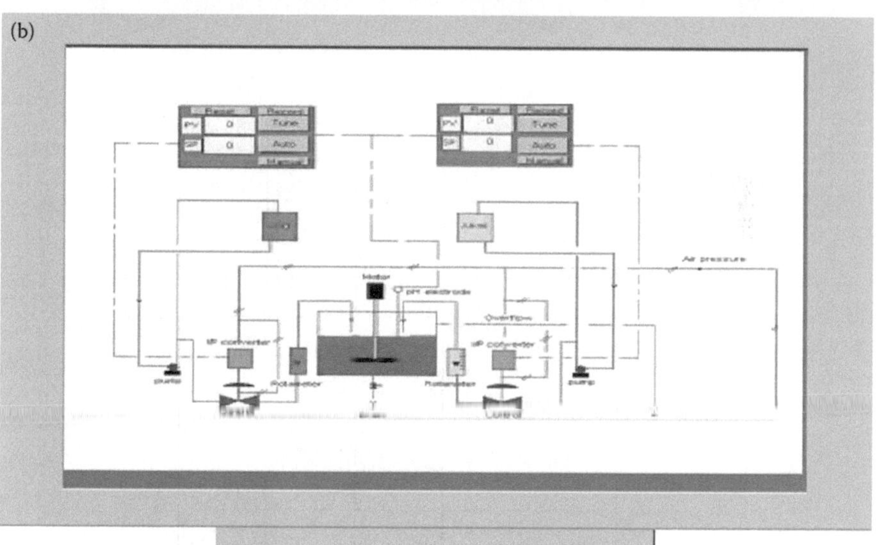

FIGURE 9.7 Supervisory computer control with automatic data acquisition in the laboratory showing (a) video camera view of the field instruments (b) computer screen view for supervisory operation.

After entering the necessary data in the window, control actions can be viewed. Separate programs are written for auto, manual, and tune modes. On clicking the "Recorder" button, the pH and output power of the alkali controller is available as shown in Figure 9.9. There is also a provision for saving the run and printing the graph. Pressing the "Reset" button will start the recording afresh.

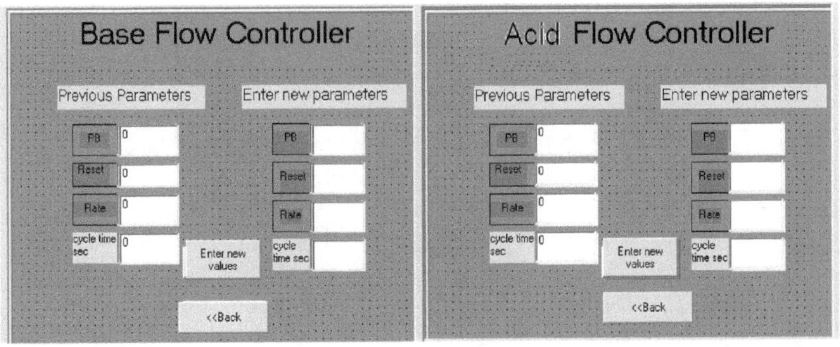

FIGURE 9.8 Data entry window pops up when the "Tune" button is pressed.

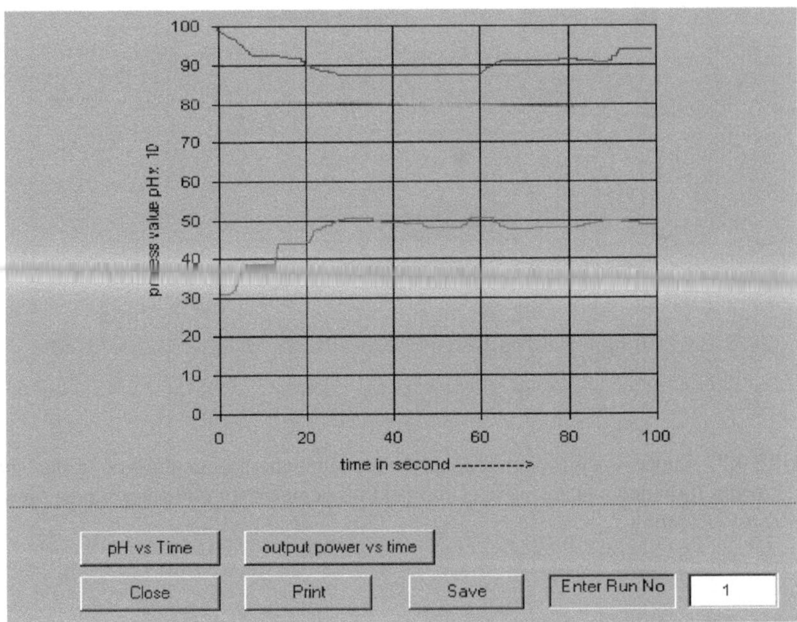

FIGURE 9.9 Recorder view of the run for the experiment showing pH and output power from the alkali controller in the field.

10 Selected Problems and Solutions of GATE Examinations from 2002 to 2012

1. (Q.1.14, 2002)

Closed-loop poles of a stable second-order system could be

(a) Both real and positive
(b) Complex conjugate with positive real parts
(c) Both real and negative
(d) One real positive and another real negative

Solution:

Answer is (c) both real and negative.

Explanation:

Poles are the roots of the characteristic equation of a closed-loop transfer function at controller gain, $K_c = 0$.

For a second-order system, the characteristic equation is

$$\tau^2 s^2 + 2\tau\xi s + 1 + K_c = 0$$

At $K_c = 0$, $\tau^2 s^2 + 2\tau\xi s + 1 = 0$.

Solving the above quadratic equation, we get roots $s = \{-\xi \pm \sqrt{(\xi^2 - 1)}\}/\tau$, which indicates that the roots of the characteristic equation may be real, imaginary, or complex, but having real parts for both the roots are negative.

For any stable system, the real parts of the roots of the characteristic equation should be negative. Hence, the answer is (c).

2. (Q.1.15, 2002)

A first-order system with unity gain and time constant τ is subjected to a sinusoidal input of frequency $\omega = 1/\tau$. The amplitude ratio for this system is

(a) 1
(b) 0.5
(c) $1/\sqrt{2}$
(d) 0.25

Solution:

First-order transfer function is $1/(\tau s + 1)$

So amplitude ratio $= 1/(\sqrt{(\tau^2\omega^2 + 1)} = 1/\sqrt{(\tau^2/\tau^2 + 1)} = 1/\sqrt{2}$

Answer is (c).

3. (Q.2.16, 2002)

The frequency response of a first-order system has a phase shift with lower and upper bounds given by

(a) $[-\infty, \pi/2]$
(b) $[-\pi/2, \pi/2]$
(c) $[-\pi/2, 0]$
(d) $[0, \pi/2]$

Solution:

Phase shift of a first-order system, $\phi = \tan^{-1}(-\omega\tau)$
For lower bound $w = 0$, $\phi = 0$
and for upper bound, $w = +\infty$, $\phi = -\pi/2$
Answer is (c).

4. (Q.CH-18, 2002)

A mercury thermometer can be used to measure body temperature by placing it either in the mouth or in the armpit of a patient. The true body temperature can be taken to be the temperature inside the mouth, which is usually higher than the temperature in the armpit by 0.5 K. Assume that the true body temperature of the patient is 312 K, and the thermometer is initially at 300 K. Also, assume that the thermometer behaves like a first-order system with a time constant of 40 seconds.

(a) Obtain a relationship for the thermometer reading $T(t)$ as a function of time in terms of its initial temperature and body temperature T_B.
(b) How long should the thermometer be placed in the patient's mouth in order to ensure that the error in the measurement is not greater than 0.05%?
(c) Because the body temperature in the armpit is less, the measurement made here using the thermometer is corrected by adding 0.5 K. How long should the thermometer be placed in the armpit in order to ensure that the error in the corrected measurement is not greater than 0.05% of the true body temperature?

Solution:

For the first-order system, we know, $Y(t) = A\{1 - \exp(-t/\tau)\}$
where $Y(t) = T(t) - T(0) = T(t) - 300$
and $A = T_B - T(0) = T_B - 300$
So

(a) $T(t) - 300 = (T_B - 300)\ \{1 - \exp(-t/40)\}$
or $T(t) = 300 + (T_B - 300)\{1 - \exp(-t/40)\}$, where t is in seconds
(b) While the thermometer is in the patient's mouth,
$T_B = 312$ K and $T(t) = (1 - 0.0005) \times 312 = 311.84$ K
So
$311.84 = 300 + (312 - 300) \times \{1 - \exp(-t/40)\}$
or $\exp(-t/40) = 1 - 11.84/12 = 0.013$
or $t = 173.77$ sec.
(c) While the thermometer is in the armpit,
Armpit temperature = 312 – 0.5 = 311.5 K and corrected temperature = $T(t) = (1 - 0.0005) \times 312 = 311.84$ K

thermometer reading +0.5 = 300 + (311.5 – 300) * (1 – exp(–t/40)) + 0.5
or 311.84 = 300 + (311.5 – 300) × (1 – exp(–t/40)) + 0.5
or exp(–t/40) = 1 – (311.84 – 300.5)/11.5 = 0.16/11.5 = 0.013913
or t = 170.10 sec.

5. (Q.CH-19, 2002)

Consider a system of two tanks in series as shown in Figure 10.1.

The level h_2 in tank II is measured and has to be controlled by manipulating the flow rate F_1. It is given that $F_2 = 0.005\ h_1$ and $F_3 = 0.0025\ h_2$, m^3/s, where h is in m. Cross-sectional area of tanks I and II are both equal to 1 m^2.

(a) Determine the transfer function of the process.
(b) Compute the time constants of the process. Is the open-loop process overdamped, underdamped, or critically damped?
(c) If proportional control is used with constant $K_c > 0$, determine the value of K_c for which the closed-loop response becomes critically damped.

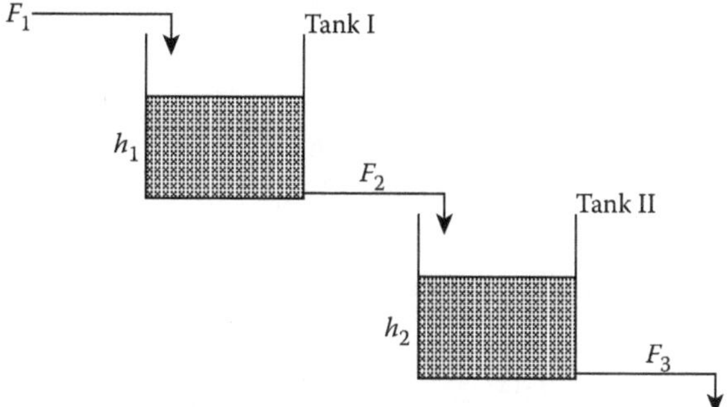

FIGURE 10.1 Two tanks in a series.

Solution:

(a) Because the level h_2 in tank II is measured and has to be controlled by manipulating the flow rate F_1, the transfer function will have to be related with h_2 as the output variable and F_1 as the input variable.

Let us take $X_1(s) = F_1(s) - F_{1s}$
and $Y_1(s) = h_1(s) - h_{1s}$, $X_2(s) = F_2(s) - F_{2s}$ and $Y_2(s) = h_2(s) - h_{2s}$
where F_{1s}, h_{1s}, F_{2s}, and h_{2s} are the steady-state values.
From the first tank

$$X_2(t) = \frac{Y_1(t)}{R_1}$$

and

$$X_1(t) - \frac{Y_1(t)}{R_1} = A_1 \frac{dY_1(t)}{dt}$$

or

$$\frac{Y_1(s)}{X_1(s)} = \frac{R_1}{(\tau_1 s + 1)}$$

and

$$\frac{X_2(s)}{X_1(s)} = \frac{1}{(\tau_1 s + 1)}$$

Similarly, from the second tank

$$\frac{Y_2(s)}{X_2(s)} = \frac{R_2}{(\tau_2 s + 1)}$$

So

$$\frac{Y_2(s)}{X_1(s)} = \frac{R_2}{(\tau_2 s + 1)(\tau_1 s + 1)}$$

Hence, the desired transfer function is

$$\frac{Y_2(s)}{X_1(s)} = \frac{400}{(200s + 1)(400s + 1)}$$

where
$R_1 = 1/0.005 = 200$ and $R_2 = 1/0.0025 = 400$
So $\tau_1 = R_1 A_1 = 1/0.005 = 200$ sec (as $A_1 = A_2 = 1$)
and $\tau_2 = R_2 A_2 = 1/0.0025 = 400$ sec

(b) The denominator of the transfer function evaluated above is

$$200 \times 400s^2 + 600s + 1$$

comparing with the second-order denominator

$$\tau^2 s^2 + 2\tau\xi s + 1$$

we get the second-order time constant of the process

$$\tau = \sqrt{(200 \times 400)} = 282.84 \text{ sec}$$

and damping coefficient, $\xi = 600/(2 \times 282.84) = 1.06$, i.e., the process (open loop) is overdamped.

Time constants of the individual tanks are already determined in part (a) above as 200 and 400 sec, respectively.

(c) When the above process is closed loop with a proportional controller, the characteristic equation becomes

$$200 \times 400s^2 + 600s + (1 + 400\ K_c) = 0$$

or

$$\frac{(8\times 10^4 s^2 + 600s)}{(1+400K_c)} + 1 = 0$$

or

$$\tau = 282.84\sqrt{\{1/(1 + 400K_c)\}}$$

and

$$2\tau\xi = 600/(1 + 400K_c).$$

To make the closed loop critically damped ($\xi = 1$) K_c is obtained by solving

$$2 \times 282.84\sqrt{\{1/(1 + 400K_c)\}} = 600/(1 + 400K_c)$$

or

$$K_c = 3.125 \times 10^{-4}$$

6. (Q.22, 2003)

Match the measured process variables with the list of the measuring devices given below,

Measured Process Variables	Measuring Devices
(P) Temperature	(1) Bourdon tube element
(Q) Pressure	(2) Orifice plates
(R) Flow	(3) Infrared analyzer
(S) Liquid level	(4) Displacer devices
(T) Composition	(5) Pyrometer

(a) P-5, Q-1, R-2, S-4, T-3
(b) P-3, Q-1, R-4, S-2, T-5
(c) P-1, Q-3, R-4, S-2, T-5
(d) P-3, Q-1, R-2, S-4, T-5

Answer:

The correct answer is (a) P-5, Q-1, R-2, S-4, T-3.

7. (Q.23, 2003)

Suppose that the gain, time constant, and dead time of a process with the following transfer function

$$G_c(s) = 10\ \exp(-0.1s)/(0.5s + 1)$$

are known with a possible error of ±20% of their values. The largest permissible gain K_c of a proportional controller needs to be calculated by taking the values of process gain, time constant, and dead time as

(a) 8, 0.6, 0.08
(b) 12, 0.6, 0.12
(c) 8, 0.6, 0.12
(d) 12, 0.4, 0.08

Solution:

Possible values of the process parameters within ±20% are obtained as

$$\text{process gain } (K_p) = 8 \text{ to } 12$$

$$\text{dead time } (\tau_d) = 0.08 \text{ to } 0.12$$

$$\text{process time constant } (\tau_p) = 0.4 \text{ to } 0.6$$

according to the Cohen–Coon, the tuning method for a proportional controller is given by

$$K_c = \frac{\tau_p}{K_p\tau_d}\left(1+\frac{\tau_d}{3\tau_p}\right)$$

(a) $K_c = \dfrac{0.6(1+0.08/1.8)}{8\times 0.08} = 0.979$

(b) $K_c = \dfrac{0.6(1+0.12/1.8)}{12\times 0.12} = 0.444$

(c) $K_c = \dfrac{0.6(1+0.12/1.8)}{8\times 0.12} = 0.67$

(d) $K_c = \dfrac{0.4(1+0.08/1.2)}{12\times 0.08} = 0.444$

Hence, the answer is (a) where the value of K_c is the maximum, i.e., 0.979

Answer: (a)

8. (Q.24, 2003)

Water is flowing through a series of four tanks and getting heated as shown in Figure 10.2. We want to design a cascade control scheme for controlling the temperature of the water leaving tank 4 as there is a disturbance

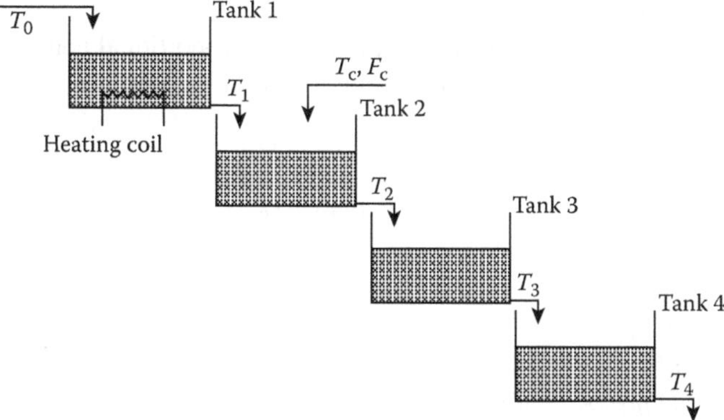

FIGURE 10.2 Process scheme of problem 8.

in the temperature of a second stream entering tank 2. Select the best place to take the secondary measurement for the secondary loop.

(a) Tank 1
(b) Tank 2
(c) Tank 3
(d) Tank 4

Solution: Answer is (b) tank 2

Explanation:

Because the temperature in tank 2 will be affected faster than in tanks 3 and 4; hence, the temperature in tank 2 will be the secondary measurement. The primary controller will set the temperature of the secondary controller (slave) in tank 2, which will manipulate the heating rate in tank 1. This cascade control scheme is shown in Figure 10.3.

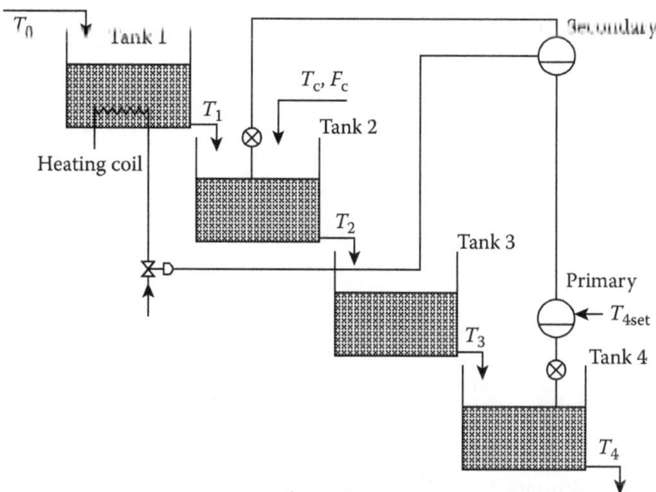

FIGURE 10.3 Cascade control scheme.

9. (Q.77, 2003)

Find the ultimate gain and frequency for a proportional controller in the case of a process having the following transfer function:

$$G_p(s) = \frac{1}{(4s+1)(2s+1)(s+1)}$$

(a) $\omega = \frac{1}{14}, K_c = \frac{45}{7\sqrt{14}}$
(b) $\omega = \sqrt{(7/6)}$, $K_c = 46/3$
(c) $\omega = 1$, $K_c = 13$
(d) $\omega = \sqrt{(7/8)}$, $K_c = 45/4$

Solution:

For a proportional controller in the control loop, the open-loop transfer function is

$$G_{ol}(s) = \frac{K_c}{(4s+1)(2s+1)(s+1)}.$$

So

$$\frac{AR}{K_c} = \frac{1}{\sqrt{\{(16\omega^2+1)(4\omega^2+1)(\omega^2+1)\}}}$$

and $\phi = \tan^{-1}(-4\omega) + \tan^{-1}(-2\omega) + \tan^{-1}(-\omega) = \alpha + \beta + \gamma$
where $\alpha = \tan^{-1}(-4\omega)$
$\beta = \tan^{-1}(-2\omega)$
$\gamma = \tan^{-1}(-\omega)$
At the crossover frequency, $\phi = -180°$
so

$$\tan(\alpha+\beta+\gamma) = \frac{7\omega - 8\omega^3}{1-14\omega^2} = \tan(-180°) = 0$$

or $\omega = \sqrt{(7/8)}$
so

$$\frac{AR}{K_c} = \frac{1}{\sqrt{(15 \times 9/2 \times 15/8)}} = 4/45.$$

Hence, ultimate $K_c = 45/4$ when the $AR = 1$.
Answer is (d).

10. (Q.78, 2003)

Match the type of controller given in group 2 that is most suitable for each application given in group 1.

Group 1	Group 2
(P) Distillation bottom level to be controlled with bottom flows	(1) P control
(Q) Distillation column pressure to be controlled by manipulating vapor flow from the top plate	(2) PI control
(R) Flow control of a liquid from a pump by positioning the valve in the line	(3) PID control
(S) Control of temperature of a CSTR with coolant flow in jacket	

(a) P-1, Q-1, R-2, S-3
(b) P-2, Q-2, R-3, S-3
(c) P-2, Q-2, R-1, S-1
(d) P-2, Q-3, R-2, S-3

Solution:

Answer is (a): P-1, Q-1, R-2, S-3.

11. (Q.79, 2003)

In the case of a feed-forward control scheme, which of the following is **NOT** true?

1. It is insensitive to modeling errors.
2. Cannot cope with unmeasured disturbances.
3. It waits until the effect of disturbance has been felt by the system before control action is taken.
4. Requires good knowledge of the process model.
5. Requires identification of all possible disturbances and their measurement.

(a) 1 and 3
(b) 1 and 4
(c) 2 and 5
(d) 3 and 4

Solution:

Answer is (a).

12. (Q.80, 2003)

Temperature control of an exothermic chemical reaction taking place in a CSTR is done with the help of cooling water flowing in the jacket around the reactor. The types of valves and controller action to be recommended are

(a) Air-to-open valve with the controller directly acting
(b) Air-to-close valve with the controller indirectly acting
(c) Air-to-open valve with the controller indirectly acting
(d) Air-to-close valve with the controller directly acting

Solution:

The correct answers are (a) and (b)

Explanation:

Directly acting means when measured value increases, the controller increases the output signal and vice versa.

Indirectly or reverse acting means when measured value increases, the controller decreases the output signal and vice versa.

An air-to-open valve means the valve is normally closed, and it starts opening with increasing air pressure.

An air-to-close valve means the valve is normally open, and it starts closing with increasing air pressure.

Thus,

Case (a) As the temperature increases, the controller (directly acting) increases the air pressure signal, causing coolant flow to increase, as the valve is air to open and vice versa. This is desired.

Case (b) As the temperature increases, the controller (indirect or reverse acting) decreases the air pressure signal, causing coolant flow to increase, as the valve is air to close and vice versa.

(Note: For safety, air-to-close valve for coolant flow should be used such that at the time of power failure the valve fully opens to cool the reactor.)

13. (Q.1, 2004)

The inverse of the Laplace Transformation of the function

$$f(s) = \frac{1}{s(1+s)}$$

is

(a) $1 + e^{t}$
(b) $1 - e^{t}$
(c) $1 + e^{-t}$
(d) $1 - e^{-t}$

Answer: (d)

14. (Q.4, 2004)

$$\frac{d^2y}{dx^2} + \sin x \frac{dy}{dx} + ye^{x} = \sin hx$$

is

(a) First order and linear
(b) First order and nonlinear
(c) Second order and linear
(d) Second order and nonlinear

Answer: (d)

15. (Q.78, 2004)

Match a first-order system with group I with the appropriate time constant in group II

Group I	Group II
(P) Thermometer	(1) $(mC_p)/(hA)$
(Q) Mixing	(2) q/V
	(3) V/q
	(4) $(hA)/(mC_p)$

(a) P-4, Q-2
(b) P-4, Q-3
(c) P-1, Q-2
(d) P-1, Q-3
Answer is: (d)

16. (Q.79, 2004)

The exponential response of the controlled variable $y(t)$ for a step change of magnitude P in the manipulated variable $x(t)$ is shown below:

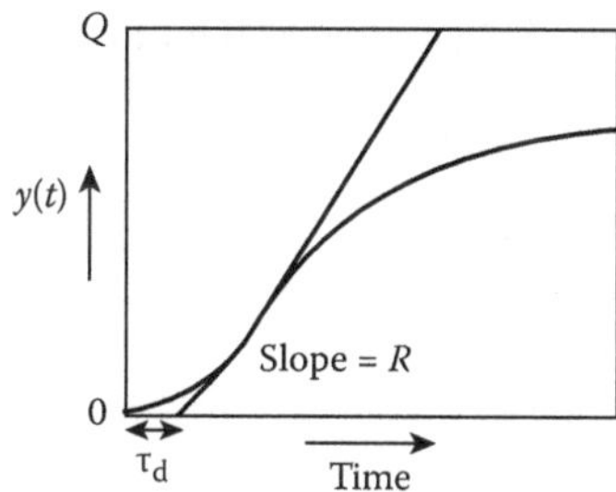

The appropriate transfer function of the process is

(a) $\dfrac{(Q/P)\,e^{-(Q/R)\,s}}{\tau_d s+1}$

(b) $\dfrac{(Q/R)\,e^{-\tau_d s}}{(Q/P)s+1}$

(c) $\dfrac{(Q/P)\,e^{-\tau_d s}}{(Q/R)s+1}$

(d) $\dfrac{(Q/R)\,e^{-(P/Q)s}}{\tau_d s+1}$

Answer is: (c)

17. (Q.80, 2004)

Consider a system with the open-loop transfer function

$$G(s) = 1/(s + 1)(2s + 1)(5s + 1).$$

Match the range of ω (frequency) in group I with the slope of the asymptote of the log (AR) versus log ω plot in group II.

Group I	Group II
(P) $0 < \omega < 0.2$	(1) −5
(Q) $\omega > 1$	(2) −3
	(3) −2
	(4) −1
	(5) zero

(a) P-5, Q-2
(b) P-4, Q-2
(c) P-5, Q-3
(d) P-4, Q-1
Answer is: (a)

$$\begin{aligned}\log(AR) &= -1/2\log\{(\omega^2+1)(4\omega^2+1)(25\omega^2+1)\}\\ &= -1/2\log\{(1)(1)(1)\} \text{ for } \omega<1\\ &= 0,\ \text{slope}=0\end{aligned}$$

and

$$\begin{aligned}\log(AR) &= -1/2\log\{(\omega^2)(4\omega^2)(25\omega^2)\} \text{ for } \omega>1\\ &= -1/2\log(100\omega^6) = -1-3\log(\omega),\ \text{hence slope} = -3\end{aligned}$$

[illegible]

The process and disturbance transfer function for a system are given by

$$G_p(s) = y(s)/m(s) = 2/\{(2s+1)(5s+1)\}$$

$$G_d(s) = y(s)/d(s) = 1/\{(2s+1)(5s+1)\}.$$

The feed-forward controller transfer function that will keep the process output constant for changes in disturbance is
(a) $2/\{(2s+1)^2(5s+1)^2\}$
(b) $\{(2s+1)^2(5s+1)^2\}/2$
(c) 1/2
(d) $(2s+1)(5s+1)$
Answer: (c)

As the feed-forward controller transfer function is

$F(s)$ = transfer function of manipulated value/transfer function of disturbance value = $m(s)/d(s) = 1/2$.

19. (Q.82, 2004)

The block diagram is shown below

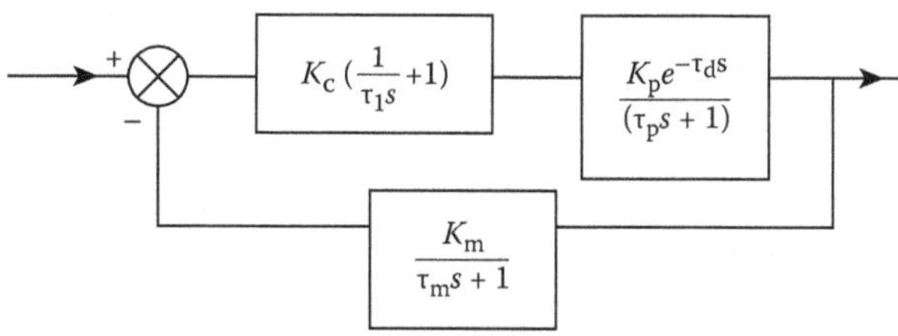

The characteristic equation is

(a) $\tau_I s(\tau_p s+1)+K_cK_p(\tau_I s+1)e^{-\tau_T s}=0$
(b) $(\tau_p s+1)(\tau_m s+1)\ K_mK_pe^{-\tau_T s}=0$
(c) $\tau_I s(\tau_p s+1)(\tau_m s+1)+K_cK_pK_m(\tau_I s+1)e^{-\tau_T s}=0$
(d) $(\tau_p s+1)(\tau_m s+1)+K_eK_pK_me^{-\tau_T s}=0$

Answer is (c)

20. (Q.8, 2005)

The unit step response of a first-order system with time constant τ and steady-state gain K_p is given by

(a) $K_p(1-e^{-t/\tau})$
(b) $K_p(1+e^{-t/\tau})$
(c) $K_p(1-e^{-2t/\tau})$
(d) $K_p\, e^{-t/\tau}/\tau$

Answer: (a) $K_p(1-e^{-t/\tau})$

21. (Q.9, 2005)

An example of an open loop, second order underdamped system is

(a) Liquid level in a tank
(b) U-tube manometer
(c) Thermocouple in a thermowell
(d) Two noninteracting first-order systems in series

Answer: (b) U-tube manometer ($\xi<1$) is an underdamped system

22. (Q.10, 2005)

Control valve characteristics are selected such that the product of process gain and the valve gain

(a) Is a linearly increasing function of the manipulated variable
(b) Is a linearly decreasing function of the manipulated variable
(c) Remains constant as the value of the manipulated variable changes
(d) Is an exponentially increasing function of the manipulated variable

Answer: (c)

23. (Q.11, 2005)

Cascade control comes under the control configuration, which uses

(a) One measurement and one manipulated variable

(b) More than one measurement and one manipulated variable
(c) One measurement and more than one manipulated variable
(d) More than one measurement and more than one manipulated variable
Answer: (b) More than one measurement and one manipulated variable

24. (Q.46, 2005)

Match the process variables (group I) given below with the measuring devices (group II)

Group I	Group II
(P) High temperature	(1) Orifice meter
(Q) Flow	(2) Chromatograph
(R) Composition	(3) Radiation pyrometer
	(4) Bi-metallic thermometer

(a) P-1, Q-2, R-3
(b) P-1, Q-3, R-2
(c) P-3, Q-1, R-2
(d) P-4, Q-2, R-1
Answer: (C)

25. (Q.47, 2005)

Given the characteristic equation below, select the number of roots that will be located to the right of the imaginary axis

$$S^4 + 5S^3 - S^2 - 17S + 12 = 0$$

(a) One
(b) Two
(c) Three
(d) Zero
Answer: (b) Because there are two sign changes from the equation, there will be two roots on the right of the imaginary axis.

26. (Q.49, 2005)

Given the process transfer function $G_p = 20/(S-2)$, and controller transfer function $G_c = K_c$, and assuming the transfer functions of all other elements in the control loop are unity, select the range of K_c for which the closed-loop response will be stable.
(a) $K_c < 1/10$
(b) $K_c < 1/100$
(c) $1/100 < K_c < 1/10$
(d) $K_c > 1/10$
Answer: (d)

The characteristic equation is $G_{OL} + 1 = 0$ or $20K_c/(S-2) + 1 = 0$ or $S - 2 + 20\ K_c = 0$.

As using the Routh–Hurwitz array of polynomials of degree one, there will be two rows and the first column elements to be positive for closed-loop stable.

$$\begin{vmatrix} 1 \\ 20K_c - 2 \end{vmatrix}.$$

So for closed-loop stable system, $20K_c - 2 > 0$ or $K_c > 1/10$.

27. The value of the ultimate period of oscillation P_u is 3 minutes, and that of the ultimate controller gain K_{cu} is 2 minutes. Select the correct set of tuning parameters (controller gain K_c, the derivative time constant τ_D in minutes, and the integral time constant τ_I in minutes) for a PID controller using Ziegler–Nichols controller settings.
(a) $K_c = 1.1, \tau_I = 2.1, \tau_D = 1.31$
(b) $K_c = 1.5, \tau_I = 1.8, \tau_D = 0.51$
(c) $K_c = 1.5, \tau_I = 1.8, \tau_D = 0.51$
(d) $K_c = 1.2, \tau_I = 1.5, \tau_D = 0.38$
Answer: (d)
In the PID mode

$$K_c = 0.6 \times 2 = 1.2, \ \tau_I = 3/2 = 1.5, \ \tau_D = 3/8 = 0.38$$

28. (Q.20, 2006)
The control valve characteristics for three types of control valves (P, Q, and R) are given in the figure. Match the control valve with the characteristics.
(a) P-quick opening, Q-linear, R-equal percentage
(b) P-linear, Q-square root, R-equal percentage
(c) P-equal percentage, Q-linear, R-quick opening
(d) P-square root, Q-quick opening, R-linear

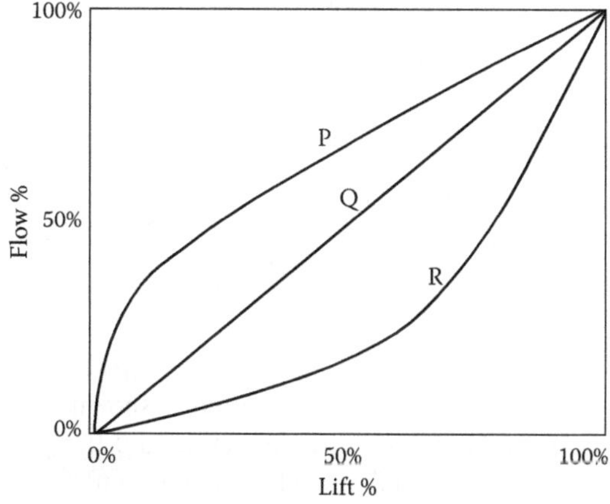

Answer: (a) P-quick opening, Q-linear, R-equal percentage.

(As for P, the flow percentage is greater than those of Q and R at the same lift percentage. For Q, the flow percentage is proportional to the lift percentage, and for R, the equal percentage flow increases exponentially as $\ln(m/m_0) = \beta x$ or $m = m_0\, e^{\beta x}$.)

29. (Q.55, 2006)

The Laplace transformation of the input function, $X(t)$, given in the figure below

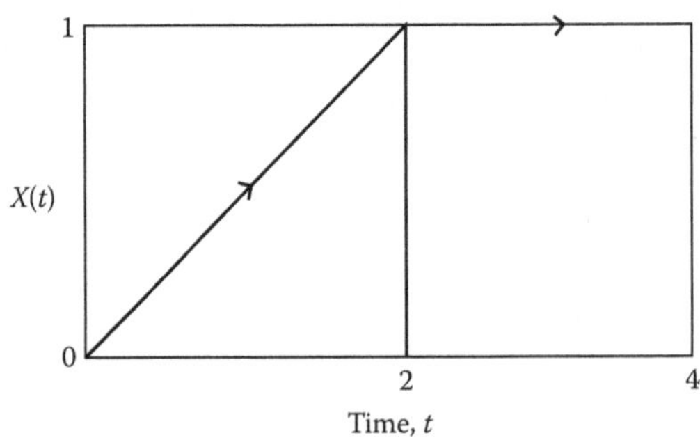

is given by

(a) $(1 - e^{-2s})/2s^2$
(b) $(1 + e^{-2s})/2s^2$
(c) $(1 + e^{2s})/s^2$
(d) $(1 - e^{-2s})/s^2$

Answer: (b)

$$(\text{as } X(t) = 1/2t\,[u(t) + u(t-2)]$$

$$= 1/2tu(t) + 1/2\, tu(t-2)\,.$$

So

$$X(s) = 1/2s^2 + 1/2\, e^{-2s}/s^2$$

$$= (1 + e^{-2s})/2s^2)$$

30. (Q.56, 2006)

A liquid level control system is configured as shown in the figure below. If the level transmitter (LT) is directly acting and the pneumatic control valve is air-to-open, what kind of control action should the controller (LC) have and why?

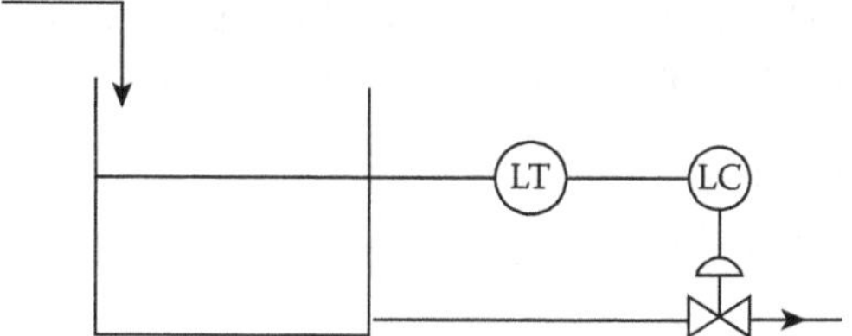

(a) Direct acting because the control valve is direct acting
(b) Reverse acting because the control valve is reverse acting
(c) Direct acting because the control valve is reverse acting
(d) Reverse acting because the control valve is direct acting
Answer: (a)

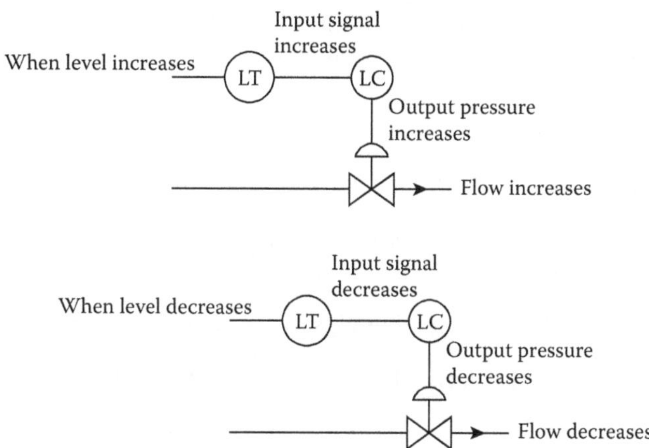

(Because the control of the level will be achieved when the level of liquid rises above the desired set point, the outflow through the control valve should be increased and vice versa. That is, overall action should be direct acting. Because the control valve opens with the increase of air pressure, the controller output is to be increased when its input (i.e., level) increases over set point and vice versa. Hence, the controller should be set at directly acting mode. These actions during increase and decrease in levels are explained in the above figure.)

31. (Q.57, 2006)

A 2-input, 2-output process can be described in the Laplace transformation domain as given below:

$$(\tau_1 s + 1)\, Y_1(s) = K_1 U_1(s) + K_2 U_2(s)$$

$$(\tau_2 s + 1)\, Y_2(s) = K_3 U_2(s) + K_4 Y_1(s)$$

where $U_1(s)$ and $U_2(s)$ are the inputs and $Y_1(s)$ and $Y_2(s)$ are the outputs. The gains of the transfer functions $Y_1(s)/U_2(s)$ and $Y_2(s)/U_2(s)$, respectively, are

(a) K_2 and K_3
(b) K_1 and $K_3 + K_2K_4$
(c) K_2 and $K_3 + K_1K_4$
(d) K_2 and $K_3 + K_2K_4$

Answer: (d)

As the ultimate value is the gain, we determine the gains as follows:

$$(\tau_1 s + 1)\ Y_1(s) = K_1U_1(s) + K_2U_2(s).$$

Consider $U_1(s) = 0$ and $U_2(s) = 1/s$ (unit step change)

$$Y_1(s) = K_2/(\tau_1 s + 1)s$$

$$\lim_{t \to \infty} Y_1(t) = \lim_{s \to 0} s\ Y(s) = K_2$$

and

$$(\tau_2 s+1)Y_2(s) = K_3U_2(s) + K_4[K_1U_1(s) + K_2U_2(s)]/(\tau_1 s+1)$$

$$= [K_3 + K_4K_2/(\tau_1 s+1)]U_2(s)$$

$$Y_2(s) = [K_3/(\tau_2 s+1) + K_4K_2/(\tau_2 s+1)(\tau_1 s+1)U_2(s)$$

$$= [K_3/(\tau_2 s+1) + K_4K_2/(\tau_2 s+1)(\tau_1 s+1)]/s$$

$\lim Y_1(t) = \lim s\ Y(s) = K_3 + K_4\,K_2$
as $t \to \infty$ as $s \to 0$

Hence, the gains of $Y_1(s)/U_2(s)$ and $Y_2(s)/U_2(s)$ are respectively given as in (d).

32. (Q.58, 2006)

A process is disturbed by a sinusoidal input, $u(t) = A \sin wt$. The resulting process output is $Y(s) = KAw/(\tau s + 1)(s^2 + w^2)$. If $y(0) = 0$, the differential equation representing the process is

(a) $dy(t)/dt + \tau y(t) = Ku(t)$
(b) $\tau\, dy(t)/dt + y(t) = KAu(t)$
(c) $\tau\, dy(t)/dt + y(t) = Ku(t)$
(d) $\tau\, [dy(t)/dt + y(t)] = KAu(t)$

Answer: (c)

As,

$$\tau\ [sy(s) - y(0)] + y(s) = KAw/(s^2 + w^2)$$

or

$$y(s) = KAw/(s^2 + w^2)(\tau s + 1) \text{ as } y(0) = 0$$

33. (Q.74 and Q.75, 2006)

The block diagram of a closed-loop control system is shown in the figure below. Y is the controlled variable, D is the disturbance, Y_{sp} is the set point, G_1, G_2, and G_3 are the transfer functions, and K_c is the proportional controller.

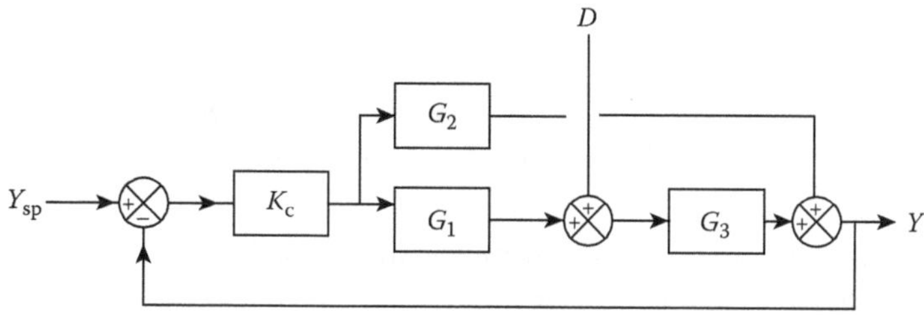

(I) The closed-loop transfer function $Y(s)/D(s)$ is given by

(a) $G_3G_1/[1 + (G_1G_3 + G_2)K_c]$

(b) $G_1/[1 + (G_1G_3 + G_2)K_c]$

(c) $G_3/[1 + (G_1 + G_2)G_3K_c]$

(d) $G_3/[1 + (G_1G_3 + G_2)K_c]$

Answer: (d)

As $\varepsilon = Y_{sp}(s) - Y(s)$

$$Y(s) = (K_c\varepsilon G_1 + D)G_3 + K_c\varepsilon G_2 = (K_cG_1G_3 + K_cG_2)\varepsilon + DG_3$$

$$= (K_cG_1G_3 + K_cG_2)[Y_{sp}(s) - Y(s)] + DG_3$$

or

$$Y(s) = [K_c(G_1G_3 + G_2)Y_{sp}(s) + DG_3]/[1 + K_c(G_1G_3 + G_2)]$$

While $Y_{sp}(s) = 0$ then, $Y(s)/D(s) = G_3/[1 + K_c(G_1G_3 + G_2)]$

(II) Let $G_1 = 1$, and $G_2 = G_3 = 1/(s + 1)$. A step change of magnitude M is made in the set point. The steady-state offset of the closed loop Y is

(a) $M/(1 + 2K_c)$

(b) $M/(1 + K_c)$

(c) $M(K_c - 1)/(1 + 2K_c)$

(d) Zero

Answer: (a)

$$\text{As } Y(s)/Y_{sp}(s) = [K_c(G_1G_3 + G_2)]/[1 + K_c(G_1G_3 + G_2)]$$

$$= [K_c2/(s+1)]/[1 + 2K_c/(s+1)]$$

And $Y_{sp}(s) = M/s$,
so $\lim Y(t) = \lim sY(s) = MK_c\, 2/[1 + 2K_c]$
as $t \to \infty$ as $s \to 0$
offset $= Y_{sp} - Y(\infty) = M - MK_c\, 2/[1 + 2K_c] = M/[1 + 2K_c]$

34. (Q.84 and Q.85, 2006)

For the system below, $G_1(s) = 1/(\tau_1 s + 1)$ and $G_2(s) = 1/(\tau_2 s + 1)$ and $\tau_2 = 2\tau_1$

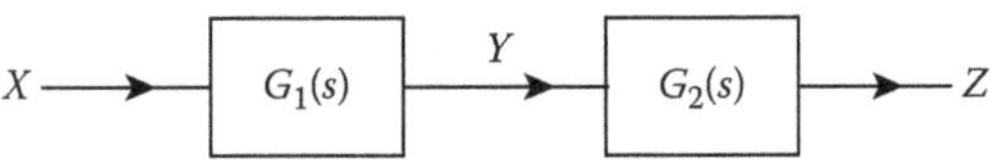

When the system is excited by the sinusoidal input $X(t) = \sin wt$, the intermediate response Y is given by

$$Y = A \sin (wt + \varphi).$$

(I) If the response of Y lags behind the input X by 45°, and $\tau_1 = 1$, then the input frequency w is
(a) 1
(b) $\pi/4$
(c) Zero
(d) –1
Answer: (a)
As $\varphi = -45° = \tan^{-1}(-w)$ for the first-order system; hence, $w = 1$.

(II) For the same input, the amplitude of the output Z will be
(a) 1.00
(b) 0.62
(c) 0.42
[illegible]
Answer: (d)
As $Z(s) = G_1 G_2 X(s) = 1/(s + 1)(2s + 1)$
So $AR = |Z(jw)| = 1/\sqrt{[(w^2 + 1)(4w^2 + 1)]} = 1/\sqrt{10} = 0.32$
Hence, the amplitude of Z is 0.32 as amplitude in the input sine is unity.

35. (Q.16, 2007)

An operator was told to control the temperature of a reactor at 60°C. The operator set the set point of the temperature controller at 60%. The scale actually indicated 0% to 100% of a temperature range of 0°C to 200°C. This caused a runaway reaction by over-pressurizing the vessel, which resulted in injury to the operator.

The actual set point temperature was
(a) 200°C
(b) 60°C
(c) 120°C

(d) 100°C

Answer: (c)

If T = temperature, P = %

$T = mP + c$

$0 = 0 + c$

$200 = 100m$

$m = 2$

So $T = 2P$; hence, when $P = 60\%$ set, the temperature is $T = 2 \times 60 = 120$°C.

36. (Q.39, 2007)

The pressure differential across a venturi meter, inclined at 45° to the vertical (as shown in the figure) is measured with the help of a manometer to estimate the flow rate of a liquid through it. If the density of the flowing fluid is ρ, and the density of the manometric fluid is ρ_m, the velocity of the fluid at the throat can be obtained from the expression

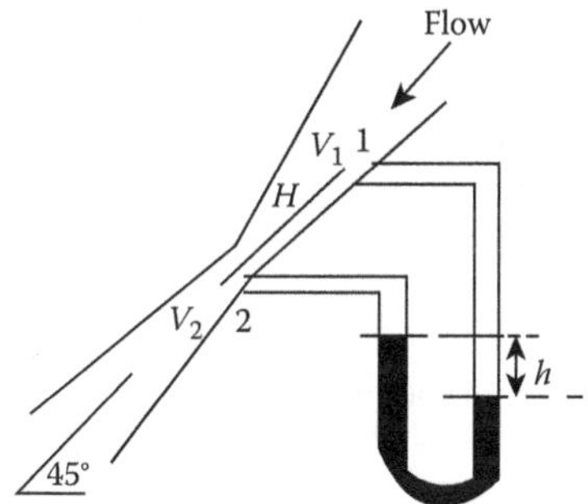

(a) $(V_2^2 - V_1^2)/2g = h(\rho_m - \rho)/\rho + H \sin 45°$

(b) $(V_2^2 - V_1^2)/2g = h\rho_m/\rho + H \sin 45°$

(c) $(V_2^2 - V_1^2)/2g = h\rho_m/\rho$

(d) $(V_2^2 - V_1^2)/2g = h(\rho_m - \rho)/\rho$

Solution:

Applying Bernoulli's theorem in Sections 1 and 2,

$$V_2^2/2g + P_2/\rho = V_1^2/2g + P_1/\rho + H \sin 45°$$

As P.E at 2 is taken to be base = 0 and the P.E at 1 is the vertical difference of head

$$H \sin 45°$$

Answer: (a)

37. (Q.60, 2007)

The dynamic model for a mixing tank open to atmosphere at its top as shown in the figure below is to be written. The objective of mixing is to cool the hot water stream entering the tank at a flow rate of q_2 and feed temperature T_s with a cold water feed stream entering the tank at a flow rate of q_1 and feed temperature T_0. A water stream is drawn from the tank bottom at a flow rate of q_4 by a pump, and the level in the tank is proposed to be controlled by drawing another water stream at a flow rate of q_3. Neglect evaporation and other heat losses from the tank.

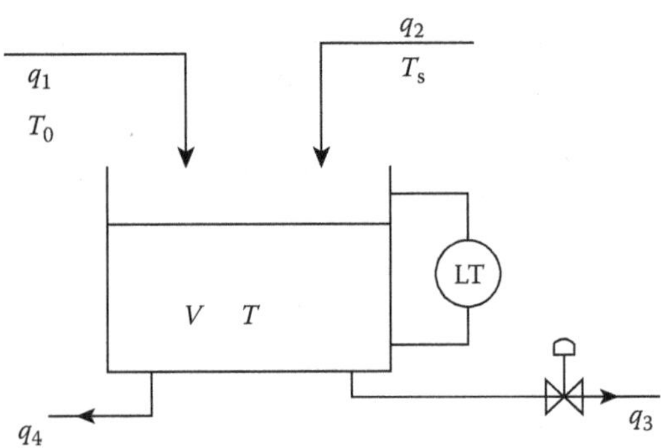

The dynamic model of the tank is given as

(a) $dV/dt = q_1 + q_2 - q_3$, $V\,dT/dt = q_1T_0 + q_2T_s - q_3T$

(b) $dV/dt = q_1 - q_4$, $d(VT)/dt = q_1T_s - q_4T$

(c) $dV/dt = q_1 + q_2 - q_4$, $d(VT)/dt = q_1T_0 + q_2T_s - q_4T$

(d) $dV/dt = q_1 + q_2 - q_3 - q_4$, $V\,dT/dt = q_1(T_0 - T) + q_2(T_s - T)$

Answer (d)

As

material balance for level,

$$dV/dt = q_1 + q_2 - q_3 - q_4$$

where q is the volumetric flow rate

Heat balance for the tank is

$$d(\rho cVT)/dt = \rho c q_1 T_0 + \rho c q_2 T_s - (q_3 + q_4)\rho c T$$

or $d(VT)/dt = q_1T_0 + q_2T_s - (q_3 + q_4)T$

or $V\,d(T)/dt + T\,dV/dt = q_1T_0 + q_2T_s - (q_3 + q_4)T$

or $V\,d(T)/dt + T(q_1 + q_2 - q_3 - q_4) = q_1T_0 + q_2T_s - (q_3 + q_4)T$

or $V\,dT/dt = q_1(T_0 - T) + q_2(T_s - T)$

38. (Q.61, 2007)

Match the transfer function with the responses to a step input in the figure.

(i) $-2.5\,(-4s + 1)/(4s^2 + 4s + 1)$
(ii) $-2e^{-10s}/(10s + 1)$
(iii) $-5/(-20s + 1)$
(iv) $-0.1/s$
(v) $(4s + 3)/(2s + 1)$

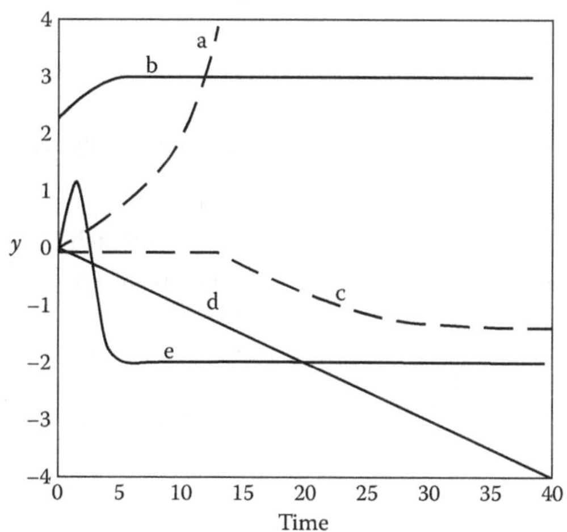

(a) i-e, ii-c, iii-a, iv-d, v-b
(b) i-a, ii-b, iii-c, iv-d, v-e
(c) i-b, ii-a, iii-c, iv-e, v-d
(d) i-e, ii-a, iii-c, iv-b, v-d

Answer: (a)

(i) $y(s) = -2.5(-4s + 1)/\{s(4s^2 + 4s + 1)\}$, so $y(t)$ is given by the final value theorem
as $t \to$ infinity, $\lim sY(s) \to -2.5$, i.e., graph e

(ii) $y(s)/x(s) = -2e^{-10s}/(10s + 1)$, the step response is sigmoidal with decreasing value with a dead lag up to 10 seconds, as in graph c

(iii) $y(s) = -5/s(-20s + 1) = -5[1/(s - 1/20) - 1/s]$, $y(t) = 5(e^{t/20} - 1)$; hence, graph a

(iv) $y(s)/x(s) = -0.1/s$, $y(s) = -0.1/s^2$, so $y(t) = -0.1t$; hence, graph d

(v) $y(s) = (4s + 3)/(2s + 1)$, $y(s) = (4s + 3)/s(2s + 1) = 2/s + 1/s - 1/(s + 1/2) = 3/s - 1/(s + 1/2)$, $y(t) = 3 - e^{-t/2}$, graph b

39. (Q.62, 2007)

Consider the following instrumentation diagram for a chemical reactor. C_{sp} represents a concentration set point.

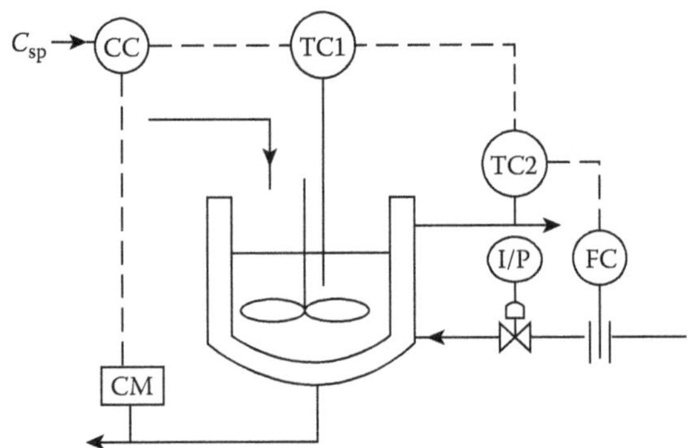

Match the items in group A with the corresponding items in group B.

Group A	Group B
(P) Control strategy	(1) Feed forward control
(Q) Primary control variable	(2) Cascade control
(R) Slowest controller	(3) Concentration in the reactor
(S) Fastest controller	(4) Reactor temperature
	(5) Jacket temperature
	(6) Concentration controller
	(7) Reactor temperature controller
	(8) Jacket temperature controller
	(9) Flow controller
	(10) Selective control

(a) P-2, Q-3, R-6, S-9
(b) P-1, Q-4, R-8, S-7
(c) P-10, Q-7, R-9, S-6
(d) P-1, Q-8, R-5, S-9

Answer: (a)

As a concentration controller will sense the concentration that will be changed only when the reactor temperature is changed. The reactor temperature will change only when jacket temperature changes. Thus, the hierarchy of the control will be in the sequence of fastness as jacket temperature > reactor temperature > concentration.

Control strategy is cascade control, CC is the master of TC1, and TC1 is the master of TC2, and TC2 is the master of FC. The primary control variable is concentration, the slowest controller is the concentration controller, and the fastest controller is the flow controller of the jacket fluid.

40. (Q.71, Q.72, and Q.73, 2007)

Common data for questions: A cascade control system for pressure control is shown in the figure given below. The pressure transmitter has a range of 0 to 6 bar(g), and the flow transmitter range is 0 to 81 nm^3/hr. The normal flow rate through the valve is 32.4 nm^3/hr corresponding to the value of set point for pressure = 1 bar(g), and to give the flow, the valve must be 40% opened. The control valve has linear characteristics and is fail-open (air to close). Error, set point, and control variable are expressed in percentage transmitter output (% TO). Proportional gain is expressed in the units of percentage controller output (CO/%TO).

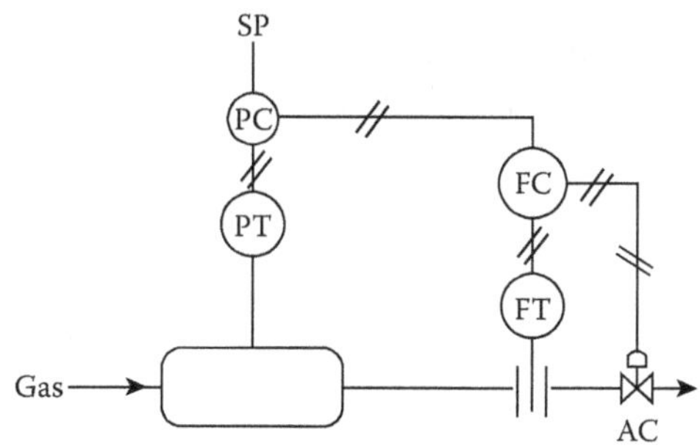

(Q.71) The types of action for the two controllers are

(a) Direct acting for the pressure control and direct acting for the flow control
(b) Indirect acting for pressure control and indirect acting for the flow control
(c) Direct acting for the pressure control and indirect acting for the flow control
(d) Indirect acting for the pressure control and direct acting for the flow control

(Q.72) The bias values for the two controllers, so that no offset occurs in either controller, are

(a) Pressure controller: 40%; Flow controller: 60%
(b) Pressure controller: 33%; Flow controller: 67%
(c) Pressure controller: 67%; Flow controller: 33%
(d) Pressure controller: 60%; Flow controller: 40%

(Q.73) Given that the actual tank pressure is 4 bar(g) and a proportional controller is employed for pressure control, the proportional band setting of the pressure controller required to obtain a set point to the flow controller to 54 nm^3/hr is

(a) 50%
(b) 100%
(c) 150%
(d) 187%

Solutions:

{Q.71. Answer is either (c) or (d)}.

In order to control pressure, the flow rate must be increased to reduce pressure to reach set point when the pressure rises above the set point. Similarly, when pressure falls below the set point, flow must be decreased. Thus, the relationship between the pressure of the tank and the flow rate should be directly acting, i.e., increase–increase, decrease–decrease. The PT transmitter is a directly acting sensor, i.e., as the pressure increases, the output of the PT will also increase and vice versa. The output of the controller PC is set to indirect mode, so the output of the controller will decrease while the pressure increases over the set point. Now the flow controller (FC) is set in direct acting mode, i.e., the output will increase as its input increases, and output will decrease as its input increases. The control valve being an air-to-open type, the flow will increase as the output from FC decreases and vice versa. The overall action will become direct acting as explained in the following diagrams.

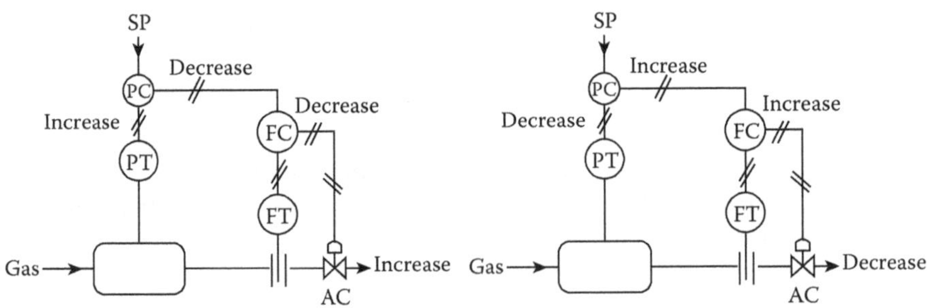

Actions when PC is in reverse acting and FC is in direct acting modes

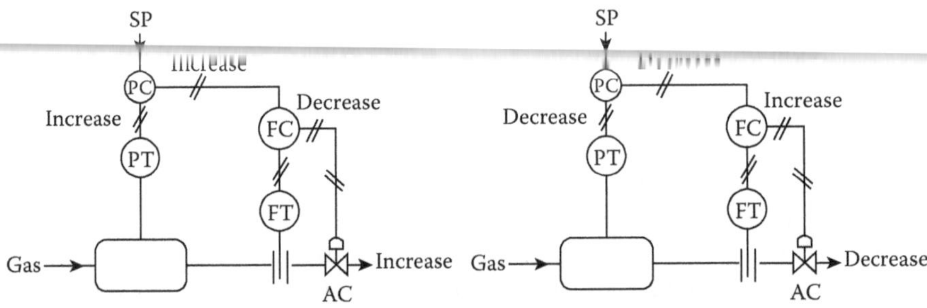

Actions when PC is in direct acting and FC is in reverse acting modes

(Q.72) For the flow 32.4 nm³/hr, FT's output = 2.4 bar(g), the steady-state output of FC is 3.6 bar(g) as calculated by simple relationships.

FT = 0 – 6 bar, for flow = 0 to 81 nm³/hr; so flow = 81/6 FT.

Control valve opening 100% to 0%, for flow = 81 to 0 nm³/hr; thus, flow = 81–81/6 FC.

Thus, when flow is 32.4 nm^3/hr, FT = 2.4 bar, 32.4 = 81–81/6 FC or FC (output from FC) = 3.6 bar. Because the output signal at the zero offset is the bias value of the signal, the bias of the FC is 3.6 bar(g). Similarly, the input signal to the FC corresponding to the 32.4 nm^3/hr flow is 2.4 bar, which must be equal to the set point to FC, i.e., input to the FC is also 2.4 bar, which is the output signal from PC. Hence, the bias value of the PC is 2.4 bar while offset is zero. This is shown in the following figure.

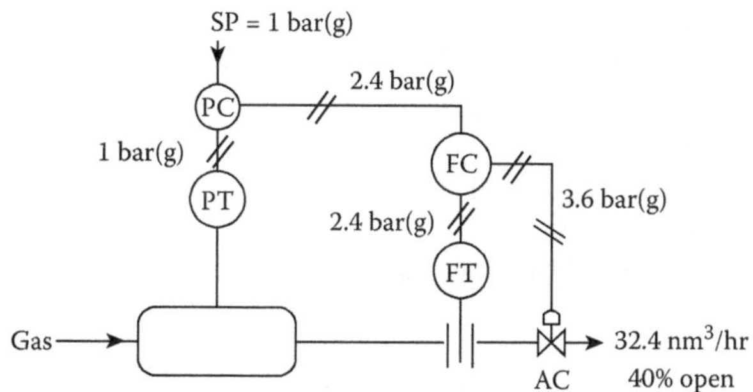

Thus, the bias value of PC = 2.4 bar = 40% of TO, and the bias value of FC = 3.6 bar = 60% of the TO.

Answer is (a).

(Q.73) The answer is (d)

As PT = 1 bar, $q = 32.4$, i.e., output of FC = 3.6 bar

As PT = 4 bar, $q = 54.0$, i.e., output of FC = (81 – 54) × 6/81 = 2.0 bar

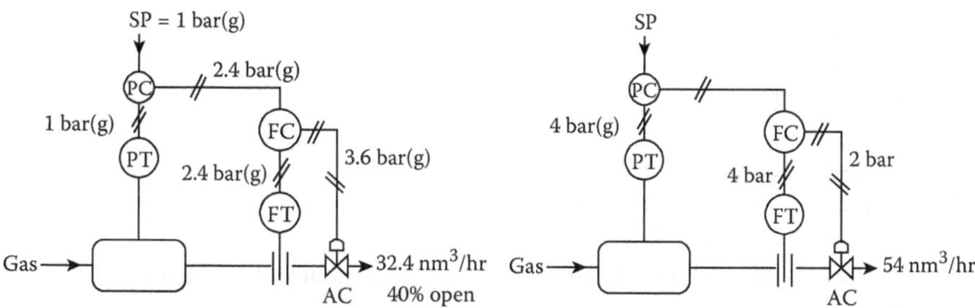

PC being a proportional controller,

Δoutput = 2.0 – 3.6 = –1.6 bar and Δinput = 4 – 1 = 3 bar

So proportional band = 3/1.6 × 100% = 187.5%

Hence, the answer is (d).

41. (Q.63, 2007)

The first two rows of Routh's tabulation of a third-order equation are

S^3	2	2
S^2	4	4

Select the correct answer from the following choices:

(a) The equation has one root in the right half of the s-plane.

(b) The equation has two roots on the j axis as $s = j$ and $-j$. The third root is in the left-half plane.

(c) The equation has two roots on the j axis at $s = 2j$ and $s = -2j$. The third root is in the left-half plane.

(d) The equation has two roots on the j axis at $s = 2j$ and $s = -2j$. The third root is in the right-half plane.

Solution: Answer is (b)

For third-order polynomials, the total number of rows of the Routh array will be 4

S^3	2	2
S^2	4	4
	0	

Because the third row is zero, there will be two roots on the right-half plane satisfying the relationship $CS^2 + D = 0$ where $C = 4$ and $D = 4$, i.e., $s = +j$ and $-j$. Third root will be on the LHP.

Answer is (b).

42. (Q.60, 2008)

The unit impulse response of a first-order process is given by $2e^{-0.5t}$. The gain and time constant of the process are, respectively,

(a) 4 and 2

(b) 2 and 2

(c) 2 and 0.5

(d) 1 and 0.5

Answer: (a)

As $y(s)/x(s) = L\{2e^{-0.5t}\} = 2/(s + 0.5)$

For $x(t)$ = unit impulse, $x(s) = 1$

So $y(s) = 2/(s + 0.5) = 4/(2s + 1)$,

which is equivalent to a first-order process with gain of 4 and time constant 2.

43. (Q.61, 2008)

A unit step input is given to a process that is represented by the transfer function $(S + 2)/(S + 5)$. The initial value ($t = 0+$) of the response of the process to step input is

(a) 0

(b) 2/5

(c) 1
(d) ∞
Answer: (c)

As, $y(s)/x(s) = (S + 2)/(S + 5)$ and $x(s) = 1/s$ because of unit step change, so initial value of $y(t) = \lim sy(s)$ as $s \to \infty$ is $(1 + 2/S)/(1 + 5/S) = 1/1 = 1$

Or, by L' Hospital's rule for undefined form, by differentiating both the numerator and denominator w.r.t. S, the same result is obtained. Hence, the answer is (c).

44. (Q.62, 2008)

A tank of volume 0.25 m^3 and height 1 m has water flowing in at 0.05 m^3/min. The outlet flow rate is governed by the relationship

$$F_{out} = 0.1h$$

where h is the height of the water in the tank in m, and F_{out} is the outlet flow rate in m^3/min.

The inlet flow rate changes suddenly from its nominal value of 0.05 m^3/min to 0.15 m^3/min and remains there. The time (in minutes) at which the tank will begin to overflow is given by

(a) 0.28
(b) 1.01
(c) 1.73
(d) ∞

Answer: (c)

As, at $t = 0$, $F_{inlet} - F_{out} = dV/dt$ or

$F_{inlet} - 0.1h = A\, dh/dt$ or $X(t) = 0.1H(t) + A\, dH/dt$ where $X(t) = F_{inlet}(t) - F_{inlet}(0)$ and $H(t) = h(t) - h(0)$

$A = 0.25/1 = 0.25$ m^2. So, $H(s)/X(s) = R/(\tau S + 1)$ where $\tau = AR = 0.25 \times 10 = 2.5$ min (as $F_{out} = 0.1h = h/R$, $R = 1/0.1 = 10$); hence, $H(s)/X(s) = 10/(2.5S + 1)$ for $X(t) = 0.15 - 0.05 = 0.10$ m^3/min as a step change,

$X(s) = 0.1/S$ and, hence, $H(s) = 10 \times 0.10/\{(2.5S + 1)S\}$ and $H(t) = (1 - e^{-t/2.5})$ where $F_{out}(0) = F_{input}(0) = 0.05 = 0.1\, h(0)$ or $h(0) = 0.5$ m.

The tank will overflow when the level becomes greater than 1 meter, i.e., $H(t) = h(t) - h(0) = 1 - 0.5 = 0.5 = (1 - e^{-t/2.5})$ or $e^{-t/2.5} = 0.5$ or $t/2.5 = 0.6931$ or $t = 1.732$ min.

Hence, the answer is (c).

45. (Q.65, 2008)

Match the following

Group 1	Group 2
(P) Ziegler–Nichols	(1) Process reaction curve
(Q) Underdamped response	(2) Decay ratio
(R) Feed-forward control	(3) Frequency response
	(4) Disturbance measurement

(a) P-3, Q-2, R-4
(b) P-4, Q-2, R-3
(c) P-3, Q-4, R-2
(d) P-1, Q-4, R-2
Answer: (a)

Ziegler–Nichols and frequency response (P & 3)
Underdamped response and decay ratio (Q & 2)
Feed-forward control and disturbance measurement (R & 4)

46. (Q.64, 2008)

Match the following

Group 1	Group 2
(P) Temperature	(1) Hot wire anemometry
(Q) Pressure	(2) Strain gauge
(R) Flow	(3) Chromatographic analyzer
	(4) Pyrometer

(a) P-1, Q-2, R-3
(b) P-4, Q-1, R-3
(c) P-1, Q-2, R-4
(d) P-4, Q-2, R-1
Answer: (d)

47. (Q.84 and Q.85, 2008)

The crossover frequency associated with a feedback loop employing a proportional controller to control the process represented by the transfer function

$$G_p(s) = 2e^{-s}/(\tau s + 1)^2, \text{ (where time is in minutes)}$$

is found to be 0.6 rad/min. Assume that the measurement and valve transfer functions are unity.

(Q.84)

The time constant, in minutes, is

(a) 1.14
(b) 1.92
(c) 3.23
(d) 5.39

Solution:

$$|G(jw)| = 2/(\tau^2 w^2 + 1) \text{ and } \theta = -w + 2\tan(-\tau w)$$

At crossover frequency, $w = 0.6$ and $\theta = -180° = -wx180/\pi + 2\tan^{-1}(-\tau w) = -0.6 \times 180/3.14 + 2\tan^{-1}(-\tau w)$

or $2\tan^{-1}(-\tau x0.6) = -180 + 34.39 = -145.6$; so $0.6 \times \tau = 3.23$ or $\tau = 5.38$ minutes

Answer is (d)

(Q.85)

If the control loop is to operate at a gain margin of 2.0, the gain of the proportional controller must equal

(a) 0.85
(b) 2.87
(c) 3.39
(d) 11.50

Answer:

For a proportional controller of gain K_c the open-loop transfer function is

$$G_{ol} = K_c G_p(s) = 2K_c e^{-s}/(\tau s + 1)^2$$

then $|G(jw)/K_c| = 2/(\tau^2 w^2 + 1)$ at the crossover frequency, $AR = |G(jw)| = 2K_c/(5.39^2\ 0.6^2 + 1) = 2K_c/(11.45)$

Gain margin = $1/AR$ at the crossover frequency = $11.45/(2K_c) = 2$ (given)

So $K_c = 11.45/4 = 2.87$

Answer: (b)

48. (Q.16, 2009)

Which one of the following sensors is used for the measurement of temperature in a combustion process ($T > 1800°C$)?

(a) Type *j* thermocouple
(b) Thermistor
(c) Resistance temperature detector
(d) Pyrometer

Answer: (d) Pyrometer

49. (Q.17, 2009)

The roots of the characteristic equation of an underdamped second-order system are

(a) Real, negative, and equal
(b) Real, negative, and unequal
(c) Real, positive, and unequal
(d) Complex conjugates

Answer: (d) complex conjugates

as the characteristic equation of a second-order system is

$$\tau^2 s^2 + 2\tau\xi s + 1 = 0$$

$$s = \{-2\tau\xi \pm \sqrt{(4\tau^2\xi^2 - 4\tau^2)}\}/2\tau^2 = \{-\xi \pm \sqrt{(\xi^2 - 1)}\}/\tau$$

For an underdamped system,

$$\xi < 1, \text{ so } s = -\xi/\tau \pm j\sqrt{(1 - \xi^2)}/\tau$$

The roots are complex conjugates with negative real part.

50. (Q.41, 2009)

The inverse Laplace transformation of $1/(2s^2 + 3s + 1)$ is

(a) $e^{-t/2} - e^{-t}$
(b) $2e^{-t/2} - e^{-t}$

(c) $e^{-t} - 2e^{-t/2}$
(d) $e^{-t} - e^{-t/2}$
Answer: (a)
Solution:

$$Y(s) = 1/(2s^2 + 3s + 1) = 1/\{(2s + 1)(s + 1)\} = 1/(s + 1/2) - 1/(s + 1)$$

$$Y(t) = e^{-t/2} - e^{-t}$$

51. (Q.42, 2009)
The characteristic equation of a closed-loop system using a proportional controller with gain K_c is

$$12s^3 + 19s^2 + 8s + 1 + K_c = 0.$$

At the onset of instability, the value of K_c is
(a) 35/3
(b) 10
(c) 25/3
(d) 20/3
Answer: (a)
From the third-order polynomial characteristic equation, the Routh array is

$$\begin{vmatrix} 12 & 8 \\ 19 & (K_c+1) \\ \{19\times 8 - 12(K_c+1)\}/19 & \\ (K_c+1) & \end{vmatrix}$$

In order to have a stable system, all the elements of the first column should be positive, i.e.,

$$\{19 \times 8 - 12(K_c + 1)\} > 0$$

or $K_c < (19 \times 8/12 - 1)$, i.e., $K_c < 35/3$

52. (Q.43, 2009)
The block diagram for a control system is shown below:

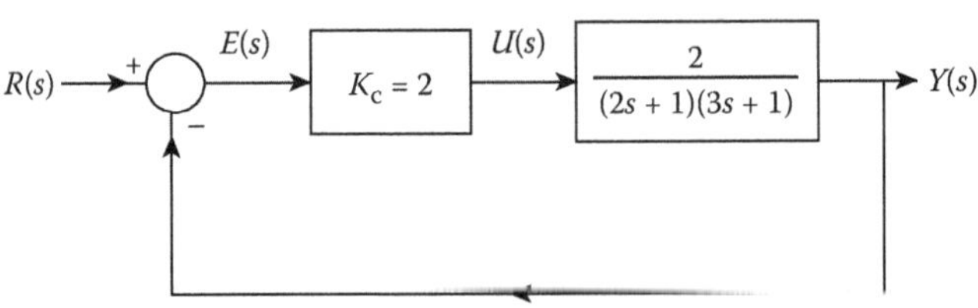

For a unit step change in the set point, $R(s)$, the steady-state offset in the output $Y(s)$ is
(a) 0.2
(b) 0.3
(c) 0.4
(d) 0.5
Answer: (a)

$$Y(s)/R(s) = \{2 \times 2/(2s + 1)(3s + 1)\}/\{1 + 2 \times 2/(2s + 1)(3s + 1)\}$$
$$= 4/\{(2s + 1)(3s + 1) + 4\}$$

or $Y(s) = 4/\{s\{(2s + 1)(3s + 1) + 4\}\}$ as $R(s) = 1/s$
or $Y(t)$ as $t\to\infty$ is $\lim sY(s)$ as $s\to 0$ is $4/\{(0 + 1)(0 + 1) + 4\} = 4/5$
offset $= R(t) - Y(t)$ as $t\to\infty$ is $1 - 4/5 = 1/5 = 0.2$

53. (Q.44, 2009)
For a tank of cross-sectional area 100 cm^2 and inlet flow rate (q_i, cm^3/s), the outlet flow rate (q_0, cm^3/s), is related to the liquid height (h, cm) as $q_0 = 3\sqrt{h}$ as shown below

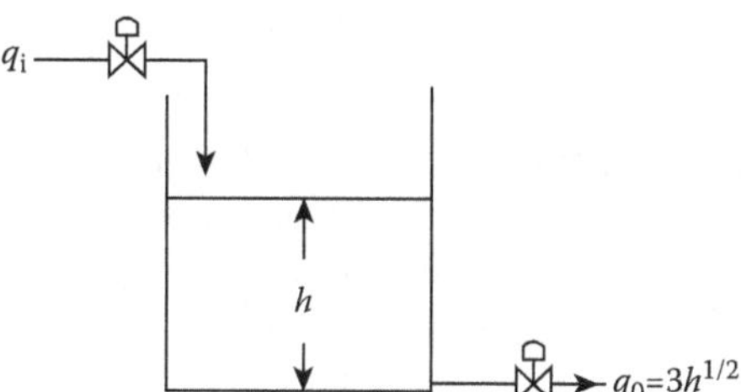

Then the transfer function $H(s)/Q_i(s)$ of the process around the steady-state point
$q_{is} = 18$ cm^3/s and $h_s = 36$ cm is
(where Q and H are q and h with respect to steady-state values)
(a) $1/(100s + 1)$
(b) $2/(200s + 1)$
(c) $3/(300s + 1)$
(d) $4/(400s + 1)$
Answer: (d)
From the diagram,
$q_0 = 3\sqrt{h}$, $q_0(t) \approx 3\sqrt{h_s} + 3/2\ (h(t) - h_s)\ \sqrt{h_s}$, (linearized)
$= q_s + 3/2\ (h(t) - h_s)\ \sqrt{h_s}$
or $Q_0(t) = 3/2\ H(t)/\sqrt{36} = H(t)/4$; hence, $R = 4$
as $A = 100$, $\tau = A \times R = 100 \times 4 = 400$

$$H(s)/Q(s) = R/(\tau s + 1) = 4/(400s + 1)$$

54. (Q.12, 2010)

The Laplace transformation of the function shown in the figure below is

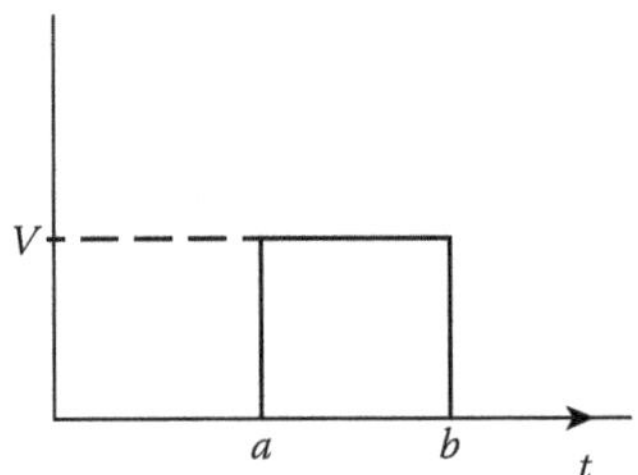

(a) $Ve^{(a-b)s}/s$
(b) $V(e^{-bs} - e^{-as})/s$
(c) $V(e^{-as} - e^{-bs})/s$
(d) $V(e^{as} - e^{bs})/s^2$
Answer: (c)

$$f(t) = Vu(t - a) - Vu(t - b)$$

$$f(s) = Ve^{-as}/s - Ve^{-bs}/s = V(e^{-as} - e^{-bs})/s$$

55. (Q.23, 2010)

Flow-measuring instruments with different specifications (zero and span) are available for an application that requires flow rate measurements in the range of 300 liters/hr to 400 liters/hr. The appropriate instrument for this application is the one whose specifications are
(a) Zero = 175 liters/hr, span = 150 liters/hr
(b) Zero = 375 liters/hr, span = 100 liters/hr
(c) Zero = 275 liters/hr, span = 150 liters/hr
(d) Zero = 475 liters/hr, span = 100 liters/hr
Answer: (c)

As per the requirement, the minimum value on the scale (zero) should be 300 liters/min., and span should be (400–300), i.e., 100 liters/min.

If (a) was selected, a minimum of 300 liters can be read, but the maximum, 400 liters/hr, cannot be read by the instrument, which will be reading the maximum at (175 + 150) 325 liters/hr. Hence, not suitable.

If (b) is selected, a minimum = 375 and maximum = 475, i.e., a minimum of 300 liters/hr cannot be read although the required maximum 400 liters/hr can be made.

If (c) was selected, a minimum = 275 and maximum = 275 + 150 = 425, i.e., both the required minimum and maximum rates at 300 and 400 liters/hr, respectively, can be read. Hence, this should be selected.

If (d) was selected, a minimum = 475 and maximum = 575 is totally out of the required range of 300 to 400. Hence, this will not show any of the desired flow rates. It cannot be selected.

56. (Q.24, 2010)

The transfer function, $G(s)$, whose asymptotic Bode diagram is shown below, is

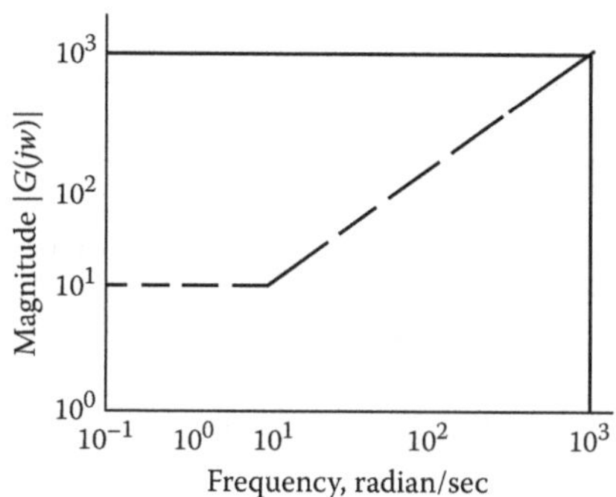

(a) $10s + 1$
(b) $s - 10$
(c) $s + 10$
(d) $10s - 1$

Answer: (c)

Solution:

(a) If $G(s) = 10s + 1$, $|G(jw)| = \sqrt{(100w^2 + 1)}$, for $w = 0.1, 1, 10$, $|G(jw)| = \sqrt{2}, \sqrt{101}, \sqrt{10^4}$, i.e., not constant at 10.

(b) If $G(s) = s - 1$, $|G(jw)| = \sqrt{(w^2 + 1)}$ with negative slope for $w = 0.1, 1, 10$, $|G(jw)| = 1, \sqrt{2}, \sqrt{101}$, i.e., not constant at 10.

(c) If $G(s) = s + 10$, $|G(jw)| = \sqrt{(w^2 + 100)}$, for $w = 0.1, 1, 10$, $|G(jw)| = \sqrt{100.01}, \sqrt{101}, \sqrt{200}$, i.e., nearly constant at 10 then at $w = 100, 1000$, with positive slope, for $w > 10$, $|G(jw)| = \sqrt{(10^4 + 100)}$, and $\sqrt{(10^6 + 100)}$ ≈ 100 and 1000, respectively, hence (c) is the answer.

(d) If $G(s) = 10s - 1$, $|G(jw)| = \sqrt{(100w^2 + 1)}$, with negative slope after $w > 10$, for $w = 0.1, 1, 10$, $|G(jw)| = \sqrt{2}, \sqrt{101}, \sqrt{10^4}$, i.e., not constant at 10.

57. (Q.42, 2010)

A block diagram for a control system is shown below.

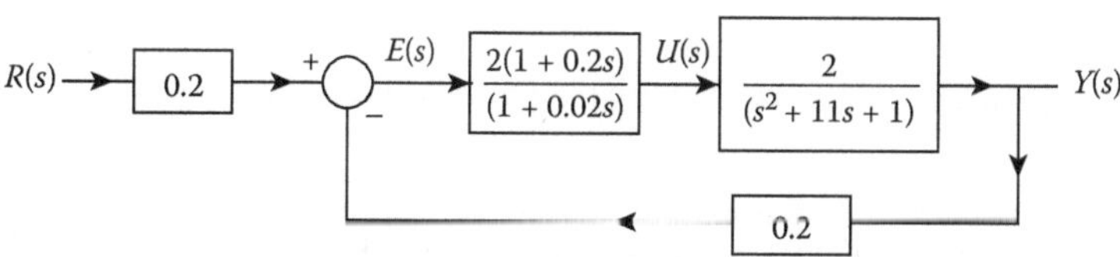

The steady-state gain of the closed-loop system between output $Y(s)$ and set point $R(s)$ is

(a) 5/9
(b) 4/9
(c) 1/3
(d) 2/9

Answer: (b)
Solution:

$$Y(s)/[0.2R(s)] = 4(1+0.2s)/[(1+0.02s)(s^2+11s+1)]/[1+0.8(1+0.2s)/\{(1+0.02s)(s^2+11s+1)\}]$$

$$= 4(1+0.2s)/[(1+0.02s)(s^2+11s+1)+0.8(1+0.2s)]$$

or $Y(s)/R(s) = 0.8(1 + 0.2s)/[(1 + 0.02s)(s^2 + 11s + 1) + 0.8(1 + 0.2s)]$

or $Y(s) = 0.8(1 + 0.2s)/[s\{(1 + 0.02s)(s^2 + 11s + 1) + 0.8(1 + 0.2s)\}]$

If $R(s) = 1/s$ because of a step change of unity, then $Y(t)$ as $t \to \infty$ is obtained by a final value theorem, i.e.,

$$Y(\infty) = \lim sY(s) \to 0.8(1 + 0)/[\{(1 + 0)(0 + 0 + 1) + 0.8(1 + 0)\}] = 0.8/1.8 = 4/9 \text{ as } s \to 0$$

So a steady-state value of $Y(t)$ is 4/9. The answer is (b)

58. (Q.43, 2010)

Consider the cascade control configuration shown in the figure below:

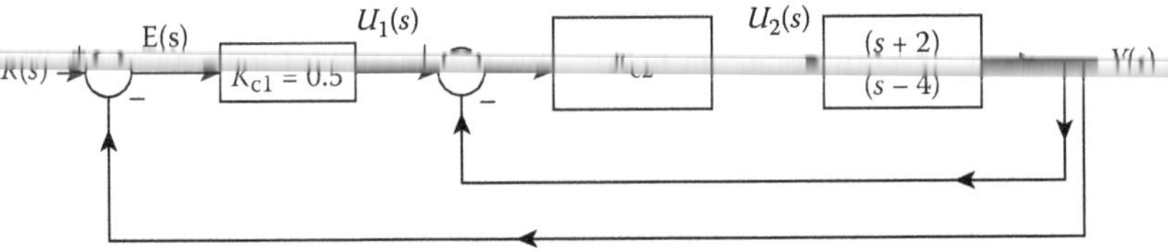

The system is stable when K_{c2} is

(a) 3/4
(b) 1
(c) 5/4
(d) 3/2

Answer: (d)
Solution:

$$G(s) = Y(s)/U_1(s) = K_{c2}(s + 2)/(s - 4)/[1 + K_{c2}(s + 2)/(s - 4)]$$

and

$$Y(s)/R(s) = K_{c1}G(s)/(1 + K_{c1}G(s)] = 0.5/(1 + 0.5G(s)]$$
$$= 0.5K_{c2}(s + 2)/[(1.5K_{c2} + 1)s + (3K_{c2} - 4)].$$

So the characteristic equation of the closed loop is

$$(1.5K_{c2} + 1)s + (3K_{c2} - 4) = 0.$$

Using the Routh array

$$1.5K_{c2} + 1$$

$$3K_{c2} - 4.$$

Hence, to have a positive value of $3K_{c2} - 4 > 0$, $K_{c2} > 4/3$.

Hence, the value of K_{c2} must be selected greater than 4/3 = 1.33. Hence, the value of K_{c2} = 3/4, 1, 5/4, and 3/2; only 3/2 is greater than 1.33. Hence, K_{c2} = 3/2 = 1.5 is selected. The answer is (d)

59. (Q.43, 2010)

Consider the process as shown below:

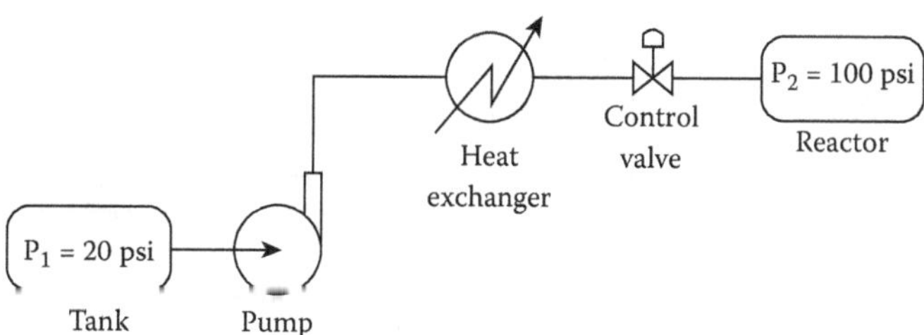

The constant head pump transfers a liquid from a tank maintained at 20 psi to a reactor operating at 100 psi through a heat exchanger and a control valve. At the design conditions, the liquid flow rate is 1000 liters/min while the pressure drop across the heat exchanger is 40 psi, and that across the control valve is 20 psi. Assume that the pressure drop across the heat exchanger varies as the square of the flow rate. If the flow is reduced to 500 liters/min., then the pressure drop across the control valve is

(a) 30 psi
(b) 50 psi
(c) 80 psi
(d) 150 psi

Answer: (b)

Solution:

Pressure drop across the heat exchanger at 500 liters/min. = $(500/1000)^2 \times 40 = 0.25 \times 40 = 10$ psi.

Discharge pressure head of the pump at 1000 liters/min. flow = 40 + 20 + 100 = 160 psi.

The pump has a constant head discharge, so the pressure at the discharge of the pump will be 160 psi when flow changed to 500 liters/min., i.e.,

$$160 = 10 + \text{pressure drop across the control valve} + 100$$

Hence, pressure drop across the control valve at the 500 liters/min. flow = 160 – 110 = 50 psi.

Answer: (b).

60. (Q.44, 2011)

The following diagram shows a CSTR with two control loops. A liquid phase, endothermic reaction is taking place in the CSTR, and the system is initially at steady state. Assume that the changes in physical properties of the system are negligible.

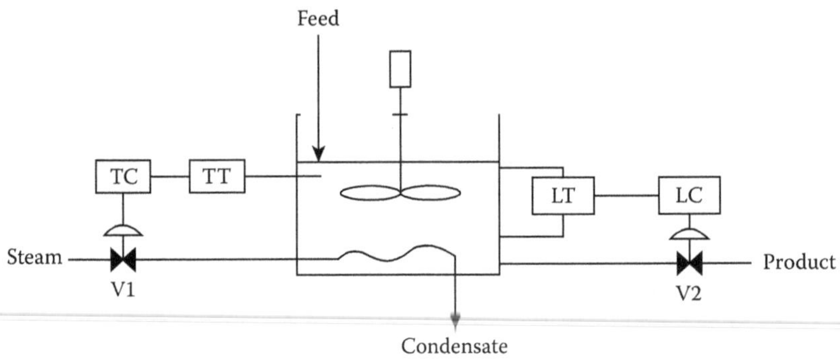

TC: Temperature controller, LC: Level controller, TT: Temperature transmitter, LT: Level transmitter, V1 and V2: Control valves.

Which one of the following statements is true?

(a) Changing the level controller set point affects the opening of V2 only.
(b) Changing the temperature controller set point affects the opening of V2 only.
(c) Changing the temperature controller set point affects the opening of both V1 and V2.
(d) Changing the level controller set point affects the opening of both V1 and V2.

Answer: (d)

As reactor overall material balance, component balance, and the heat balance involve V (i.e., Level × Area), change in set point of the level controller will affect both the level and the temperature away from the existing process values; hence, both the valves will be actuated for achieving control. So the answer is (d).

61. (Q.54 and Q.55, 2011)

A PID controller output $p(t)$ in the time domain is given by

$$P(t) = 30 + 5e(t) + 1.25\int_0^t e(t)\,dt + 15\frac{de(t)}{dt}$$

where $e(t)$ is the error at time t. The transfer function of the process to be controlled is

$G_p(s)$ = 10/(200s + 1). The measurement of the controlled variable is instantaneous and accurate.

The transfer function of the controller is

(a) $5(12s^2 + 4s + 1)/3s$
(b) $5(12s^2 + 3s + 1)/3s$
(c) $5(12s^2 + 4s + 1)/4s$
(d) $5(12s^2 + 3s + 1)/4s$

Answer: (c)

The controller equation involving parameters K_c, τ_i, and τ_D is

$$P(t) = A + K_c\,e(t) + \frac{K_c}{\tau_i}\int_0^t e(t)\,dt + K_c\tau_D\frac{de(t)}{dt}.$$

Comparing this relationship,

$$\text{Bias} = A = 30,\ K_c = 5,\ \tau_i = 4,\ \text{and}\ \tau_D = 3.$$

The transfer function of the controller is

$$K_c\,(1 + 1/(\tau_i s) + \tau_D s) = 5(1 + 1/4s + 3s) = 5(4s + 1 + 12s^2)/4s.$$

Hence, the answer is (c).

For the same problem above, which one is the characteristic equation of the closed loop?

(a) $6s^2 + 102s + 1 = 0$
(b) $700s^2 + 102s + 25 = 0$
(c) $100s^2 - 196s - 25 = 0$
(d) $240s^3 + 812s^2 + 204s + 1 = 0$

Answer: (b)

$1 + G_p\, G_c = 0$, where the transfer functions of valve and the measuring element are taken as unity.

So the characteristic equation of the closed loop is

$$1 + 10/(200s + 1) \times 5(12s^2 + 4s + 1)/4s = 0$$

or

$$4s(200s + 1) + 50(12s^2 + 4s + 1) = 0$$

or

$$1400s^2 + 204s + 50 = 0$$

or

$$700s^2 + 102s + 25 = 0$$

Hence, the answer is (b).

62. (Q.8, 2011)

The range of the standard current signal in process instruments is 4 to 20 mA. Which one of the following is the reason for choosing the minimum signal as 4 mA instead of 0 mA?

(a) To minimize resistive heating in instruments.
(b) To distinguish between signal failure and minimum signal condition.
(c) To ensure a smaller difference between maximum and minimum signal.
(d) To ensure compatibility with other instruments.

Answer: (b)

If the minimum signal was 0 mA, it would be difficult to understand whether the instruments were damaged or open as at those events too no signal would be generated. Whereas if the minimum signal was chosen as 4 mA, the breakdown of the instrument or disconnection (open) of the instrument would be detected when no signal would be generated or communicated.

Hence, the answer is (b).

63. (Q.24, 2012)

A thermometer initially at 100°C is dipped at $t = 0$ into an oil bath maintained at 150°C. If the recorded temperature is 130°C after 1 minute, then the time constant of the thermometer (in min) is

(a) 1.98
(b) 1.35
(c) 1.26
(d) 1.09

Solution:

A thermometer is considered to be a first-order system; hence, the response or thermometer reading is related to the bath temperatures as a function of time and is given as

$$Y(t) = A(1 - e^{t/\tau})$$

where $T(t)$ is the instantaneous thermometer reading = 130°C at t = 1 min, $T(0)$ is the initial reading equal to 100°C, and A is the difference between the new and previous bath temperatures = 150 – 100 = 50°C

$$Y(t) = T(t) - T(0) = 130 - 100 = 30°C.$$

Hence, $30 = 50(1 - e^{-1/\tau})$

Solving, $\tau = 1.09$; hence, the answer is (d)

64. (Q.25, 2012)

The Bode stability criterion is applicable when

(a) Gain and phase curves decrease continuously with frequency
(b) Gain curve increases and phase curve decreases with frequency
(c) Gain curve and phase curve both increase with frequency
(d) Gain curve decreases and phase curve increases with frequency

Solution: Answer is (a)

65. (Q.39, 2012)

A thermometer having a linear relationship between 0°C and 350°C shows an EMF of zero and 30.5 mV, respectively, at these two temperatures. If the cold junction temperature is shifted from 0°C to 30°C, then the EMF correction (in mV) is

(a) 3.13
(b) 2.92
(c) 2.61
(d) 2.02

Solution:

Because the linear relationship is valid

$$E = aT + b$$

where E is the EMF, T = hot junction temperature, and a and b are constants.

When $T = 0$, $E = 0$, i.e., $b = 0$, and when $T = 360$, $E = 30.5$, $a = 30.5/350$ while the cold junction temperature is 0. Hence, $E = 30.5\ T/350$.

The EMF correction because of the shift of the cold junction temperature from 0 to 30°C is $E = 30.5 \times 30/350 = 2.61$ mV, i.e., this corrected EMF is to be added to get the correct reading of any temperature measured.

This can be said in the other way while the EMF is expressed in terms of hot and cold junction T and T_c,

$E = 30.5/350 \times (T - T_c)$; thus, at $T = 350$ and $Tc = 30$, $E = 30.5 \times 320/350 =$ 27.88 mV, which is lower than 30.5 mV while the cold junction was at 0°C. Hence, the correction is 30.5 – 27.88 = 2.61 mV.

Hence, the answer is (c)

66. (Q.40, 2012)

The characteristics equation for a system is

$$S^3 + 9S^2 + 26S + 12(2 + K_c) = 0$$

Using the Routh test, the value of K_c that will keep the system on the verge of instability is

(a) 20.9
(b) 18.4
(c) 17.5
(d) 15.3

Solution: The Routh array is given below with four rows:

$$\begin{vmatrix} 1 & 26 \\ 9 & 12(2+K_c) \\ \dfrac{9\times 26-12(2+K_c)}{9} & \\ 12(2+K_c) & \end{vmatrix}$$

According to the Routh stability criterion, all the elements of the first column should be positive and nonzero. To satisfy this condition, it is required that $12(2 + K_c) > 9 \times 26$.

Or the maximum value of $K_c = 17.5$.

Hence, the answer is (c).

67. (Q.42, 2012)

The block diagram of a system with a proportional controller is shown below:

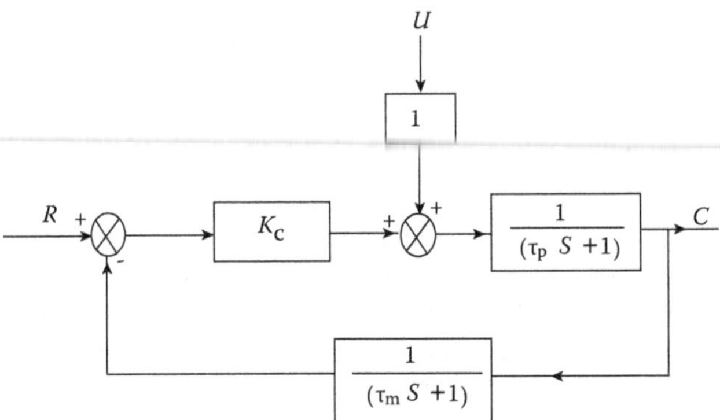

A unit step input is introduced in the set point. The value of K_c to provide a critically damped response for $U = 0$, $\tau_p = 8$, and $\tau_m = 1$ is

(a) 3.34
(b) 2.58
(c) 1.53
(d) 1.12

Solution:

The denominator of the closed-loop transfer function $C(s)/R(s)$ is

$$\frac{R(s)}{C(s)} = \frac{\dfrac{K_c}{(\tau_p S+1)}}{1+\dfrac{K_c}{(\tau_p S+1)(\tau_m S+1)}} = \frac{K_c(\tau_m S+1)}{(\tau_p S+1)(\tau_m S+1)+K_c}$$

$$= \frac{K_c(\tau_m S+1)}{\tau_p\tau_m S^2+(\tau_p+\tau_m)S+(1+K_c)}$$

$$= \frac{K_c(\tau_m S+1)/(1+K_c)}{\dfrac{\tau_p\tau_m S^2}{(1+K_c)}+\dfrac{(\tau_p+\tau_m)S}{(1+K_c)}+1}$$

Comparing with the second-order denominator, $(\tau^2 + 2\tau\xi + 1)$
Hence,

$$\tau^2 = \tau_p\tau_m/(1 + K_c) \text{ and } 2\tau\xi = (\tau_p + \tau_m)/(1 + K_c)$$

or

$$\tau = \sqrt{(8/(1 + K_c)} \text{ and } \xi = 9/\{2\sqrt{8(1 + K_c)}\}$$

for

$$\xi = 1,\ 1 + K_c = 81/32 \text{ or } K_c = 1.531$$

Hence, the answer is (c).

Index

H

I

L

M

N

O

P

Q

R

S

T

U

V

Z